AF829063

Springer Geography

The Springer Geography series seeks to publish a broad portfolio of scientific books, aiming at researchers, students, and everyone interested in geographical research. The series includes peer-reviewed monographs, edited volumes, text-books, and conference proceedings. It covers the major topics in geography and geographical sciences including, but not limited to; Economic Geography, Landscape and Urban Planning, Urban Geography, Physical Geography and Environmental Geography.

More information about this series at http://www.springer.com/series/10180

Régis Darques

Mapping Versatile Boundaries

Understanding the Balkans

Méditerranée
revue géographique des pays méditerranéens
Journal of Mediterranean geography

Springer

Régis Darques
UMR 5281 ART-Dev
CNRS
Montpellier
France

ISSN 2194-315X ISSN 2194-3168 (electronic)
Springer Geography
ISBN 978-3-319-40924-5 ISBN 978-3-319-40925-2 (eBook)
DOI 10.1007/978-3-319-40925-2

Library of Congress Control Number: 2016943440

© Springer International Publishing Switzerland 2017
This work is subject to copyright. All rights are reserved by the Publisher, whether the whole or part of the material is concerned, specifically the rights of translation, reprinting, reuse of illustrations, recitation, broadcasting, reproduction on microfilms or in any other physical way, and transmission or information storage and retrieval, electronic adaptation, computer software, or by similar or dissimilar methodology now known or hereafter developed.
The use of general descriptive names, registered names, trademarks, service marks, etc. in this publication does not imply, even in the absence of a specific statement, that such names are exempt from the relevant protective laws and regulations and therefore free for general use.
The publisher, the authors and the editors are safe to assume that the advice and information in this book are believed to be true and accurate at the date of publication. Neither the publisher nor the authors or the editors give a warranty, express or implied, with respect to the material contained herein or for any errors or omissions that may have been made.

Printed on acid-free paper

This Springer imprint is published by Springer Nature
The registered company is Springer International Publishing AG Switzerland

To my children, Théodora and Timothée

The Tonghu who were neighbours of the Hsiung-nu, asked Mao-tun to give them his best horse and then they asked him for his favorite wife. Contrary to the advice of his followers Mao-tun did as he had been asked for he wanted relations between the two peoples to remain peaceful. But when the Tonghu asked the Hsiung-nu for territory which lay fallow and Mao-tun's followers advised him to grant that wish, Mao-tun was angry and said:

"Land is the foundation of the state, so how can we give it away?"

And he executed all those who had wanted him to relinquish land.

Mete Han or Baghatur (Modu Chanyu or Mao-tun), founder of the Xiongnu (Hsiung-nu) nomadic Empire, 2nd century BC. Supposed ancestor of Attila the Hun.

Taken from the *Shiji* of Sima Qian and quoted by Anatoly M. Khazanov (Khazanov 1994), p. 138.

Preface

The area of *Balkan Studies* has been too long neglected in recent history. The dispersed microcommunities who have an interest in the topic coexist more than they collaborate. The main body of literature is made of works written in foreign *exotic* languages such as French, German, Greek, Bulgarian, Albanian, Macedonian, Romanian, Serbo-Croatian, and Turkish. Edited books take usually precedence over journal papers. Although the author has drawn much from such a rich corpus of local knowledge, English is far from being the most preferred language of scholars who deal with the Balkans, especially in the area of geography. It would have been useless to mention a long list of references that are unintelligible to most people—and even unreadable (four different alphabets).

Three other corpus of documents widely used for preparing and building the GIS, remote sensing, and statistical databases will be provided with references only in specific cases (facsimile reproduction, sources of figures or tables). Details will not be included because of the obvious overload that it would entail:

- Topographic or historical maps (about 400 documents consulted and/or used);
- Statistical books edited by the many Balkan statistical offices over two centuries (dozens of references);
- The body of literature relevant to the scope of geodesy, advanced GIS studies, and remote sensing techniques does not appear either (numerous references, mostly journal articles).

Such choices are aimed at helping readers not to get lost within a broad range of documents that have no straightforward incidence on the analysis. Concentrating on the essential is a priority. The Balkan issue is fairly complicated and confusing. It makes no sense to add fuel to the fire.

Accordingly, the general principle may be that nonessential information should be curtailed. This is made all the more easy as geography has long been the poor relation of research initiatives in the Balkan area. Historical and geopolitical approaches are more popular, although deprived of any spatial dimension. Within this limited corpus, the border issue is even more randomly addressed. This together with the primary goal of producing an extensive body of original information has led to limiting the conventional review of literature. I express my sincere regrets to all respected authors whose works are missing. After more than twenty years of applied research, this "background" bibliographic apparatus remains of the utmost importance.

Aknowledgements

I would like to express my sincere thanks to my dearest friends, scattered throughout France and Europe, whose support to this long-term project has been invaluable, both materially and morally. Nothing could have been possible without them. Special acknowledgment goes to Michel Bouillet, Georges Sidiropoulos, Emile Kolodny, Guy and Galia Burgel, and Christophe Morhange whose encouragements have contributed to the achievement of this work. The commitment of Dr. Jay Gatrell (Bellarmine University, Louisville, Kentucky, USA) has been essential in helping me find the most appropriate publishing medium. His wise advice have largely contributed to the promotion of this work. I would like to thank the anonymous readers for their useful comments on the original manuscript, the entire Springer editorial team whose competencies are recognized internationally, and especially Dr. Robert Doe. His guidance, support and assistance were critical to the preparation of this book.

My gratitude goes to the UMR 5281 ART-Dev, of which I have the privilege to be a member. The *Association des Amis de la Revue Méditerranée*, strongly active in the area of Mediterranean studies, and its team of passionate scientists have also agreed to accompany and support this project despite its seemingly unrealistic initial objectives.

I want to pay heartful tribute to the very large list of anonymous people that I have had the pleasure of meeting on a number of occasions during my travels. The information supplied in lively conversations is just irreplaceable. Such encounters have provided opportunities for building profound and long-lasting friendships.

Last but not least, I wish to extend my gratitude to my wife, kids, and family for their patience and understanding. A charming smile often goes a long way.

Contents

List of Figures

List of Tables

Abstract

Some people consider that international boundaries are lines drawn on maps without identifying characteristics. Conceptual more than material entities, they divide lands and states, but their existence is always understood in reference to the central areas they encompass and define. Borderlands are rarely regarded per se, i.e., as spatial objects, as if geopolitics had succeeded in depriving geography of any content. That is at least what conventional maps suggest—those often found in newspapers. Through this and other means, mapping applications distort reality and artificially create *non-places*—lands without substance, disembodied spaces.

In a seemingly contradictory way, the dominant political culture presents valuable arguments in favor of a *borderless* world. *Supermodernity* defined as the alteration of spatial awareness and *globalization* as the reign of powerful *transnational* dynamic obviously provide useful tools for helping make sense of the modern world. However, there is mounting evidence to show that the defeat of classical geography is far from being achieved. Modern economic and military *empires* still have to deal with tiny states that resist inclusion in the *global sphere* and refuse to accept the loss of their sovereignty despite international pressure.

The Balkans and Eastern Europe in general always seem to swim against the tide. "There," everything is complex, out of control, and unpredictable. The inviolability of borders—a dogma broadly accepted throughout the *developed* world—is regularly challenged. States that may never exceed the size of a county are not afraid to fight for, and gain, their independence against the greatest powers. Recently, the fall of communism has brought back revolution, war, ethnic cleansing, economic collapse, and a powerful revival of all mafias. One might thus believe that the region has become what it never ceased to be: a "no-rights" area.

Such instability, deeply rooted in history, has far-reaching consequences. In an attempt to disentangle the issue, this book aims at addressing the Balkans on a systematic transnational basis. Attaining this goal supposes to bypass major military and administrative restrictions. A research methodology has been developed, which combines field inquiries, GIS, and remote sensing analyses. This has been achieved by virtue of long-term investigations carried out over the course of twenty years across Greece, former Yugoslavia, Bulgaria, Albania, and Romania.

The enormous volume of data collected and processed results in an extensive body of maps and figures that re-evaluates the manner in which the Balkans are seen. Since the emergence of nation-states at the beginning of the nineteenth century, the region has been torn up by hundreds of territorial demarcations. The census process demonstrates that *frontiers* are everywhere. As a consequence, boundaries can no longer be considered *external* geographic objects because they directly determine the nature of *heartlands*. As can be seen from research, the general confrontation between states, autonomous districts, self-declared republics, and other empires evidences a spatial variability whose dimensions exceed all expectations. The powerful return of applied geography seriously questions the idea of a borderless world that freely submits to the domination of few leading powers and dominant territories.

Chapter 1
Introduction

Abstract Among the issues closest to people's hearts, one of the most important is that of land confines. In the Balkans, borders and boundaries shape landscapes and impregnate mindsets to such an extent that obsession and phobia constitute two faces of the same prism. In this first section, attention focuses on the conceptual framework that will guide analysis. A series of pictorial representations puts emphasis on how mapping can help address this much disputed issue. Complex geographical systems must be approached with due regard to intermediate spaces whose significance cannot be underestimated.

Keywords Land conflicts · Border geography · Geopolitics · Transnationalism · Nation-states · Mapmaking · GIS · Complexity · Graphic modeling

Visiting borderlands? Don't even think about it!

My friends always granted me with the same feedback when I advised them of my goals for the weekend. What reason would take a French traveler to spend time wandering around unlikely spaces while other "frontiers" are so tantalizing? Aren't the Greek or Croatian beaches more appealing than remote mountains peopled by anonymous peasants? Reactions such as these are easily understandable. Exploring "never-before-seen" places is not as much rewarding as usually expected. Balkan borderlands are endless countryside covered by trees, agricultural areas, fallow lands, or rocky soils. Imagining more dreary landscapes is a real challenge, especially in winter.

I still wonder today why the Balkans have early drawn my attention. I started to travel throughout the Peninsula just after the communist collapse, at the beginning of the 1990s. With the implosion of Yugoslavia, the border issue had reached the proportions of an international concern. From the former Federation, a myriad of dwarf states had come into being. Although Balkan history resembles a long chain of political crises, insurrectional movements, and wars, the implications of the ongoing changes seemed grossly out of proportion. Nonsense had been made a general rule regulating daily life. How important are boundaries when, to survive, each microstate has to establish close relationships with the neighbors it was fighting just before, on the eve of its independence?

© Springer International Publishing Switzerland 2017

R. Darques, *Mapping Versatile Boundaries*,
Springer Geography, DOI 10.1007/978-3-319-40925-2_1

Fig. 1.1 Russian soldier facing Edirne, Selimiye Mosque, Russo-Turkish War of 1877–1878. *Source* www.lostbulgaria.com

The second capital of the Ottoman Empire became one of the largest border cities of the Balkans after 1912. Edirne was the second Ottoman city of Rumelia by its population at the end of the nineteenth century, behind Istanbul. Locked between Greece, Bulgaria and Turkey, it stands as a major geopolitical *hot point* of the Peninsula up to now.

From such experiences, the lesson to remember was clear: Borders and boundaries permeate all aspects of society. Not only do they define landscapes, but they also impregnate mindsets (Fig. 1.1). Of course, the main central areas try to maintain some distance with the "hard materiality" of lasting boundaries. However, in outlying regions, *borders* impose themselves as geographic objects governing all aspects of land development. Everyday conversations cannot avoid this underlying subject. Was it even possible for a geographer to escape addressing the issue?

1.1 Borderlands: A Bad Name

Bunkers, mined areas, surveillance towers, barbed wire fences, barking dogs, and armed officers are common references associated with the idea of frontiers as limes of identity and places of confrontation. Land confines have a poor reputation. Boundaries are relics of wars, last witnesses of countless victims. They are geographic souvenirs left behind them by retreating soldiers as a reminder of all sacrifices agreed to protect homelands. In southeastern Europe, no ground has been gained without bitter and deadly fights. Negotiations come later, once passions have reached a climax and all belligerents are exhausted. Everywhere the earth bears signs of hard battles where countless fighters sacrificed their lives hoping to hand down a peaceful future to their families. Borderlands have a well-deserved reputation. Are they only *doomed areas* where the number of graves is always higher than that of living people?

Many fiction writers and moviemakers have embraced the border issue with the intent to depict the *Balkan soul*. Ismail Kadare, Ivo Andrić, and Aris Fakinos are not the least of them. In 1991, Theo Angelopoulos presents *The Suspended Step of the Stork* as his latest film. The movie provides a fascinating insight into the *Balkan heartlands*. While disregarding sunny beaches and wide-open sea horizons, the filmmaker fixes his attention on remote places the true colors of which he brings to the fore. Heavy skies and gloomy atmosphere draw the line of the work throughout the journey of foreign journalists lost in the middle of *nowhere*. It was on the eve of the Yugoslav Wars. The step of the stork is a metaphor: The foot of a soldier kept suspended in time and space, not allowed to hit ground across a borderline—geographic artifact dividing lands but also breaching men.

In his novel *Surveillance Zone* published in 1972, Aris Fakinos pictures other anonymous village communities of northern Greece, close to the Rhodope mountain range, caught in the middle of the Cold War. The author makes every effort to avoid describing the borderline and its forbidden zone through any landscape description. Deprived of materiality, the boundary creates a general atmosphere while managing people's lives silently. Without geographic *body*, it becomes a second cloth or skin for local residents. Villagers and other civilians are barely authorized to stay nearby. Kept under constant control, they are strangers to their homeland. Police and military officers are in charge of any decision.

> -Here is the surveillance zone!... Seven hundred and thirty-nine villages... Four hundred and twenty in Macedonia and Thrace, two hundred in Thesprotia, hundred around Jannina! [...] Twenty in Corfu! Seventy-seven thousand square kilometers, Master. that we know meter by meter, fellow by fellow, the same way you care for the children of your school. Everyone is registered, weighted and sorted, from the grandfather to the grandson. Even your mother doesn't know you as well as this cabinet. (Fakinos 1972), p. 199.

> For you who live inside the surveillance zone, illegality, oppression and arbitrariness are rooted like the thistles of your mountains. [...] You don't have the right to leave your village, to travel, to buy or sell any field; for that, the approval of military authorities is required. To visit your children who live outside the surveillance zone, you must obtain the approval of the division. The population of your villages decreases; slowly but surely they are deserted. Before the war, there were three hundred families in Trahoni; a week ago, there were only thirty. In Syli, there are eight persons left and three of them have already asked the permission for moving to Kavala. Map in hands, we have been seeking for Vitovo, Thisavro, Retsinitsa and Kaltsovo: we only found ruined houses, fallen walls, streets invaded by weeds. Maps hanging on classroom walls are lying. Thirty-seven villages have disappeared a long time ago. In one of them, we picked up a rotten sign of the security battalions: "The Barbarians won't go through". (Fakinos 1972), pp. 209–210.

The curse emerges from the combination of strategic issues and competing political ambitions. Preventing and potentially stopping any outside threat to the motherland are absolute priorities. The price to be paid is a heavy one. Border districts are condemned to decline. Bowing under the pressure of considerable forces, people are caught dreaming of exile; this will lead them to abandon home and fields. “Anywhere else better than here” is instituted as a credo. Generations of migrants move abroad hoping for a better future in America, Australia, Germany, France, Italy, and even Russia. The major capitals of the world are particularly coveted destinations.

From the shores of the Danube Delta to the high Carpathian Mountains, or across the Adriatic coast, the Thracian plain, or the Pindus mountain range, the Balkans offer so varied and breathtaking landscapes they hardly seem compatible with such degraded presentation. Deeply rooted in the Cold War era, the words of Aris Fakinos are clear. Aren’t they loose with the truth? Each year, millions of tourists visit the outstanding natural beauties of the Aegean and Adriatic coasts and appreciate the advantages of flourishing local traditions. Would they be enticed with wrong promises? Even during the Cold War era, the crystal blue waters and sandy beaches of the Mediterranean never ceased to draw massive flows of tourists. However, few people know that, for centuries, some Aegean islands were used as penitentiary camps reserved for common law and political prisoners, or as insane asylums. Many islets locked out of ferry routes are places where obtaining medical assistance is still a problem today. Attractive and repellent spaces are not established per se. This is what some authors call the performativity of space phenomena (Green 2010). Most dangerous lands are also fascinating destinations. Geography is full of contradictory statements.

Fortunately, humans have the capacity to delete embarrassing information from their memories. Battlefields might become resort places, and holiday centers might easily replace abandoned military barracks. Erasing the physical traces of the past is not the most difficult business people can handle. Nature adopts the same tactics: In just a few short years, bare soils can change into flourishing ecologic reserves, and vice versa, erosion and climate aggressions may ruin any human landscape in no time. By the end of 1991, in a couple of weeks, Mostar and Dubrovnik renounced their tourist leadership and engulfed in a nightmare scenario while crushed under a hail of bombs. Nothing is definitively established.

Gaining insight into the border issue is a good way of understanding the ins and outs of a complex Balkan system. Over the past two centuries, no period has escaped the quagmire of long-term conflicts between neighboring countries. All regions have been marked by several episodes of violence, with devastating effects on local life. As a result of these events, new territorialities emerge.

In his novel, Aris Fakinos depicts a typical fellow that most travelers meet when spending some time around tables: the *old fighter*. Thrassos is already an experienced *Capetanios* in the 1960s. His story is worth telling rather than living. During the Balkan Wars (1912–1913), as a young man he first quits his wife to reach his village brigade and chase some Bulgarians away from Greece. Between 1914 and 1918, he joins the regular Greek troops and loses a hand in a trench around Skra, at the Yugoslav border. In 1922, he follows the Greek army in Asia Minor with his eldest son, Mitsos. After the defeat, he comes back home with a broken face, but Mitsos is shot dead. His second son Spyros dies in 1940 in Southern Albania. Thrassos learns the death of his last son Constantine while fighting against the

Germans in the maquis. Even in his old age, he is not spared from the outcomes of the Greek Civil War and Cold War (p. 167). His military service never came to its end. In Greece, as well as in other Balkan countries, land and people are constantly under fire. Each citizen is first and foremost a soldier.

In contrast, anyone submitted to "public distrust" is reserved a dreadful fate. Nationalist fevers promote martyrdom on all sides. Political opponents and ethnic minorities will not be tolerated; they have to comply. When too much pressure is applied, or when "rebels" prefer to keep standing positions, exile is the only alternative to detention. Just like land, families are hurt and divided.

The situation of Thrassos is all the more saddening that it is not uncommon. The same rule has been applied throughout the nineteenth century. The emergence of nations and the gradual collapse of the Austrian and Ottoman Empires not only took place through "regular" interstate wars. Each village used to manage its "resistance" thanks to voluntary armed groups ready to fight against any threat. Gunmen, smugglers, and bandits under the conduct of *Captains* control the Balkan countryside. They may rely on ethnic allegiances to create large insurgencies, adopt extreme political positions, or change side when necessary. They may even "vanish" within rural communities. Disputed boundaries produce new types of "heroes."

1.2 A Map-Based Approach of Geography

Geography as a scientific discipline has a very bad press in the Balkans perhaps because cartographers have been almost exclusively employed by military authorities. Such an applied knowledge likely to be misused had to be kept under control. Accordingly, the development of any form of *civilian* geography has been prohibited until lately, i.e., the end of the Cold War. Why such a discreet existence?

Among other purposes, the present volume aims to release geography from fixed national frameworks and to address borders and boundaries as ordinary objects. Minor subjects often reveal chief questions. National political staff members are often keen to react against outside observers who would be interested in studying border issues. They would argue that awakening ghosts might be risky. Silence and mystery are considered to be better solutions for maintaining peace. Against such allegations, one might argue back that the Yugoslav Wars and other resurging conflicts of the past few years would not have erupted if that were true. Researchers are commonly endowed with wrong powers.

So many scientists have brought support to political aims during the last wars that defending the neutrality of scholars does not work either. And conversely, the rationale of general suspicion toward researchers is an inherently limited posture. Once accepted without critics, this reasoning leads to a mind-numbing effect. The best way of transforming risky topics into disputable questions is to implement strong methods of analysis, which produces admissible evidences later.

Frontiers are geographers' obsession (Foucher 2012) for long. In 1938, on the eve of WWII, in a context of global political breakdown, when all territorial ambitions of the states were to be released, one of the most influential scientists of his generation publishes a book about border geography (Ancel 1938). Jacques Ancel is well known for his work in Macedonia during the 1920s. Having acquired substantial experience in field operations, he is a founding member of *geopolitics* as an emerging discipline (Ó. Tuathail 1996). Former captain of the French Army's Oriental Corps, attached to its headquarters, his approach to the border question is unique. His clear understanding directly comes into conflict with the monolithic vision the German *Geopolitik* would like to impose to the world—such determinism is inherited from Friedrich Ratzel. Ancel embodies a tradition of foreign geographers who used to study the Balkans "behind the back" of faltering Empires at the turn of the nineteenth and twentieth centuries. WWII and the upcoming Cold War would put a term to this tradition later. Deprived of any transnational perspective, interested scientists will no longer escape the binding framework of state territories during 50 years onward. Schizophrenia had seized geography, making it focus on central places and forget the in-betweens that had drawn the attention of so many pioneers of science.

The work of Jacques Ancel allows us to discuss a few points that will stand for basic interrogations and milestones to the analysis:

1. Ancel suggests that boundaries are "political isobars" that determine the balance terms between two pressures. Therefore, they cannot be studied per se, but relatively to the territorial bodies they divide and connect. Boundaries are mobile objects that vary in space and in time, constantly. They are versatile. International treaties, when discussed and agreed, are papers to be revoked at any time.
2. In old times, boundaries and frontiers were abstract entities whose existence was based on formal maps. Most pastoral societies that used to travel from "winter homelands" to other "summer homelands" had no other choice but to ignore them. With their elected chiefs, they belonged to no one and were granted with factual *autonomy*. Sarakatsani and Vlachs were famous among them, but their way of life was already set in jeopardy in the interwar period.
3. The *fixity* of state boundaries is a tricky point. Throughout history, many states have promoted alternative forms of territoriality. When nations are not fully established and confirmed, defining external boundary lines is for them the least concern. In such cases, state confines are areas where people are authorized and encouraged to stay. The nature of frontiers depends on the dynamics of the states that define them. Growing political bodies do not require fixed frames. Besides, Empires that agree to the idea of their own "universality" virtually have no limits (Ateş 2013). Instead, nations that feel they have achieved their goal—when state limits match with the nation's space—are inclined to settle down stable boundaries. Only then would their confines be changed into human deserts.
4. Agrarian societies and pastoral groups think differently about their frontiers. Ancel opposes "forest states" and "steppe states." The first category is surrounded by green deserts created through deforestation. Their outside faces have clear physical limits. The second group is made of "dromocraties": Control is effective through communication means. Land is not thoroughly mastered because states do not depend on a rooted territoriality. Sovereignty is absolute,

but its application is not deeply anchored in space. The *country* "follows" the nomadic group to which it is attached across its journeys (Khazanov 1994).

5. Fixed lines drawn on paper are the application of a recent concept. Political maps are intrinsically lying because they create artificial delineations from scratch. Linear boundaries are creations of draughtsmen of the Renaissance who expressed a need for simplification. According to Ancel, all mapmakers overcome their rights when they draw lines. This is widely echoed in recent debates concerning the so-called territorial trap (Agnew 1994, 2010; Paasi 2009).

The above remarks are comments rather than words taken from Ancel's hand. The position of the author is brilliant: He mixes general statements rooted in his personal culture with knowledge grounded in practical experiences. The cocktail is relevant, even though the approach suffers from different methodological weaknesses.[1] The probative power of the speech is limited. Gathering examples from the "origin of the world" to most recent times, putting together, and comparing countries scattered around the world are debatable options. As a founder of *geopolitics*, the author already unveils what shall become the Achilles heel of his emerging discipline:

- The word *territory* expresses a complex interaction between people and land (Elden 2010; Gottmann 1973; Forsberg 1996). The concept is often reduced to its simplest political dimension. However, political sciences and geopolitics/geography should not be confused. Geographers see limits everywhere, not only on the external face of political bodies.
- Feelings of suspicion toward maps are rationally grounded and result from a specific historical context: Isn't the interwar period a time of absolute submission of cartography to planned military ambitions? The distrust of the author is understandable. His book lacks any iconography, as if geography had to break with mapmaking to become geopolitics. The forewords of a future division between both disciplines are already settled—a position we shall deny (Black 1997; Harley and Laxton 2001; Pickles 2004).

We wish here to adopt a definition of geography that does not consider maps as optional inputs/outputs of knowledge production systems[2] (Dodge et al. 2011; Perkins 2004; Boria 2013). Too many works are written on the basis of exclusive textual approaches, even when discussing spatial issues. What do maps promote? For geographers, quantitative analyses, GIS, and other remote-sensing techniques

[1]The author and his heirs shall forgive the anachronistic smell of this statement.

[2]This standpoint is clearly away from Denis Wood who considers any map as a pure political product throughout history (Wood 2010)—what is undoubtedly true for *official* mapping and political counter mapping.

For our part, we shall assume a maieutic approach to cartography as a science that may help *some truth* to emerge from the surrounding chaos. In this regard, science is not only a knowledge system manipulated by governments with the aim of enforcing their authority—this is what D. Wood rightly turns down. Cartography will be considered as an expression of hermeneutics. As a consequence, a map tends to be all the more so authoritative (in the worst sense of the word) that it is deprived of scientific *weight*.

are effective means designed to address land issues. They help scientists not to make laws from exceptional cases. They introduce complexity where everyone would follow simple ideas—a sure way to defeat ideologic stances. When data mining and processing are properly mastered, the distinction between opinion and evidence becomes clearer. This position contrasts with the widely shared assumption in postmodern geography that maps are *only* cultural products or social constructions among others (Crampton 2010). When Gunnar Olsson explains that "we find our way in the unknown by drawing on invisible maps of the invisible and by following cultural compasses whose needle points us not to the physical existence of the magnetic north pole but to the social subsistence of the culturally taken-for-granted" (Olsson 2007; Abrahamsson and Gren 2012), he focuses the debate on a specific aspect of map "production/consumption" systems. Fortunately, other exact uses of maps exist (Robinson and Petchenik 1976). A large community of geographers is inclined to believe that the sophisticated digital tools available today have introduced an unprecedented scientific and technologic revolution in the process of knowledge production.

Not everyone shares this way of conducting applied research. In modern geography, worlds confront each other.[3] *Geosciences* and *geocomputation* now use a large panel of digital tools that make it possible to process masses of data and produce original graphics (Longley et al. 2010). Facing rough competition with "computer geeks," some "old-fashioned" workers survive who ignore geomatics engineering and adopt sociologic/ethnographic approaches. For them, space is nothing more than a research category among others. Against this background, the present work lies at the interface of different scientific disciplines. Innovations often emerge from the combination of different approaches and techniques (Crampton 2011).

Methodological questioning and long-term personal commitment are required to achieve the targets. The study had to meet several criteria:

1. Overcoming a so-baffling issue implies processing a great volume of original data. Research supposes to bypass global restrictions on access to information, whether imposed by military or civilian authorities.
2. The most advanced techniques in geoscience bring a solution to the first point. Learning photogrammetry, statistics, geodesy, remote sensing, GIS theory, and application is not a matter of a few months. It is a lifetime challenge, considering that geosciences move fast, especially through the advent of innovative algorithms and softwares.
3. Mapping could not be just about manipulating linear features issued from GIS databases. Producing new images, maps, and figures is a sure way to legitimize the existence of *unknown* lands. Had Jacques Ancel discovered *fuzzy logics* in remote sensing and GIS applications, his opinion about cartography would have been different.

[3]See the heated debate on whether geography is committed to supporting epistemology or ontology—critical geographical theory versus GIScience (Crampton 2009; Leszczynski 2009a, 2009b; Schuurman 2000; Schuurman 2006; Goodchild 2006).

4. Obviously, boundaries need to acquire *some* spatial extent to become geographic objects in the true sense of the word and get out from under the umbrella of political sciences. The existence of interfaces is based on *surfaces*, not lines. Maps do not have to bend down territory.
5. Field studies play a central role in avoiding the production of a "fake" vision of borderlands. The human dimension of geography is never acquired through remote sensing engineering. "There is no science but concrete," Ancel used to say. Intentionally, theoretic considerations will be kept in a second place. *Borderless worlds, hyper borders, border performances, cross-border cooperation, bordering and debordering processes, refugees and border security, green belts, and transboundary conservation* are among the latest fashion themes and fields of expertise. Most often issued from assumed ideologic stances, new concepts and neologisms collide[4] (Andersen et al. 2012; Sassen 2013; Ruggie 1993; Brenner et al. 2003; Herrschel 2011; Houtum et al. 2005; Kolossov 2005). To avoid any ambiguity, the word "border" shall not be used metaphorically in this work. It will refer to spatial objects that shape land.[5]

---

[4]Most discussions about *territory*, *borders*, and *geography* are led today by other sciences—sociology, political science, anthropology—that pay little or no attention to cartography. The list of related references is endless. From my perspective, these stimulating debates are not fully relevant and do not dare to go the whole way because lands and territories are seen as abstractions rather than material objects to study through innovative methodologies and dedicated tools. Many conceptions of what geography is should be collide. This work supports the position adopted by some "old-school" geographers for whom geography needs to be *rooted* in some spatial reality, as opposed to other *non-earth* approaches.

[5]As everyone knows, the terms *border*, *boundary*, and *frontier* have different meanings in British and American English. Depending on the cultural background, a word may have opposite meanings. *Frontier* may thus be interpreted as a political frontier (UK) or as a frontier of settlement (US). The polysemy of words is a fascinating subject. Linguists know that no exact equivalent can be found between terms when translated from one language to another. The meaning of words also varies in time and space. The works of D. Power and N. Standen (Power and Standen 1999), P. Garde in French (Garde 2004), S. Ateş (Ateş 2013), and K. Rankin and R. Schofield (Rankin and Schofield 2004) perfectly illustrate the linguistic problem any author has to face on this subject. Assuming here that no *modus vivendi* might be reached between Greek, Turkish, Albanian, Bulgarian, Serbo-Croatian, Romanian, and other French, English, or German senses of the words, and to avoid the greatest confusions, we will adopt the following three basic definitions:

- *Boundary* will refer to a political limit with hampered interactions across a defined line.
- *Border* will be used as a generic term referring to the ends of a territory, without necessarily attaching the barrier connotation implied in British English. As will become obvious below, Balkan *borderlands* may be seen as a "land of opportunity" (American English). The "challenge of wilderness" and personal freedom also infuse mental representations in the Balkans. The question is disputable.
- *Frontier* will apply to a border area deprived of strict linear division. Spatial fuzziness implies multiple interactions.

Many wise readers shall feel unsatisfied by such definitions—so shall we. To minimize confusions, these words will be explained by an appropriate use of epithets when necessary.

The forthcoming figures illustrate different ways of thinking interfaces and borders Figs. 1.2, 1.3, 1.4, 1.5, 1.6, 1.7 and 1.8. These are graphic transcriptions of a conceptual framework that will serve as a reference for further spatial analyses. Unexperienced readers should feel at ease: Graphic signifiers provide *instant* access to the signified (Olsson 1991; Schuurman 2006).

1.3 In-Betweens and Dead-Ends

"To find one's way, one has to get lost," says the adage. The precept is relevant for geographers who wish to find original information. Exoticism is just at the street corner, provided that our valuable GPS is left home. Visiting beaches or archeological sites in Dalmatia, Greece, or Bulgaria is certainly a fine way to relax and enlarge anyone's culture. However, the *Balkans* are much more than that.

Although they receive little flows, backcountries broaden the scope of possibilities for the boldest of tourists. Traveling conditions are of course problematic in remote districts, as always. However, the deep malaise of abandonment associated with those areas is caused not just by the perception of unequal levels of economic development. From Baranja to Dobrudja, across Epirus and Rhodope, many worthwhile destinations are swept under the rug of tour operators and vanish at the same time out of public view. After the human mind, geography is granted with conscious and unconscious areas. And we know that the unconscious is reluctant to any external intrusion.

Throughout the Balkans, scientists have to face unwilling administrations when they try to gather information. To obtain satisfaction, they focus on capitals at first —it is a mere sin to confess—because central governing offices are usually more inclined to share data. The farther the land confines, the harder the access to information. Years ago, when I began to study Salonika (Thessaloniki), the second largest city of Greece, both active and passive resistance already impeded my investigations. It was nothing compared to the challenges I had to face while studying the Albanian confines. In outlying places and along borderlands, closed lips follow silent landscapes.

Understanding borders and boundaries implies studying the relationship between a territorial unit and its comprising parts. Margins exist in reference to identified centers. The theme is as old as geography (Smith 1977). Such a categorization determines hierarchical positions whose power balance is not static. The Balkans are an area where land hierarchy has been turned upside down many times over the past centuries. New states regularly upset older spatial balances. When this happens, anonymous settlements emerge from the shadows and attain the rank of major cities. Instead, wealthy cities and rural lands considered relics of the past change into semi deserts. Under such a scenario, the status of centrality and remoteness is extremely versatile.

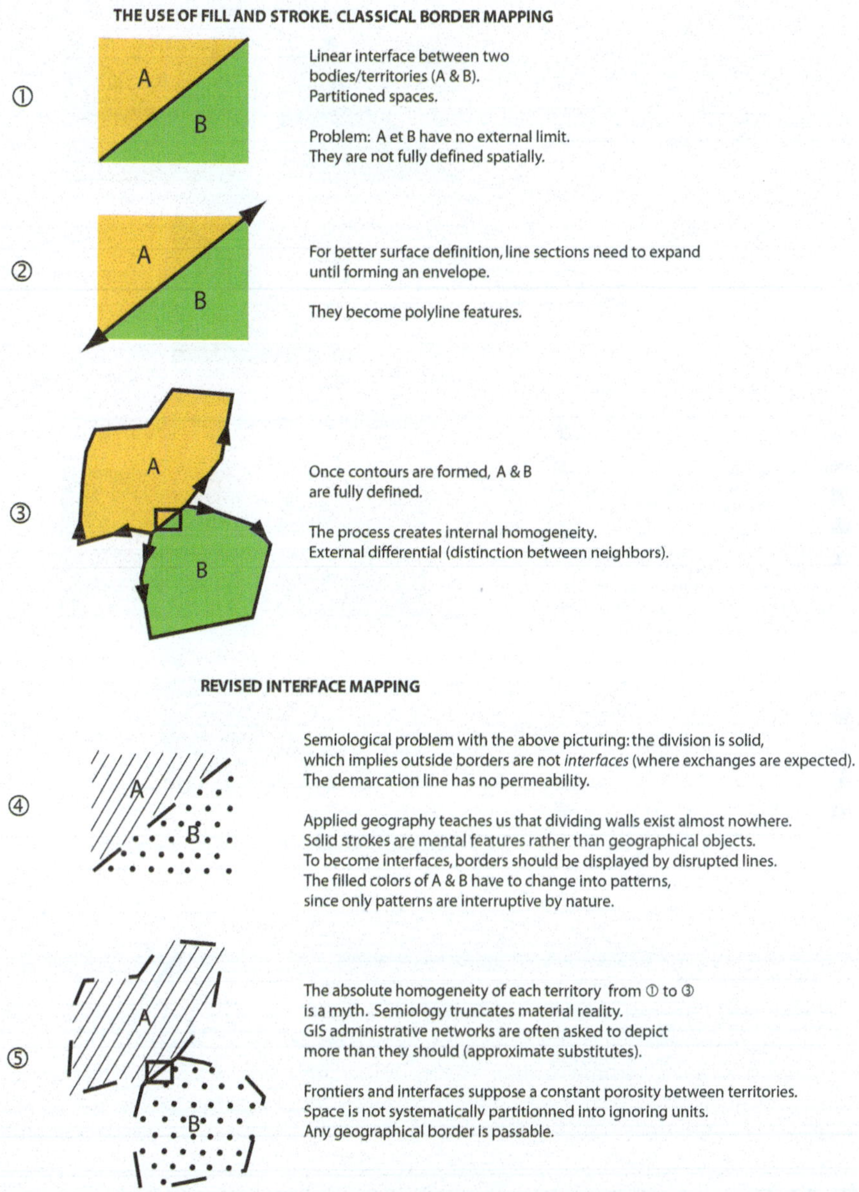

Fig. 1.2 Territories defined by interfaces. Mapping mistakes about fill and stroke. *Source* R. Darques, 2015

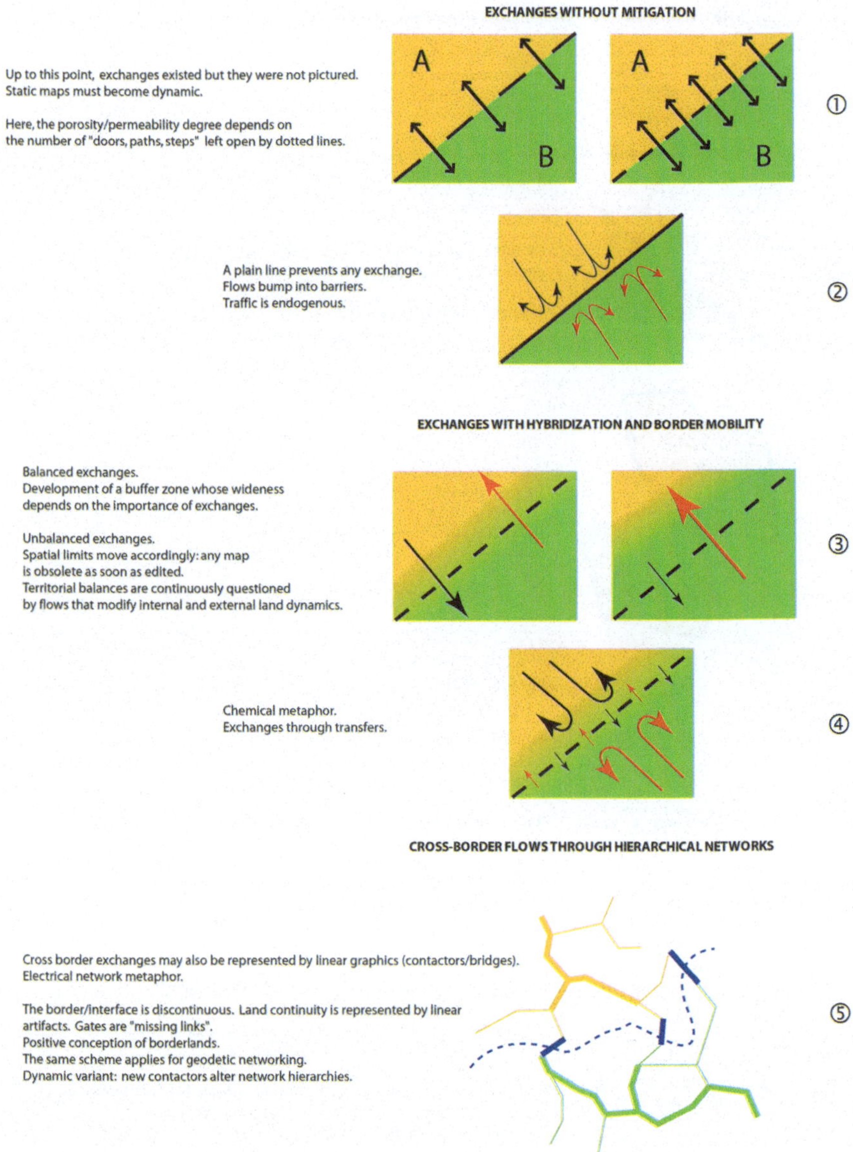

Fig. 1.3 Flows and interactions. Dynamic picturing. *Source* R. Darques, 2015

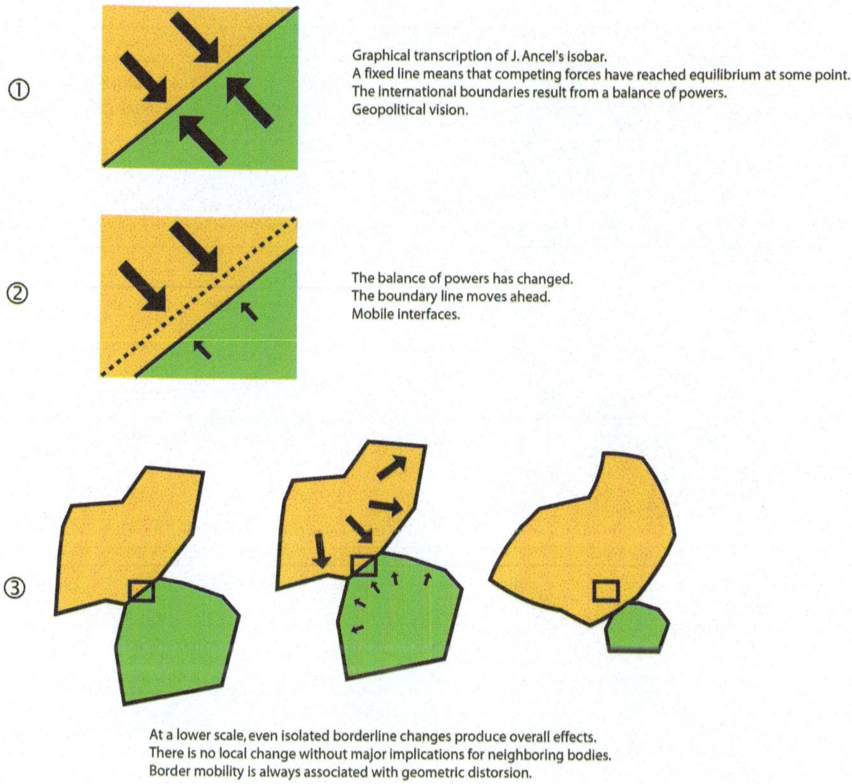

Fig. 1.4 Pressures and moving boundaries. *Source* R. Darques, 2015

Two cases may exemplify the varying condition of "land confines":

1. The *Pashalık* of Ali, the notorious governor of Jannina (present Greece) born in Tepelen (today's Albania), is a historical and geographic oddity. At the beginning of the nineteenth century, Ali manages to obtain by force and opportunistic maneuvering the "autonomy" of Epirus—an area placed under his governorship by the Sultan. Such "autonomy" had already resulted elsewhere in the establishment of (semi-)independent nation-states at the expense of the Ottoman Empire. Unfortunately (for Ali), after a series of withdrawals, the Sultan decides not to drop his arms in front of these ambitions. He sends his army to bring back Epirus into his domain and orders to get him the Pasha's head. The military defeat of Ali in 1822 leaves Jannina almost destroyed. The disaster explains a long-term decline and division of the region whose implications are still tangible today. In other places, such colorful adventures have produced a whole bunch of heroes obsessed with the idea of liberating lands and nations from the oppression of "foreign occupying forces." There is no map left of Ali's dominion. It seems he never worried about how to settle landmarks to his self-declared "domain."

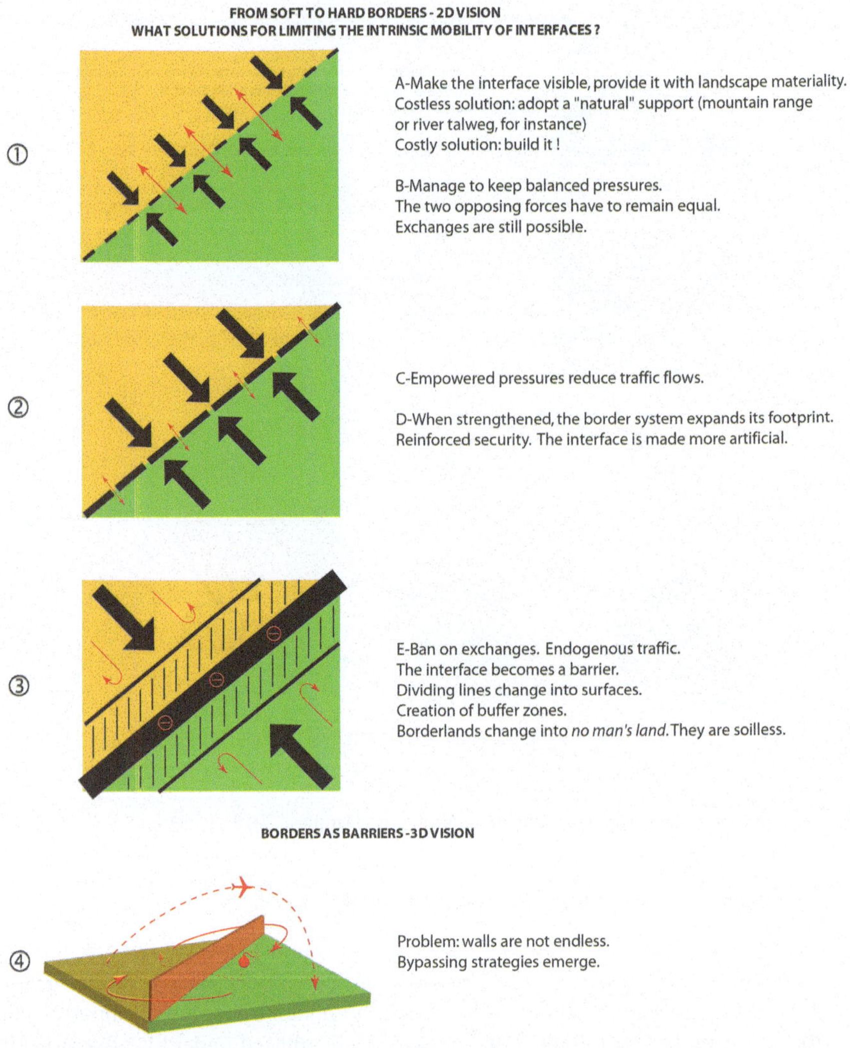

Fig. 1.5 Fixing, strengthening, and forbidding borders. *Source* R. Darques, 2015

2. The second case is a reverse example. Mount Athos is anything but the short-lived *undefined* political body of Ali. The boundaries of this religious institution have been legally fixed a long time ago. Nobody can deny the remoteness of Mount Athos. The isolation of the "Republic of the monks" is the outcome of a planned program. The third *finger* of Halkidiki is protected by a 1000 years' boundary that has survived all invasions. The sovereign Peninsula has witnessed the birth and death of dozens of states, Empires, and despotates, successfully preserving its privileges and self-governorship. Local selfgoverning bodies can thus discreetly

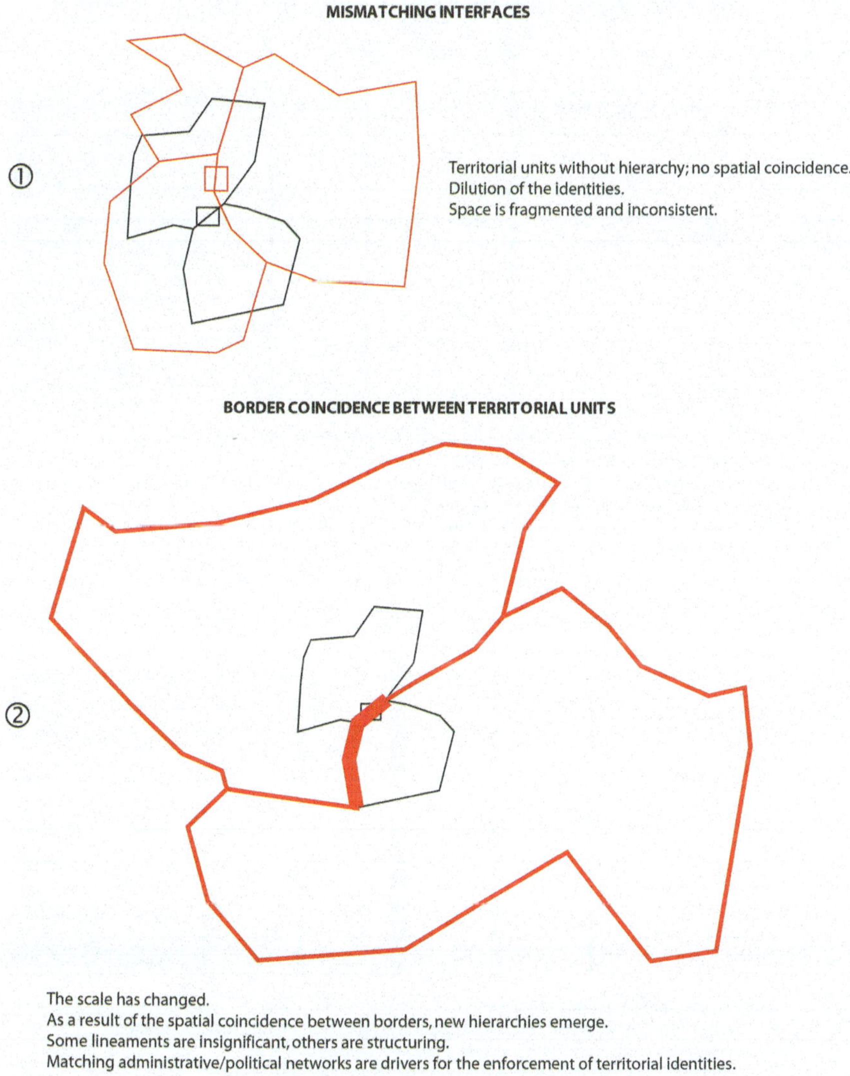

Fig. 1.6 Overlapping and coinciding borders. *Source* R. Darques, 2015

overcome powerful states. Remoteness often leads to decay; it sometimes produces "outlandish," yet resilient territories.

Focusing on such geopolitical oddities changes the too-common view of lonesome conflicting nations. Land confines add deepness—in every sense of the word—to geography. Borderlands are "young" areas. Gained through deadly fights and established as step-forward acquisitions, the farthest provinces have always been opposed to the "old" central regions where core nations settle down first, closely

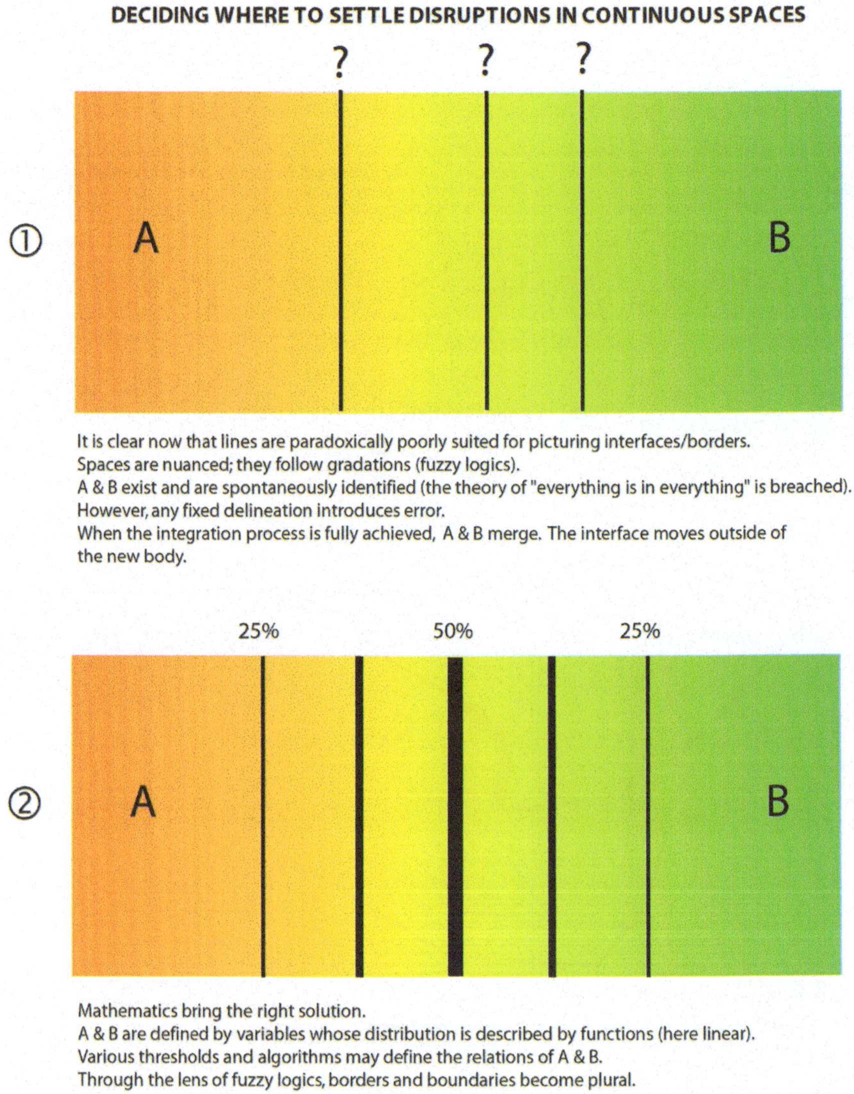

Fig. 1.7 Dilution and permeation. When arithmetics meet geometry. *Source* R. Darques, 2015

organized around their capitals. Young territories are wild. They are often accused of resisting the "central" authorities, whatever imperial or national these might be. The *Liberation* of land and people is the official way of qualifying the forced introduction of a new social and political order. In the face of higher-level authorities, local communities have no other choice but to submit themselves to external jurisdictions.

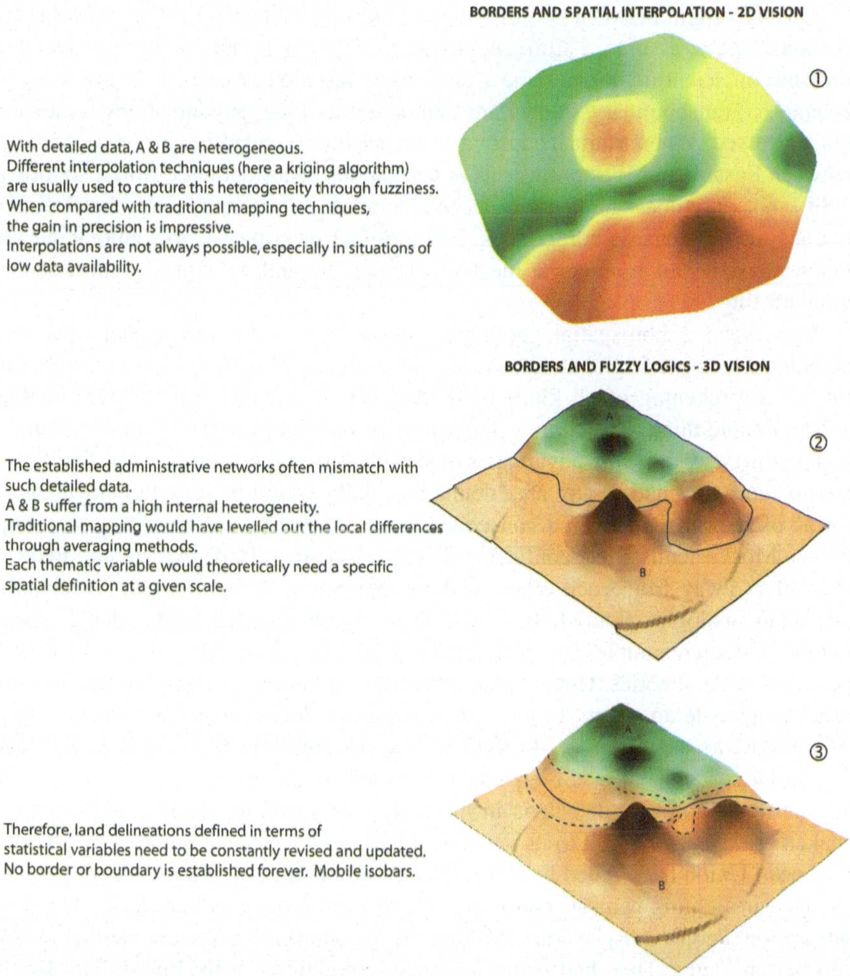

Fig. 1.8 Thematic mapping, interpolated boundaries, and fuzzy logics. *Source* R. Darques, 2015

Army troops have the prime responsibility for imposing law and order before stepping aside in favor of civil administrators. This territorial integration often meets an economic collapse of all gained provinces—the process is supported by mass population movements. Out of their *youth*, the Balkan countryside lends its remaining forces to cannibalizing metropolitan areas and slowly falls into early *old age*. Borderlands are quiet. They keep secrets, untold stories, and cultivate unlikely landscapes. As a key tool for land control, mapping is closely monitored by central authorities whose spatial perception is selective. Once over-represented when they were a goal of territorial ambitions, land "confines" sink into oblivion when placed under the authority of new governments.

Usually caught between two investigation levels, national and supranational, the *Balkan Studies* assume a difficult position. National frames apply considerable pressure on scientific communities. They force research projects to follow buoyed channels. Transboundary studies are authorized as long as contentious issues are not addressed. When Maria Todorova examines the so-called *bridges* or *transitions* between Eastern and Western worlds in her famous historiographic work (Todorova 2009), not only does she deconstruct the Orientalism associated with the *Balkans*, but she also disregards borderlands. Few original or critical works embracing the Peninsula dare put frontiers at the focus center. Should we consider the issue as uninteresting?

Nowadays, a non-spatial geography subscribes to the widespread—but disputable—concept of "globalized world" and embraces the idea of a vast movement toward a weakening of all kinds of borders (Ohmae 2005; Sassen 2015; Ohmae 1995). People think that *daddy's geography* has stretched to its limits: No country could praise any more the advantages of self-identity and self-awareness. The *extra-territoriality* attached to any area defined spatially would be a myth denied by all forms of hybridization. As a consequence, geography would have become ubiquitous. Most scientists feel confident when they compare lands that are hundreds of kilometers away from each other, without distinction. Long-distance correlations and snap analogies flourish here and there, ignoring what makes local places unique. Counterexamples are disregarded when they contradict well-established, yet often futile, theories. Understanding the Balkan borderlands implies adopting an exactly opposite approach. In these writings, *insignificant* anonymous places shall be brought to the forefront. Geography has to concentrate on in-betweens and interstitial spaces whose self-existence is denied by all sides.

The implosion of Yugoslavia in the early 1990s and its "local" conflicts have shaken the civilized world to its very foundations. Normalized members of an open European Union have failed to understand the stubbornness of these "tiny" peoples engaged in fighting against each other. There are so many examples of successful integration. People must be crazy! Putting fire to countries for obscure identity claims and barren lands? They had better accepted international help; Eurodollars would build bridges between ethnic groups one day soon. Is the Eurovision Song Contest not a good way of reconciling hostile communities? Folklore exhibitions and cross-border cultural cooperative efforts provide the opportunity to set up so fair events.

Against the so-called *balkanization* process, the European integration is presented as an overall solution. Countless programs are achieved by Brussels: FEDER, PHARE, SAPARD, and other LEADER are now proud logos people easily recognize. Field observers know that most of these initiatives may produce useful public amenities and help the development of remote districts. However, the host countries are almost unknown to foreign backers, as if space was a blank page. To "fill in the gaps" and encourage any transboundary "cooperation," decision-makers prefer starting from zero thus reducing land planning to a varnishing work.

The making of *transnational regions* is obviously a challenging project, although the Peninsula still gives rise to feelings of contempt or reject (Todorova 2009; Garde 2004; Prévélakis 1994; Bartov and Weitz 2013). M. Todorova mentions a few ideas that come together in people's mind when referring to the *Balkans*: "exasperation before complexity, marginality, unredeeming backwardness, passivity and fatalism, cultural narcissism, fragmentation and war ad nauseam." The most recent years have also brought soft communism, unachieved market economy, semi-democracy, and illegal trade traffics. This makes for an explosive cocktail. The present volume precisely aims to bypass such caricatures. Fears and mistakes often find their origin into misunderstandings. Unlike all other official spokesmen, scientists have to circumvent public conventions.

The past two decades have changed the way in which our field studies are conducted. Renewed geospatial tools now allow land survey and categorization where "no information" was the inevitable answer any researcher could get from his conversation partners. "No map, no question, no trespassing" used to be the rule, but times have changed. Filling in the gaps left by decades of government and self-censorship and saving entire regions from the oblivion to which they had been confined supposes to develop unorthodox research strategy. Out of this, hopefully we can expect to help change stereotypes.

References

Abrahamsson C, Gren M (eds) (2012) GO: on the geographies of Gunnar Olsson. Burlington, Ashgate, Farnham

Agnew J (1994) The territorial trap: the geographical assumptions of international relations theory. Rev Int Polit Econ 1(1):53–80

Agnew J (2010) Still trapped in territory? Geopolitics 15(4):779–784

Ancel J (1938) Géographie des frontières. Gallimard, Paris

Andersen DJ, Klatt M, Sandberg M (eds) (2012) The border multiple: the practicing of borders between public policy and everyday life in a re-scaling Europe. Ashgate, Farnham

Ateş S (2013) The Ottoman-Iranian borderlands. Making a boundary, 1843–1914. Cambridge University Press, Cambridge

Bartov O, Weitz ED (eds) (2013) Shatterzone of empires. Coexistence and violence in the German, Habsburg, Russian, and Ottoman Borderlands. Indiana University Press, Bloomington

Black J (1997) Maps and politics. University of Chicago Press, Chicago

Boria E (2013) Geographers and maps: a relationship in Crisis. L'Espace Politique 21(3)

Brenner N, Jessop B, Jones M, MacLeod G (eds) (2003) State/space: a reader. Blackwell Pub, Malden, MA

Crampton JW (2009) Being ontological: Response to postructuralism and GIS: is there a disconnect? Environ Plann D: Soc Space 27(4):603–608

Crampton JW (2010) Mapping: a critical introduction to cartography and GIS. Malden, MA, Wiley-Blackwell, Chichester, U.K

Crampton JW (2011) Cartographic calculations of territory. Prog Hum Geogr 35(1):92–103. doi:10.1177/0309132509358474

Dodge M, Kitchin R, Perkins CR (2011) The map reader: theories of mapping practice and cartographic representation. Hoboken, NJ, Wiley, Chichester, West Sussex

Elden S (2010) Land, terrain, territory. Prog Hum Geogr 34(6):799–817

Fakinos A (1972) Zone de surveillance. Seuil, Paris

Forsberg T (1996) Beyond Sovereignty, within territoriality: mapping the space of late-modern (geo)politics. Coop Confl 31(4):355–386

Foucher M (2012) L'obsession des frontières. Perrin, Paris
Garde P (2004) Le discours balkanique: des mots et des hommes. Fayard, Paris
Goodchild MF (2006) GIScience ten years after 'Ground Truth'. Trans GIS 10(5):687–692
Gottmann J (1973) The significance of territory. University of Virginia Press, Charlottesville
Green SF (2010) Performing border in the Aegean. On relocating political, economic and social relations. J Cult Econ 3(2):261–278
Harley JB, Laxton P (2001) The new nature of maps: essays in the history of cartography. Johns Hopkins University Press, Baltimore, Md
Herrschel T (2011) Borders in post-socialist Europe: territory, scale, society. Ashgate, Burlington
Hv Houtum, Kramsch OT, Zierhofer W (eds) (2005) B/ordering space. Burlington, VT, Ashgate, Aldershot, Hants, England
Khazanov AM (1994) Nomads and the outside world. University of Wisconsin Press, Madison
Kolossov V (2005) Border studies: changing perspectives and theoretical approaches. Geopolitics 10(4):606–632
Leszczynski A (2009a) Poststructuralism and GIS: is there a 'disconnect'? Environ Plann D Soc Space 27:581–602
Leszczynski A (2009b) Quantitative limits to qualitative discussions: GIS, its critics, and the philosophical divide. Prof Geogr 61(3):350–365
Longley P, Goodchild M, Maguire D, Rhind D (2010) Geographical information systems and science. Wiley, Chichester
Ohmae K (1995) The end of the nation state: the rise of regional economies. Free Press, New York
Ohmae K (2005) The next global stage: challenges and opportunities in our borderless world. Wharton School Pub, Upper Saddle River, N.J
Olsson G (1991) Invisible maps: a prospectus. Geogr Ann 73(1):85–92
Olsson G (2007) Abysmal: a critique of cartographic reason. University of Chicago Press, Chicago
Paasi A (2009) Bounded spaces in a 'borderless world': border studies, power and the anatomy of territory. J Power 2(2):213–234
Perkins C (2004) Cartography—cultures of mapping: power in practice. Prog Hum Geogr 28 (3):381–391
Pickles J (2004) A history of spaces: cartographic reason. Mapping and the Geo-Coded World, Routledge, London
Power D, Standen N (1999) Frontiers in question: Eurasian Borderlands, 700–1700. Palgrave Macmillan
Prévélakis G (1994) Les Balkans. Cultures et géopolitique, Nathan, Paris
Rankin KJ, Schofield RN (2004) The troubled historiography of classical boundary terminology. University College Dublin, Dublin, Institute for British-Irish Studies
Robinson AH, Petchenik BB (1976) The nature of maps: essays toward understanding maps and mapping. University of Chicago Press, Chicago
Ruggie JG (1993) Territoriality and beyond: problematizing modernity in international relations. Int Org 47(1):139–174
Sassen S (2013) When territory deborders territoriality. Territory Politics Gov 1(1):21–45
Sassen S (2015) Losing control?. Columbia University Press, New York, Sovereignty in the Age of Globalization
Schuurman N (2000) Trouble in the heartland: GIS and its critics in the 1990s. Prog Hum Geogr 24(4):569–590
Schuurman N (2006) Formalization matters: critical GIS and ontology. Ann Assoc Am Geogr 96 (4):726–739
Smith DM (1977) Human geography, a welfare approach. Edward Arnold, London
Todorova M (2009) Imagining the Balkans. Oxford University Press, New-York
Tuathail ÓG (1996) Critical geopolitics. Routledge, London
Wood D (2010) Rethinking the power of maps. Guilford Press, New-York

Chapter 2
Hidden and Showcased Boundaries: A Field Approach

Abstract This chapter places special emphasis on demonstrating the growing number of boundaries and checkpoints throughout the Balkans over the past 20 years. The Peninsula has abandoned enclosure models inherited from the Cold War while converting itself to a seemingly open space facilitated by the expansion of the EU. Most national borderlands are no longer the repulsive spaces they used to be. However, the scheme is not that simple, as evidenced by many field experiences. Have these remote areas succeeded in becoming proper districts?

Keywords Border crossings · Traffic routes · Connectivity · Remoteness · Underdevelopment · Field experiences · Landscapes

Various means of transport may be used to cross boundaries: train, bus, boat, plane, and car—or, more simply, foot. I have personally experienced all these means during my travels in the Balkans. Boats and planes tend to undermine the idea of frontier, both literally and figuratively. Instead, land facilities make it much easier to keep contact with some "grounded" geography. With their dedicated equipment, custom posts are outdoor theaters dropped by distant authorities lost in the middle of *nowhere*.

Usually, people have limited contact with the surrounding landscapes. This is obvious for geographers. Hundreds of times people drive on the same routes, paying less and less attention to their environment, once stuck in a boring routine. Markings and signs invade roadsides and make drivers focus on selected targets Putting this phenomenon into mapping would show that, during their daily trips, 95 % of all travelers practice up to 5 % of the area they theoretically "occupy." For practical and behavioral reasons, man is a gregarious animal. Within cities and across open fields, invisible walls surround us and get rid of our exploration instincts. Have you ever compared the distances covered during mountain bike excursions, even if you practice this sport irregularly, to those covered by other daily journeys? A short off-road distance lets you discover more landscapes than ever. Frequency and variability of paths are here key concepts in understanding the effective limits of mobility.

© Springer International Publishing Switzerland 2017
R. Darques, *Mapping Versatile Boundaries*,
Springer Geography, DOI 10.1007/978-3-319-40925-2_2

The same restraints apply to boundaries. Most travelers gather around a limited number of checkpoints open to traffic and follow funnels peacefully. Aligned trucks and cars waiting for their transit permission are a perfect illustration of this phenomenon. Border crossings are intimate places. The extent of open spaces is limited, at least where Schengen regulations do not yet apply. Out of these, authorized boundary "practices" feed wrong feelings among anonymous travelers. Such assessment makes it necessary for geographers to distance themselves from the idealized pictures peddled in everyday conversations and the media. External perceptions are obviously biased.

2.1 Crossing Frontiers: *Balkan* Times and Distances

Exploring the Balkans is the best possible way of overcoming myths and achieving demystification. Practical experiences quickly demonstrate that time and space have different meanings in Eastern Europe and in the most *developed* parts of the world. To the open and frenetic global world, the Balkans oppose a territorial rootedness whose qualities are best expressed in slowness and duration.

How long would it take to travel from Salonika to Istanbul in the late 1990s? About 14 h are necessary to cover the 600 km between both cities by rail; buses would lessen the penalty to 12 h. Besides, we must not forget the waiting time at borders. When a dozen or so Turkish "microbusiness" women fill up the bus holds and corridors with large colorful handbags, negotiating customs duties turns out to be a time-consuming process (Egbert 2006). Let us add 4 h of tough bargaining. This latency leaves time to consider *another* Turkey, visiting back and forth tumbledown buildings. Did you think you would have enough time to choose a comfortable place to stay, once arrived in Istanbul? Forget it: Leaving the Macedonian city in the early morning, the trip ends in the middle of the night. No alternative remains but to jump into the bed of the first crumbling hotel or taste the seats of the bus station. Welcome to the Balkans!

What about taking a break for the weekend and spending some good time on the Black Sea coast on the occasion of another stay in the Greek Macedonian metropolis? The best solution would be to reach the Bulgarian capital, Sofia, and then to catch a connection with Burgas. The first part of the journey is easy: A special bus company is contracted to the international itinerary. This enables travelers to avoid the rival and wheezy train of the same line (300 km, 6 h). The trip is only slowed down by a 1-h pause at the Bulgarian border. Nothing to be afraid of in comparison with the 3 or 4 km of aligned trucks waiting for their permission to cross the boundary. Most passengers of the bus face complicated personal and professional situations that need to pay special attention to their visas. Once arrived in Sofia, connecting the train station is easy. Antiquated wagons offer hospitality to an already-tired traveler. Burgas should have been reached in 6 or 7 h. But this is without counting on the failure of the locomotive engine, which results in the train being stuck in the middle of the Balkan Range for three more hours. Aren't travels supposed to shape youth?

Using personal vehicles would not reduce the inconvenience. Until recently, is public transports brought substantial benefits. They were slow but costless. The connected network was extensive, whether considering a "post-communist" bus or its "liberal" counterpart in Greece or Turkey.[1] The situation has changed after the early 2000s. The number of private vehicles in circulation has soared, showing all too clearly the limitations of outdated road networks. The experience is painful in summer in the tourist areas of Dalmatia, when huge crowds from northern Europe attempt to reach the astounding parks, cities, and beaches of Croatia behind the wheel of their heavy sedans. *Highway* is here an overestimated name—the same applies for *speedway*. The courtesy shown by Croatian and Slovenian custom officers, usually less nitpicking than their Albanian or Bulgarian colleagues, is a small compensation.

What about a bus trip to Tirana starting from Athens or Jannina? This stands in contrast to the traffic congestion observed in Dalmatia. Few vehicles used to travel these roads until the late 1990s. As a result of heavy European investments, the road to Kakavia (main Greek–Albanian entrance) has been improved and widened. Until recently, taxis, old buses, and *furgona* used to be standard travel means in Southern Albania, a region almost entirely abandoned because of mass emigration. Twenty four hours were needed to cover the 800 km between Athens and Tirana. The bus route was submitted to various police controls and, in line with the customs' charge, the transit at Kakavia could be very time-consuming. During the Albanian revolt of 1997, the itinerary was no longer secure, and traffic happened to be closed—a measure whose consequences weighed on the Albanian working migrants who used these bus services.

There would be many other personal experiences to relate. Traveling in the Balkans is not that easy. Far from improving local communication systems, the growing number of boundaries has made them worse over the past two decades, especially in the Western Balkans. Not surprisingly, the fastest access roads (Table 2.1) concentrate along the lines between Zagreb-Belgrade and Sofia-Istanbul. These road sections accounted for the bulk of investment on infrastructures. In contrast, the emergence of new boundaries and the relative abandonment of a poor secondary road network have made day-to-day travels more and more problematic.

Tirana is a special case. The Albanian capital city is certainly one of the worst places to reach by car or bus from any other Balkan capital. Already deprived of long-distance train lines, the city is also affected by ill-adapted road networks. Covering the 300 km from Skopje may require 9–11 h, which is indeed a physical challenge. The Skopje–Tirana connection is made through the Radožda–Cafasan border post, far to the south of the country, next to the Ohrid Lake.

[1] In terms of comfort, there was no significant difference on both sides of the Iron Curtain. Most bus fleets were made of old units manufactured in the 1970s–1980s.

Table 2.1 Border crossing: time–distance between Balkan capital cities in 2012

Trip	Train	Bus	Car/taxi	Distance (km)	**Bus time/distance**
Zagreb–Belgrade	7 h 13	6 h	3 h 53	396	**0.91**
Sofia–Istanbul	12 h 55	9 h	6 h 15	563	**0.96**
Athens–Skopje	12 h	11 h 30	7 h 21	700	**0.99**
Athens–Sofia	18 h	14 h	9 h 33	791	**1.06**
Sofia–Bucharest	9 h 34	7 h	5 h 45	388	**1.08**
Skopje–Belgrade	9 h	8 h	4 h 30	429	**1.12**
Prishtina–Belgrade	–	7 h	3 h 48	352	**1.19**
Athens–Istanbul	19 h 30	22 h	11 h 54	1094	**1.21**
Sarajevo–Zagreb	9 h 27	8 h 20	5 h 22	414	**1.21**
Zagreb–Ljubljana	2 h 20	3 h	1 h 46	143	**1.26**
Prishtina–Podgorica	–	5 h 30	3 h 35	262	**1.26**
Athens–Tirana (via Ioannina)	–	17 h	11 h 21	803	**1.27**
Podgorica–Zagreb (via Dubrovnik)	–	15 h	9 h 37	694	**1.30**
Podgorica–Belgrade	8 h 40	9 h	5 h 40	416	**1.30**
Sarajevo–Belgrade	10 h 10	7 h	3 h 58	305	**1.38**
Skopje–Prishtina	3 h	2 h	1 h 09	85	**1.41**
Sofia–Skopje	9 h 47	6 h	3 h 30	220	**1.64**
Belgrade–Bucharest (via Timişoara)	13 h 17	17 h	8 h 23	615	**1.66**
Podgorica–Sarajevo	–	7 h	3 h 42	240	**1.75**
Tirana–Prishtina	–	11 h	4 h 54	347	**1.90**
Skopje–Tirana	–	11 h	4 h 08	299	**2.21**
Tirana–Podgorica	–	–	2 h 25	181	–

Time–distance between Balkan capitals (correspondences and customs time not included, traffic and road conditions idem)
Single border crossing by line. Very hypothetic and approximate timetables. Some information is passed: The Greek railway network discontinued its international service in 2011
Source Train and bus public and private agencies, *ViaMichelin* time and distance calculator

The time/distance ratio is one of the worst in Europe. The problem arises mainly from the low charge and state of the road. Along this route, “international” journeys look more like backcountry tours: Fortunately, villages and other rural landscapes are particularly worth seeing. Under these circumstances, crossing the boundary has to be somewhat “dramatized”: Modern buildings emerge from the countryside as cathedrals consecrated to national independence. No matter if unlikely shepherds are the only observers of the show. Doors have to be impressive and solemn. Here, states operate.

The “semi-independence” of Kosovo and the independence of Montenegro have clearly worsened the bilateral communications. Beyond the tensions, clashes, and militarization processes that typically accompany the creation of new "international" boundaries, one of the first practical effects of such institution building structures is a growing impediment to all exchanges. Between Kosovo and Serbia, new "gates" have been built along local roads under the form of workmen's huts. Installed in 2001, the custom point of Merdare has changed the way of life of many former Yugoslav citizens. Everything or everyone is now submitted to frequent inspections: Specialized officers proceed to the clearance of goods; they check working licenses and vehicles' papers, luggage, and collect taxes. Space has shortened, and time has lengthened. This is not the least contradiction in the Balkans to watch the simultaneous settlement of new customs and the implementation of road/highway projects, as if both initiatives were just completing each other. The Morinë–Merdare highway that shall join the Albanian boundary to Serbia through Prishtinë is an example of such confusion, considering transport efficiency at a middle-scale level. The trip from Belgrade to Prishtinë would eventually take 7 h by bus: not the worst time–distance ratio indeed, despite obstructions.

In sum, the time–distance analysis reveals the discrepancy between the effects of an enhanced Balkan fragmentation and the development objectives assigned by the European Union. Fortunately, thanks to national and international funding, major road sections have supported significant improvements, such as the Tirana-Prizren connection. This journey has undergone a precious time shortening: The former 10 h drive has decreased to 4 h. Planning policies, international relations, and territorial development are closely linked, especially in these fragmented, yet contiguous Balkan districts.

From the 1990s onward, significant improvements were made to the Balkan communication system. Apart from the specific case of former Yugoslavia, the paranoid atmosphere that permeated the Cold War period now belongs to the past. Barb wires, antitank mines, military exclusion zones, and bunkers have gradually vanished from the landscapes they used to cover for half a century. In interstate relations, the remaining weaknesses of transportation facilities are insignificant when compared to this heavy past.

2.2 Closed Versus Open Balkans: Political Interfaces and Circulation Spaces

Between confined and open space, the Balkan Peninsula has long been consumed. Starting in the 1940s, the region has been mired in ignorance, mistrust, and prejudice. For 50 years, with few exceptions, the centralized powers have managed to strengthen their outer influence while creating extended and underdeveloped, although secured, backcountries. Despite appearances, this forced "peaceful" period does not contravene the rules of a troubled history. This fictitious interlude has simply constrained interstate opposing forces to remain under cover.

The current shape of the Balkan political map results from two key events that have thoroughly altered local political frameworks. The first major disrupt goes back to the Balkan Wars and WWI. In a few short years, the Ottoman and Habsburg Empires[2] yield in front of the allied forces. State nations impose their independence and achieve to create sovereign perimeters. The final defeat of the Empires in Vienna and Istanbul against the league of "local" forces puts an end to a century of uninterrupted political upheavals. Between 1912 and 1918, two new states came into being: Albania and the Kingdom of Serbs, Croats, and Slovenes. Montenegro, one of the oldest provinces engaged in the fight for its independence, unexpectedly gives up its status of kingdom and joins the First Yugoslavia. Such a renouncement is exceptional. Meanwhile, Serbia, Greece, Bulgaria, and Romania achieve significant territorial gains. In the end, the Peninsula faces a complete restructuring.

The second major turnaround takes place one century later. Between 1990 and 2000, the Yugoslav Wars and communist collapse "unfreeze" the balance of forces inherited from the Cold War. The Yugoslav Federation and the Eastern Bloc resign themselves to accepting another range of profound changes. Out of former Yugoslavia, in just a few years, seven additional states (if we include Kosovo whose status is still unclear) find their way to independence. Across the region, on the Black Sea coast, two Republics return to full sovereignty: Moldavia and Ukraine—plus the Pridnestrovian Republic whose issue is also debated.

Unraveling the Ottoman and Russian legacies is thus a lengthy process. Up to now, the Balkans are in the thrall to the prejudice of an endless fragmentation. With their 11 000 and 14 000 km^2, Kosovo and Montenegro hardly reach the size of English or US counties. They match two French *departements*. We will deliberately ignore Transnistria (4 000 km^2) and the Bosnian *entities* that are even smaller. Between imperial mastodons and microstates, no balance point could be found.

As a consequence, boundaries proliferate. Wide-open imperial lands are gradually replaced by a spatial confinement, which in turn produces a patchwork of isolated geographic units. Referencing and mapping border crossings is a work that shall exemplify the transition from open to closed models. Few original sources enable the production of such documents.

The mapping strategy applied hereunder (Figs. 2.1, 2.3 and 2.4) relies on a simple principle. A broad definition of the *Balkan area* has been adopted, which includes Moldavia and leaves Turkey aside. The survey area is set to match the present political map. Within this perimeter, the aim is to identify the existing border crossings.

In 1908, the overall number of border crossings reaches 54 units. By the end of the 1990s, the amount increases to 73—an acceptable change. In 2013, the area receives 273 road crossing points, without considering the "unofficial" posts on the external face of Kosovo and Transnistria, and leaving aside 15 other scheduled gates. Obviously, the Balkans have faced a silent revolution that has deeply altered its transportation model, leading to thorough changes of border management systems.

Several reading keys may be forwarded. The differing quality of map sources is a set parameter to exclude. Usually, the road networks pictured on old map sheets

---

[2]A third candidate could be added to the list: the Russian Empire.

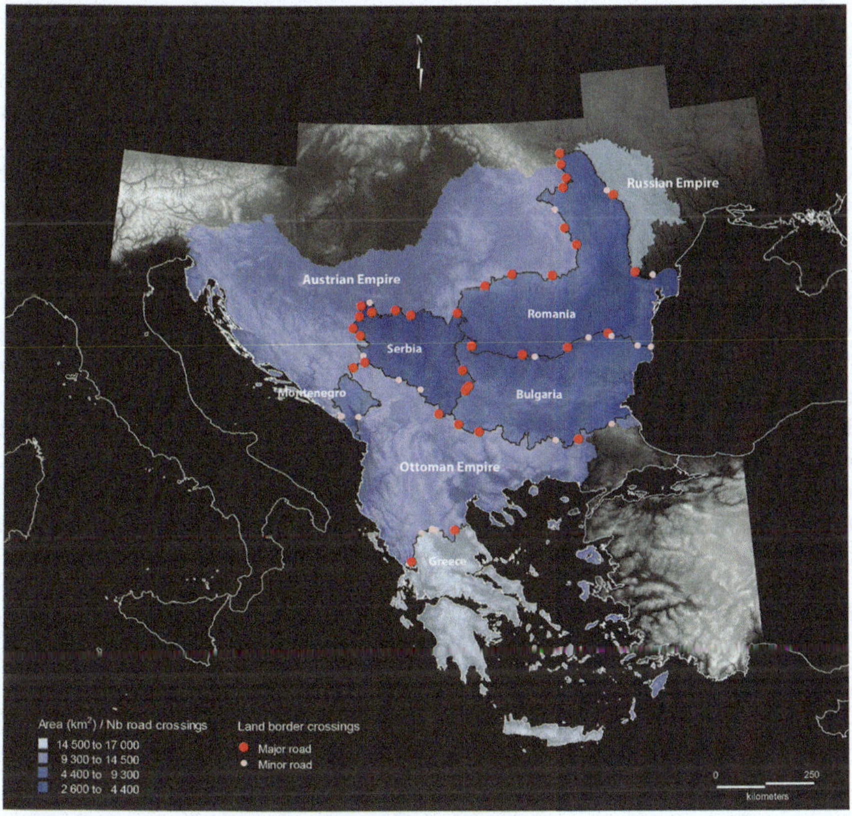

Fig. 2.1 Road border crossings in 1908. *Source* Vogel, C., *Die Balkan Halbinsel*, 1908, 1/500.000, Harvard Digital Library, DTM NASA SRTM program, author's own GIS databases

have to be taken with every precaution for various reasons, including data harmonization and production process. At the same time, few documents cover the whole Peninsula until Moldavia. Finding even and reliable data that would include the independent nation-states to the list of Austrian, Ottoman, and Russian Empires is a real challenge. Fortunately, the 1908 mapping of C. Vogel is exceptional (Vogel 1908). This source thoroughly identifies railway lines and main road ways.[3] For later sources, the information is weighed down by a high fragmentation, but details have been cross-checked using different map providers to avoid undesirable mistakes. Which lesson should we drive from this large and recent increase in the number of border crossings?

[3]The location of bridges is, however, difficult to determine because no symbol indicates these artworks. Data are less reliable on this point. Ferry lines are also disregarded. They can be activated or closed quickly depending on political or economic conditions. They usually follow seasonal demand.

The last Ottoman and Austrian imperial maps present what would be considered *unauthorized view* of the Balkan area in contemporary terms. In 1908, the transportation system is already upside-down: The railway revolution has helped many places to overcome their isolation. New transport networks come along with modernized administrations. Agricultural outputs find renewed export markets, and other leading manufacturing activities develop throughout the backcountry on both (Fig. 2.2) faces of the Empires. Besides, remote mountain communities often demonstrate a high demographic vitality.

Are the 54 road checkpoints of 1908 an effective indicator of the state of international relations? Not really. In the Central Balkans, the first motorized vehicles will be abandoned by the retreat of foreign military troops at the end of WWI, with few exceptions. The road network inherited from the Ottoman administration is almost missing. The Balkan countryside is a kingdom of the mule. Few vehicles find limited intercity road circuits. At least two-thirds of the rural districts are sidelined from the transport revolution of the time. As a result, most military reports written during WWI prefer to measure distances in time. Reaching the fighting fields of the Eastern Front implies that the French and Austrian armies

Fig. 2.2 A time-honored tradition: the *uncontrolled* fighting groups. Paramilitary militias: Albanian in Kosovo, 2001; Bulgarian Cheta (Komitadji) around Kastoria, 1902. *Sources* www.lostbulgaria.com, www.hoxha.at

build their own carriage ways. In terms of military control, sovereignty, traffic management, and taxation, settling international boundaries is a difficult task. Fifty-four checkpoints are insignificant compared with countless unreferenced ways unsuitable for modern vehicles. The fence model whose ultimate enforcement will take place during the Cold War finds no application here. However, endowed with full political and legal capacity, the imperial borderlands follow local rules that hardly comply with the injunctions of remote authorities such as Vienna and Istanbul. Smuggling not only plays a significant role in the local economy, but is also a way of living (Fig. 2.2).

According to the 1908 map, Western Montenegro is deprived of any mappable road that might connect the country to the Austrian and Ottoman Empires. Similarly, hundreds of kilometers of watercourses are deprived of bridges, a situation that should not hinder the importance of other river crossing means, randomly controlled by the authorities. When no barbed wires hinder the mobility of people and when newly settled boundaries are supposed to divide villages linked by centuries of organic ties, official checkpoints play a fairly symbolic role.

With 17 entrance gates, Serbia is centrally located. No map could describe more efficiently the strategic place of this country on the eve of WWI than Fig. 2.1. Serbia is at that time the only landlocked state of the Peninsula. At mid-distance from Istanbul and Vienna, it pays its crossroad position by plentiful transit points. On the northern side of the Dinaric range, the kingdom controls most routes connecting the heart of both Empires. Comparatively, the total number of road gates in the remaining Ottoman Rumelia only climbs to 16 points for an area three times larger than Serbia.

Obviously, the process of expanding boundaries directly impacts the newly established nation-states whose main policy is to increase the number of checkpoints to improve their financial situation. These emerging political–territorial entities also have to maintain/enforce their relationships with the remaining parts of the Empires where many "fellow citizens" still live. This provides for the collection of necessary taxes devoted to the establishment of a fully operational public administrative system. The Empires used to live under the watchword "good fences make good neighbors." The forthcoming nations build their independence on a reverse principle. The attitude might seem schizophrenic: Wars instigated with the intent to fix *hard* boundaries against outer competing enemies would drive afterward to an increased border porosity? Although surprising, this observation is definitely validated by the analysis.

The border model imposed after WWII (Fig. 2.3) is different. Tens of Russian military maps of the late 1970s to enable us to observe the transition from an "organized porosity" to a full confinement. Of course, these data are not readily comparable to the previous ones. The Russian topographic coverage stems from aerial photography, thereby allowing a more detailed examination. With a time lag of 60 years, almost no correlation between border crossings can be found. There has been a wholesale change.

As usual, observing the earth from the sky enables to look at it differently. All border crossings were picked up and classified according to their importance

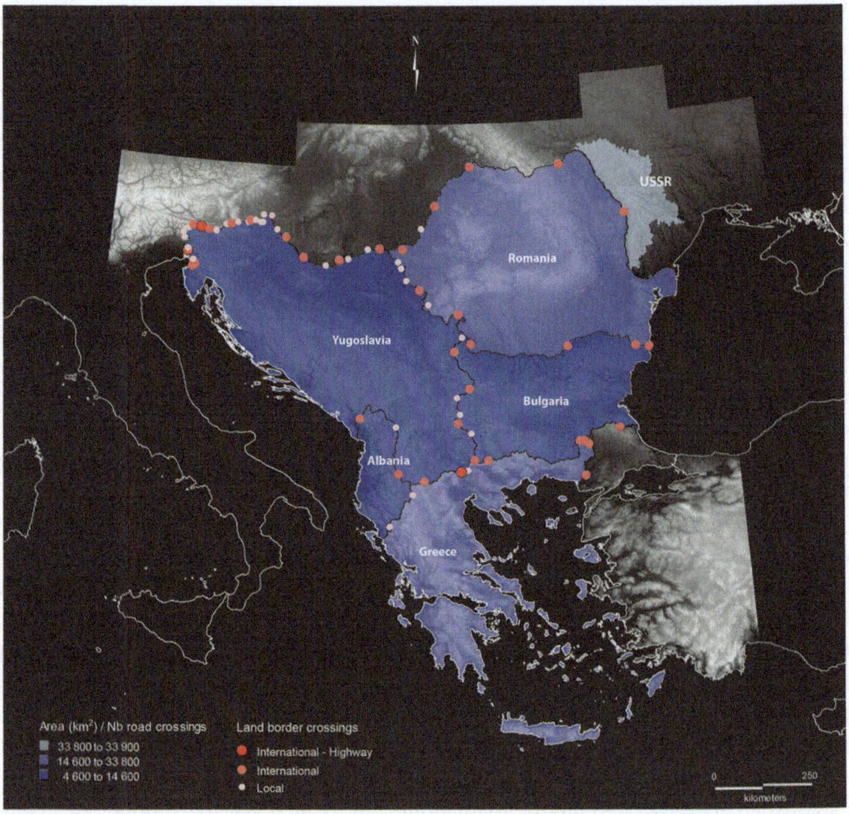

Fig. 2.3 Road border crossings in 1990. *Source* Michelin, Hallwag and other national travel maps, late 1980s, 1/700 000 and 1/1 000 000, DTM NASA SRTM program, author's own GIS databases, ESRI, Russian military maps, late 1970s, 1/500.000

(width, road surface, and hierarchical rank). This led us to identify about 300 points placed across roads and railway lines. The existence of such facilities does not mean that all of them were in use and proper working condition. Few of them were actually in service as a consequence of the Cold War. The old system is reversed. Most transport networks are neutralized to channel people and goods toward a limited number of custom points.

Roughly speaking, the Iron Curtain and the paranoid atmosphere attached to it find different applications that depend on local contexts. No global rule prevails. Each boundary has characteristics of its own. For instance, the Greek–Albanian border covers about 250 km and intersects two asphalted roads and four tracks closed to traffic. International trade and exchange are reduced to a minimum. In contrast, 24 connections punctuate the 300 km of borderline between Vojvodina and Romania: 6-11-7? cart tracks. Is this disparity caused by specific topographic conditions (lowlands vs. mountains)? Probably not, or as a secondary reason.

Both cases involve past imperial lands living for long in close proximity and submitted to a complex population mix. The economic policy applied to these outer spaces is a determining factor. Always denounced by travelers and scientists of the late nineteenth century, despite the Tanzimat reforms, the Ottoman Empire suffered from an acute underdevelopment in front of its Austrian competitor. The train could reach a very limited number of places in remote Ottoman districts. On the other hand, the Pannonian Plain is covered with railway stations that duplicate an already-developed road network. In addition, human densities are not even comparable. The first Yugoslav–Romanian boundary settled down at the end of WWI cuts into pieces neighboring communities distant from only few hundred meters. It sometimes divides urban districts. Roads and railways cross many times the same boundary as if mapmakers had ignored these networks when they were urged to define new state limits.

Between both cases, local and environmental contexts differ greatly. The Greek–Albanian borderlands never turned their backs on underdevelopment, whereas cross-border relations in Vojvodina kept thriving on past achievements. The picture would not be complete without mentioning two points:

- In contrast to other land sections, river boundaries suffer from a limited number of bridges. Danube, Drava, and Sava still play a dividing role they never gave up.
- The main transport lines stretch in an east-west pattern. They avoid crossing rivers and appreciate the natural corridors determined by topography.

Figure 2.3 clearly highlights the specific position of Yugoslavia. Unlike Greece, Albania, Bulgaria, Romania, and Turkey, which maintain distant relations in keeping active two or three checkpoints with each neighbor scattered among hundreds of kilometers of barbed wires, Yugoslavia shows a remarkable openness. The whole Federation appears as a transit platform between Western and Eastern Blocs. The Northern Slovenian border with Austria and Italy is very specific: Twenty official checkpoints are registered there in the late 1980s.

On the contrary, the highest fences and *coldest* frontiers lurk in the least expected places. Albania is almost a full isolate—a predictable condition indeed. Although tied to the Eastern Bloc and surrounded by military allies of the Warsaw Pact, Romania only keeps three road crossings with Moldavia and Ukraine along a 1 200-km interface. Similarly, Greece barely maintains two paths along the Evros River, despite its common membership with Turkey to NATO. Cold War fences are thus simple inputs to a complex border system. The ins and outs of this system are to be studied further on. As mentioned earlier, the existence of a checkpoint does not necessarily imply a permanent use. Mapping flows would reveal major disparities between posts. Some of them have been kept virtually closed to traffic over decades.

Once again, Fig. 2.4 shows an altered pattern. The boundaries' length increases as if submitted to inflationary pressures. The communist collapse and economic transition; the European integration of Bulgaria, Romania, Hungary, and Slovenia; the Yugoslav Wars; and Moldavian upheavals have produced cumulative effects in a few short years. These events have changed the face of the area. In contrast to previous periods, a clear setback can be seen. Until then, Serbia/Yugoslavia played

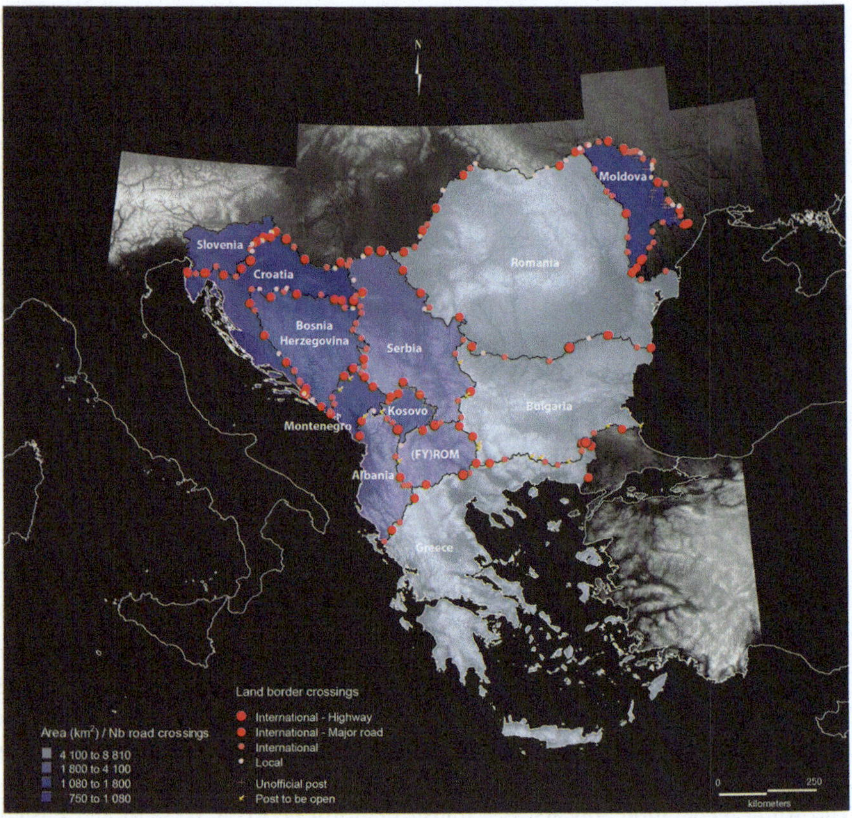

Fig. 2.4 Road border crossings in 2013. *Source* Freytag & Berndt and other national travel maps, 2004-2005, 1/400 000 and 1/500 000, updated with Google Earth, DTM NASA SRTM program, author's own GIS databases, ESRI, Home Affairs Ministries

a key role in the area. For the first time, the most *open* countries are located on the northern edge: Moldavia, Croatia, and Slovenia are perforated with regularly spaced holes. The independence of Moldavia, for instance, has turned many former internal roads of the USSR into international highways—the increasing number of border crossings thus expresses a maintained degree of territorial integration. In Slovenia, a similar process leads to an opposite result: In 2007, Slovenia and Hungary enter the Schengen Area and withdraw their corresponding custom posts. The same would eventually happen north of Croatia. In such cases, the rising number of border crossings is a preliminary step before a "linear control" resignation.

In Bosnia, Montenegro, Kosovo, and Serbia, the set of circumstances is different. The current international (or assimilated) boundaries divide lands that used to maintain close relationships. As in Moldavia, many gates are remnants of an integrated transport network whose strings were cut off. Paradoxically, the communication breakdown produces a growing number of paths. In response to its serious internal problems, Bosnia faces a similar *revolution*. The former Federative Republic went through the twentieth century without experiencing any border control on its edges. Even though, in 2013, the country was watched by 42 official road crossing points, without considering the inter-entity boundaries that are not classified as *international,* whose paths are not consequently mentioned on the map. Surprisingly, despite worsened relationships, Fig. 2.4 clearly emphasizes the long-term links between the Federation of Bosnia/Croatia and Republika Srpska/Serbia.

The Balkans are a particularly tricky issue. Any state independence initiates an immediate border opening that fosters a "vital breath" to the newborn country, as if political divides and military struggles did not compromise territorial versus economic neighborly relations. In 2013, the Ministry of Internal Affairs of Kosovo officially referenced thirteen standing and two temporary checkpoints on the outer face of its tiny "autonomous region."

In the Southern Balkans, other conditions prevail. Until recently, Greece, Albania, Bulgaria, and Turkey had to advertise their desire for integration while fulfilling the European candidate countries' criteria. Unlike what happened in former Yugoslavia, most neighboring *partners* had maintained (very) limited relationships. These confines had been deserted for long. To overcome the past monopolistic position of the military authorities, other revolutions had to occur in silence. Since the 1990s, the number of border crossings never stopped growing, thanks to European funds made available for the establishment of new offices and access facilities. Heavy foreign investments are therefore necessary, because no significant transport networks existed before.

In 1990, Greece enlisted only nine checkpoints on its northern boundaries. Half of them were almost free from any traffic. Twenty years later, six modern gates have been built in the middle of nowhere. Three extra ways are scheduled to serve Bulgaria. Between this last country and Turkey, in Eastern Thrace, three custom offices are to be created, preparing a real revolution: Traditionally, the gates of Kapitan Andreevo and Malko Trnovo concentrated the bulk of traffic. Three local connections aim to initiate renewed neighboring relationships between the divided Bulgarian and Turkish communities. The action is more symbolic than effective—for the moment.

This first approach to the Balkan complexity may end up with a rough classification of today's borderlines:

1. The old national boundaries are currently submitted to a salutary opening. After one century of limited exchanges, they receive expensive facilities that deeply alter closed landscapes—high mountains surrounded by green *corridors*.
2. The former "internal" borders of the Federations (Yugoslavia, USSR) suffer from a painful enclosure artificially limited by the creation of many checkpoints. However, fostered by abundant road crossings, neighborly relations experience severe withdrawal symptoms.

2.3 A Four-Seasons' Approach to the Balkan Borderlands

Frontiers are not only places to cross, they are also districts to stay, forests to manage, perimeters to observe and control, needy villages and communities to support, or trade centers to develop. Surprisingly, the most common image associated with borderlands is that of deserts. When people think about them, they instinctively slip from dense central cities to lazy countryside until they finally imagine empty spaces far from anywhere. In this sense, curious onlookers are strange fellows who fix their attention on places that would appreciate remaining anonymous—if we could impugn motives to them. Many—not to say all—outer districts are "hurt places" by centuries of continuous conflicts. And fighting armies usually leave barren soils behind them.

Such portrayals obscure many aspects of spatial marginality (Green 2005). Most observers focus on what is collectively considered essential. The human mind finds easy ways to shrug off unnecessary information. Geography as a scientific approach to land understanding is thus a discipline that has to bridge knowledge gaps and investigate spatial confines and mind frontiers on a systematic basis. Representations and measured facts have to confront. The account of some personal field experiences may illustrate the wide variety of local situations beyond clichés, which allows us to discover a confined world of migrants, soldiers, traders, and tourists (Gelbman and Timothy 2011; Wachowiak 2016). Readers shall excuse the futility of recalled personal memories. These experiences gathered year after year are, however, relevant for introducing our theme. They have an irreplaceable value to our understanding of local atmospheres.

As far as I remember, I never stopped hanging around borderlands. Even when I thought I was far distant, they found ways of being in the world through crooked means. I have begun to travel the Balkans when the area was at a crossroads of its history, at the beginning of the 1990s. Each country of southeast Europe in its turn made world headlines after decades of *quiet* existence. The former communist countries were undertaking their revolutions. They revealed Europe still had neighborhoods beyond the Iron Curtain. Unfortunately, economic and political collapse allowed crawling struggles to resurface.

2.3.1 *June–July 2000: Summer Memories of Dalmatia and Bosnia*

Despite the media frenzy over the Yugoslav Wars, little information had fused out of the country about the deep geographic changes happening in the background of the fights. For a decade, unless they belonged to military forces, diplomatic corps, or humanitarian organizations, external investigators were invited to postpone their works to better times. A few reporters bravely tried to make up for the heavy military silence that would have prevailed without them. Beyond the confusion, the internal restructuring of the area was so thorough that years would be necessary to issue a valid report on the ongoing spatial dynamics. A world was disappearing.

The end of the Kosovo War is dated June 1999. One year later, an opportunity presents itself to spend lengthy periods in Croatia and Bosnia. Seasonal tourist attractions and depressed feelings of people in the backcountry help to breed a strange atmosphere. In this hot summer of 2000, the inner Dalmatia combines the beauty and solitude of crop fields and natural spaces to the striking relics of human violence. The end of the Bosnian Wars goes back 4 years ago. Enough time has passed to let *Mother Nature* replenish, but the time frame is too short to hide the scars left on building facades and to help casualties of the fighting rest in peace. Along roads, gravestones and oratories mark places of painful losses. The Croatian authorities are obviously struggling to hide war relics.

Crossing the boundaries of Slovenia, Croatia, and Bosnia is unproblematic. Around Trieste, Slovenia—and especially Istria—appears as a close neighborhood of the European Union. Easy controls allow for a fluid transit of tourist flows to southern destinations. Entering Bosnia is made quite untroubled. At Metković, police officers carry out an ordinary inspection. They show little concern. Their colleagues are much more interested by frequent speed controls that provide hard currencies and taxes. The ticket amount is negotiable. Obviously, despite recent army fights, the former Yugoslavian states have not decided to build high fences around them. The situation is weird, as if free traffics had survived the wars, as a lasting legacy of the previous regime. To the driving traveler, few landmarks indicate the demarcation line between Bosnia and Croatia. Asphalted and dirt roads lazily linger along *undefined* borderlines.

The strange cohabitation of careless tourists and "hurt populations" reaches its acme in Sarajevo and Mostar. Everywhere warfare has left permanent traces. Apartment buildings ruined by heavy artillery still stand beside entrenched military bases. A kind of outrageous voyeurism makes shop traders sell videos of the past struggles. Although both cities stand away from the new international borderlines, land and human divisions appear here most clearly. Demobilized soldiers and roaming groups of young men pace the streets of Sarajevo. Bosnian military patrols cross the way of peacekeeping troops, showing obvious provocation. Tensions are running high enough to let anybody feel uncomfortable, years after the events. When the air can be cut with a knife, the borders trade their conceptual dimensions against a full practical meaning.

Croatian and Bosnian landscapes are rich in contrasts. Dalmatia appears under a shining sun, while the Adriatic Riviera receives hundreds of thousands of people running away from their rainy homelands. During this season, the mineral landscapes of the Adriatic coast offer coppery glints owing to red soils and crimson vegetation. Scattered along the coastal strip, charming cities standing on white walls and covered with red-tiled roofs melt with mineral environments. Everyone spontaneously understands that water supply is here a critical issue. The karst maintains a high photosynthetic activity in the internal basins until early July, then crops come to maturity and, unless being watered, all fields sink into drowsiness until the first rains of autumn.

The Plitvica Lakes are an unspoilt island made of preserved Mediterranean forest. With around one million visitors each year, this old National Park is among the most renowned resort places of Croatia, after coastal cities. The area highlights the inconsistency of post-Yugoslavian borderlands. A few kilometers west of the Bihać enclave, almost no sign of the boundary is visible at first sight. A closer look reveals the backsides of this tourist resort. Eight kilometers west of the Plitvica Lakes stands the ruined aerial base of Željava, destroyed in 1992—one of the biggest military bases of Yugoslavia. The first years of the Yugoslav Wars have determined the fate of the region. Following the downfall of the Republic of Serbian Krajina in 1995, the area has endured a complete renewal of its population with the departure/expulsion of most Serbs and the reverse installation of Croats from Bosnia. This is at least what statistics report. In fact, inland villages oppose an almost deserted face to the overcrowded Riviera. Many expatriated workers take the opportunity of few marriages to visit what remains of home and family.

The siege of Bihać has lasted almost 3 years (1992–1995) and has left extensive minefields as a legacy. At a crossroad between the Pannonian Plain to the north and the Adriatic coast to the south, surrounded by Serbian, Croatian, and Muslim Bosniaks villages, the area mixes all of the ingredients and wears the full complexity of land confines. Both Ottoman and Austrian Empires had made this region a solid demarcation area. Deep forests owe their existence to a former military status that has shaped the landscapes (Magris 1999).

2.3.2 August 2006. The Pannonian Plain Around Pécs and Vukovar

Luck would have chosen me to visit the northern Balkans in late spring and summer. Even during the hot season, the confluence zone of the Drava, Sava, Tisa, Danube, Mureş, south of the Pannonian Basin, is *another world*. Croatia is such a country that shows bipolar face. To the bare rocky soils of Dalmatia and varied mountain slopes of the Central Dinara succeeds an endless plain where each rising terrain might be considered a considerable summit. Nowadays, the area gathers most productive forces of four countries that share this piece of land. Large urban areas, industrial complexes, transport means: Everything essential finds here place to develop and participate in retaining high population densities. The Great Danubian Plain already smells like *Mitteleuropa*: endless horizons, continental climate, trees, and water competing for land mastering and, above all, dreary crop fields as far as the eye can see. To the extreme variety of the Southern Balkans, this marchland opposes the same boring landscapes over hundreds of kilometers.

Because of this contrast, time seems to follow another pace. Distances are covered much faster. Developed transportation networks ensure rapid connections and trips. Locked behind his car or train window, the observer soon finds advantages to losing himself in a book rather than comparing the beauty of succeeding wheat and corn fields. The Habsburgs must have felt at ease around here, doubtlessly.

The overvalorization of low-lying areas (Pannonian Plain) and the reverse underdevelopment of the uplands (foothills of the Dinaric Alps) is quite a recent geographic reality. Until late—the eighteenth or nineteenth century—depending on local conditions, regularly flooded areas, countless tributary streams of uncontrolled rivers, dead meanders, and infested ponds had an extremely repellent effect on human presence and activities (Timár et al. 2008). The most repulsive swamp area —before its general improvement—is caught between Danube and Drava, north of Osijek, amidst today's Croatia, Serbia, and Hungary. Rambling rivers are everywhere, despite facilities that would stabilize streams. Water mastering is highly risky: By definition, quaternary sediments are difficult to control, especially when water tables reach ground levels. Adopting river thalwegs for defining international boundaries is obviously a venture.

Two parks preserve today the remains of an old natural paradise/human *nightmare*—the so-called Baranja. South of Mohács, the Duna-Drava Park is at the crossroad of Croatia, Hungary, and Romania. It extends over 50 000 ha. The Kopački Rit Park covers 18 000 ha a few kilometers downstream. As for the Plitvica reserve, it belonged to the Republic of Serbian Krajina and, until the signature of the Erdut Agreement in 1995, the Serbs claimed their right on it with a large strip of land that includes Vukovar.

Beyond the quietness of fauna and flora reserves peacefully paced by summer tourists in search of "authenticity"—and good wines— fierce fights have opposed the Croatian and Serbian forces in the first half of the 1990s, leading to general ethnic cleansing. Almost totally destroyed by bombings in 1991, Vukovar is a worldwide symbol of the first crimes committed in former Yugoslavia. Ruined buildings and stolen lives are common destiny in borderlands. Restored fields cleared from mines have quickly erased any sign of the turmoil. Crumbled houses take a longer time to be eliminated.

That a battle took place in Eastern Slavonia is a small wonder. So many fights between the Ottoman and Austrian troops had happened here. The early and late Byzantine Emperors had already made the Danube River and its surroundings a contact zone to manage cautiously. The interdiction applied to any farming activity was a way to secure frontiers from external attacks. Such politics have produced large unpeopled areas along the Lower Danube and remarkably forested zones upstream (Power and Standen 1999; Panzac 1999; Fuchs et al. 2015). Nowadays, after centuries of Ottoman–Austrian "cross-Danubian" troubled relationships, natural parks seem to be the modern expression of an old function. An appropriate protection provides new purposes to repellent districts. Centuries of struggles, riots, army fights, and mutual plunder have facilitated the process of overall depopulation better than any surreptitious and devastating flood.

The region was in 2006 at a turning point in its history. Hungary had joined the EU in 2004. Many former communist countries of Eastern and Central Europe had finally found a successful conclusion to their European application. Romania would enter a year later. Hungary would also join the Schengenland in 2007. A European flag flapping in the wind at a checkpoint in Mohács was a true revolution. People easily forget the obstructions this region had faced during the Cold War, when contacts between neighboring lands and human beings were strictly limited (Schubert 2011). The event brings us back a century ago. The last time Baranja had enjoyed such freedom goes up to the 1910s, in the last years of the Austrian Empire.

For now, both national parks "freeze" a contested *tri-border* between Croatia, Serbia, and Hungary. West of Bezdan, as the legacy of an undetermined past, the boundary rambles east and west of the Danube course, playing a hide-and-seek game never to find a coincidence with the actual mainstream. "To bargain: hundreds of hectares of wild animals, preserved pieces of forest, dozens of water ponds for leisure fishing. Don't forget your favorite mosquito lotion (if you are lucky, an unexpected water pollution would replace it for a while). Please make an offer." Why make things easy when they can be complicated? River mainstreams and borderlines do not match accurately. More to the east, Szeged and its other trigon show an absolute discordance between boundary and thalwegs.

2.3.3 1998–2006: December in Epirus and Southern Albania

Most people discover the Balkans during summer holidays. Memories usually associate sunny islands and sandy beaches with other blue lagoons where multilingual crowds casually gather. On the contrary, as a researcher and land rover, avoiding summer rushes has always been essential for me to carry out my work in good conditions. Borrowing in early July an overcrowded KTEL bus in Greece may be a painful experience. It often happened to me to take such buses during 7 or 8 hours, breathing thick clouds of cigar smoke without air-conditioning, by around 40 °C. Exhausted by the drive, travelers quickly learn to program their trips off-season. The Balkan mainland produces different overall impressions in late fall and winter.

Epirus and Southern Albania are places where I have spent months between 1998 and 2006. The late 1990s were a tough period of economic collapse in all post-communist countries. Albania, (FY)ROM, and Bulgaria were at a low ebb. The debate about (FY)ROM's use of the name *Macedonia* and the *Sun of Vergina* fixed on its flag had led to an economic embargo applied by Greece to the former Yugoslav Republic. Albania and Bulgaria were managing some industrial reconstruction and vast land reforms. The *transition* revealed the existence of exhausted administrative systems. The emergence of new political parties and leaders had been long delayed. Renewed opposition forces were developing powerful positions—including mafias. Year after year, people had to leave their countries en masse toward Italy, Greece, or other European destinations hoping to enter a promising labor market. Those unable or unwilling to quit homelands—many elderly people among them—were facing hard living conditions. Change was not a short-term *evil* but a program meant to last.

In 1997, Albania experienced global troubles that have made much an impact on people's thinking. The *Pyramid crisis* had caused massive unrest, mainly in the south of the country. It revealed the corruption degree of an entire political class and the economic potential of mafias. The population had been brought on a course of suffering and despair for having placed its confidence into financial "devices," in the hope of restoring some purchasing capacity. The speculation fever had adopted a rough primary *East European* form, not so far from the unrest that would catch later the entire continent and lead to the great depression of 2008.

Chaos is the appropriate term to use as the designation for Albania's condition after the pyramids' fall. *Civil War* is an exaggerated word, considering the limited number of human casualties. However, entire cities had fallen. Hunger strikes, destruction of public buildings, state emergency, SHIK troops fighting against students, evacuation of foreign citizens, military barracks plundered by rebels: For a time, the situation had escaped any control. In a few weeks, a weakened but quiet country had slipped into anarchy. Even Tirana was touched. Its population keeps in mind the raids few armed groups used to organize on the capital as a challenge against incapable government authorities. Foreign officials did not leave Tirana without armored vehicles. A few months later, the same did not understand how an unknown *Frenchie* could cross Albania as anyone else by taxi, bus, or *furgon*.

Against such confusion, the neighboring countries had to take protective decisions and close their boundaries in order to avoid any contagion and secure outer districts. The great crisis of 1997 demonstrates the power of armed gangs able to run entire regions in full view of legal authorities. Thousands of weapons had vanished into the air that would be used later in Kosovo and (FY)ROM. The revival of local gang leaders who managed to impose their own law in the streets and, behind the threat, to rake ever-greater profits from illegal trafficking is not far from bringing us back one century ago, just before the fall of the Ottoman Empire. When states totter, other pre-existing and discrete powers take the advantage.

The Iron Curtain had survived its own ashes for 10 years. The political instability of Albania and the semi-anarchy of its southern regions, sanctioned by a terrible human and economic decay (Darques 2004), had made the northern Greek borderlands sensitive areas to survey. Few intrusions of Greek paramilitary units claiming the *liberation* of Northern Epirus have been reported in southern Albania. Cross-border incidents never completely ceased between both countries after the Balkan Wars. Tens of thousands of Albanian migrants left their homeland each year hoping to reach Greece and enter its job market, keen on using cheap and regardless workers (Sintès 2010). The flow had to be mastered *a minima*. Regularly, for political reasons, policemen used to carry out mass arrests within the Hellenic cities and in the countryside and chartered buses with illegal residents. As soon as expelled at Kakavia, the latter immediately tried their luck in crossing back the boundary.

March 1998. From Jannina, I plan to visit the small town of Konitsa, 8 km away from the Albanian boundary. The Balkan spring is late. A clouded sky makes me seek shelters to run in at the next rainfall. The KTEL bus brings few pensioners to their homes, with packages brought from the main city of Epirus. Heading a flat plain surrounded by forested mountains, Konitsa is a motionless village, trapped in time. The Ottoman place is still here, although heavily damaged by the elements. Ruined houses, bathes, and mosques left by Muslim Ottomans and/or Albanians scatter the upper city. The buildings are quite bulky, outward signs of some past wealth. Obviously, the *Athenian miracle* (Burgel 2002) has not reached these confines. Low fields stretching downwards let us imagine the past fertile *çiftliks* of the Ottoman dignitaries and the host of landless workers placed under their jurisdiction. From this past world, few things have survived. The main winner of this century-long experience is undoubtedly the forest. The *kafenion* itself is not delightful. I discuss with an old peasant woman who brings back home large faggots for firewood. I had better go back to Jannina before the night.

I did not think the return would be so colorful. KTEL buses stop at every station, even in the countryside. Dribs and drabs, people continue getting on the bus—mostly young men without luggage. All but a handful of empty seats of the outward journey are now occupied, until two consecutive police controls restore the balance. Between seat ranges, everybody stays calm. Those who cannot produce official papers are invited to get off by a head movement. The policeman puts one only question: "Where are you from?" When he reaches my seat, he asks me to show three different ID papers. Unbelievable! A French citizen in Konitsa during March. To avoid deploying too many forces, the officers observe any movement along the main road axes. People who cross the border on foot shall in the end meet these roads hoping to reach their final destination. Once leaving mobile controls behind them and entering the "heartland," their trips will be safer. Weariness is on all faces: Ceaselessly dodging around and trying to outwit each other are exhausting tasks, on both sides.

December 2001. A seizing coldness gains the capital of Epirus where I am staying for weeks. Athens knows urban chaos because of sudden snowfalls. In Jannina, these are not unexpected. The Pamvotida Lake is half frozen and heavy snowflakes hide the Clock Tower. Higher up Pindus, Metsovo is covered by one-meter-high white coat. To avoid traveling under dangerous conditions, I prefer to make a pause during the weekend and visit few places along Thesprotia. Some friends encourage me to plan a trip. The heating system of my old car can hardly warm up the passenger cell. But the day is bright. Waiting for the western part of the *Nea Egnatia* highway to be completed, I proceed on the old national road which leads to Igoumenitsa, roughly following the Kalamas River. Epirus is one of those mountain regions that have supported large rural depopulation after WWII. The Civil War did great damages, insofar that most villages never fully recovered from the loss. The general downfall made cultivated fields and pasturelands come back to wild forest over a few years. Those who succeeded in surviving fights and famine and managed not to run away had come to settle in Jannina. Present Epirus thus inherits the overwhelming weight of its regional metropolis (around 60 000 inhabitants), beside an almost deserted countryside.

When entering Thesprotia, watchful eyes may observe singular landscapes. More than half of all villages are ruined. Few people, especially tourists, imagine crossing a wounded countryside. Vegetation has enwrapped the demolished buildings for long. Almost all Albanian-speaking and Muslim Greek citizens of the region have been forced to leave home in 1944. Thus, Chameria had broken up with a famous ethnic-confessional intermixing (Hart 1999). One rural–urban migration later, most former arable lands are still left on their own. No land reform has occurred in the meantime, and no official cadastre has ever been implemented. The only places resisting decay are populated by return migration Greeks that spend their savings in German or neohellenic-style private houses. Generous benefactors from Australia or America take special pride at financing public monuments. Despite this, hundreds of remote hamlets deprived of any electrical connection or decent road access have disappeared.

Its market place has avoided a slump to Filiates. The old Ottoman town is more active than other *kaza* centers transformed into head cities of *eparhies*. Igoumenitsa has stolen its leading position to Filiates, thanks to its port facilities and access to Corfu. Albania is not far away. As a reminder of some past competition across the Iron Curtain, Albanian radios easily reach the Hellenic coast and conversely. A few years ago, listening to subversive Western music on a basic HF radio could be heavily punished on the other side of the boundary. Times have changed. For now, Filiates is asleep, but as soon as nature recovers its rights in spring, the Albanian migrants will return to their jobs in farming and tourism.

One week later, under equivalent weather, I come back to visit the village of Tsamanda, north of Filiates, in direct contact with the Albanian boundary. On the narrow road to this *end of the world*, surrounded by wild vegetation, Tsamanda tries to survive with little means. It welcomes an ethnographic museum that relates rural life as it used to be "before." Conversations in the local *kafenio* are going well. Tsamanda counted three hundred houses in old times. It lives today with only thirty to fifty permanent residents. The endless Civil War has blocked the course of development more tangibly than any boundary. However, border paths closed by years of military rule have left bitter memories. Despondent people think to the closest "Albanian" village, Lesinitsa-Leshnicë, within a 1-h walk. They have provided the museum of Tsamanda with curiosities and objects of rural life. People proudly say that Leshnicë was the true head-village of the region, and that Filiates was almost a no-place before. Who remembers today the mountain slopes and valley bottoms covered by fruit farming and wheat fields, the fat sheep flocks grazing the upper summits? Everybody has left a long time ago. Only two buses climb every week to keep ageless people in contact with modernity.

Night is coming soon in winter. Back to Jannina, I am once more stopped by police officers. Papers are useless this time. Questions come from all sides to unsettle. "Where are you from? What are you doing here? Why do you speak Greek? Where do you go?" My car is submitted to a close search. "Didn't you cross the Kalamas Bridge this afternoon"? Obviously, my trip had been watched. Even during the cold season, or because of my unexpected presence, border controls are tough. Other stays shall see me accidentally take part to man hunting operations in the nearby forests.

Winter in Gjirokastër is also worth living. In 2004, coming back from a stay in Tirana, I decide to take a break and spend a few days in the second largest city of the former Ottoman *Vilayet* of Jannina. An exhausting bus trip leaves me with a single idea in mind: Finding a bed and getting some sleep. As usual, my decisions are not programmed. No booking was made; all hotels appear to be closed. I have to make my choice quickly. At five o'clock in the afternoon, the sun falls below the horizon and coldness settles down. Fortunately, a sign placed in the upper corner of an anonymous building brings me some hope. I will eventually stay in military barracks. A few leks handed out to a young soldier, and tomorrow will be another day.

The night was eventful. After preparing myself a bed, I realized the common room appointed to me did not have heating. Albania made a routine out of it. People are accustomed to being subject to power cuts. In particular vulnerable to low incomes, people have learned to live in unheated houses. In the middle of the night, a group of young men joins the dormitory. I will get some sleep another night.

The morning sunlight finally delivers the city from its darkness and brings the hope of a comforting thaw. The place awakes with the muezzin's call to pray and bell rings of the clock tower. Colorful traditional houses appear. A small shop offers me a good breakfast and a delicious "cappuccino" with cinnamon. A walk in the upper town helps me to get warm. Gjirokastër and its surrounding settlements are asleep. A few years ago, the Greek consulate just down the castle was besieged every day by emigration candidates and those who did not manage to get legal permission forced the boundaries on foot to escape Albania at all costs. The result is now obvious. Gjirokastër is a quiet city. Compared with its countryside, it keeps minimal activity, thanks to its museum-place status and a few inherited administrative responsibilities. Maybe some lost tourists still end up here? All remains of the former regime are down: industrial plants, collective farms, middle-scale workshops, and almost all public facilities are so heavily damaged that no traces of this communist past will stand visible for long.

The entire Drino Valley fights the elements alone. The river and its tributary streams bring high amounts of sediments that overwhelm low fields almost everywhere. For a long time now, nobody matters with this. Maybe reacting to the desertion of the valley and to the disastrous consequences of land reforms, Gjirokastër firmly hangs over its rocky slopes. A new district is under construction, north of the city. Private houses and tiny apartment buildings emerge from a chaos of concrete, in the middle of open-air waste fields and cratered roads. The market is not too far, where people gather on this Sunday morning.

Back to Jannina, a few days later. Unlike its Albanian equivalent, the city of Ali Pasha is surrounded by commercial areas that stretch over kilometers: gas stations, supermarkets, taverns, construction materials, farm machinery, and other miscellaneous shops spread out as far as the eye can see. Fortunately, the place enjoys a shimmering lake, the island of Nisi, and a lovely fortified castle. Jannina is not on the UNESCO world heritage list, but it is worth visiting. The citadel was almost destroyed in the early 1820s, as a punishment inflicted by the Ottoman troops sent by the Sultan to defeat Ali of Tepelen.

I have spent enough time in Jannina. I could walk along its streets with blindfolded eyes. My old car is still waiting for me to end my usual inquiries among the Greek administrations. To breathe fresh air, I plan to visit the southern suburbs of the metropolis. Anonymous neohellenic villages scatter around the lake, absorbed by farming and residential functions. Koloniati is a settlement located on top of a rock. It gathers twenty families. Not worth lingering here.

Unexpectedly, this winter afternoon would bequeath to me lasting memories. Driving downhill, I slow down to avoid an old man wrapped up warmly in a thick coat, walking along the way a wood stick in his hand. Hearing my arrival, all of a sudden, the man turns back and steps sideways until the middle of the road, arms wide open to stop me. "People are crazy around here," I say. Dumbstruck, I lower the car window. Hidden behind a wool hood, I do not see his face. I simply notice his large stunning white mustache, poking out of the cloth.

"Where are you going?", he asks.

I did not remember where I was exactly. How should I know? Almost accidentally—it was the last name I saw on a road sign—I answer:

"Bafra"
"So do I!"

Without any invitation, the old man climbs into the car.

"Good evening first," he says, removing his hood with politeness. "Did you visit the church?"
"Yes, I did."

While showing me the way, a kind conversation begins.

"Where do you come from?"
"I'm French, from Marseille"

A French citizen who speaks Greek seems natural to him. Usually people respond by a kind "ah, ah, Massalia, the little Greek village on the seaside!" He doesn't.

"Do you see this. It's my plot," he says proudly. "There, you have a source at the bottom of the plane-tree." "Stop here. I'm going to show you the old church they are digging."

I had found the perfect guide, the one who knows the region like the back of his hand.

A few minutes and many back roads later, we reach our final destination: the *kafenio* of Bafra, where my new friend seems to be a regular. The old man, keen eye, sits down near a cautiously chosen table and invites me to join him. Immediately, he offers me a *hellenico* (Greek coffee).

"Look, I have a French identity card." He pulls out of his pocket a folded paper where his name is written. Panayiotis L., born in 1915. "You see, I'm a war veteran, a partisan, that's why it's written in French."

From his 87 years old, the *youngster* shows remarkable simplicity.

"Do you have any children?" I ask.
"Four. Two boys and two girls. The first one lives in Koloniati, the second one is a teacher at Katsika,[4] the others are not far away."
"None of them moved away?"
"Oh, on the farm, with two hundred sheep and three hundred pigs there was work for everyone. If only you had listened to that! Real music."
"And you, where are you from?"

Few surrounding villages were peopled by refugees after the *Great Catastrophe* of 1923.

"Of course I'm from here, of Greek descent."
"Did you know Jannina before. What was it like, a little city?"
"Very little. It has considerably changed. Before, there was much water in the *kambos*. Now they have dug many channels."
"And your parents?," I dare ask. Questioning an old man about his parents brings us back to the *Turcocratia*, when this part of Greece was held by the Ottomans…
"My father was a shepherd. At that time, life was hard. The entire *kambos* was in the hands of two or three owners—the *çiftliks*. With Venizelos in 1923, they distributed all fields in divisions, one plot here, one plot there. My family bought a few sheep heads to improve its condition. I was young when I left them to join the army."
"When was it?"
"In 1936. I got married in 1942. I completed my duty in 1951. Fifteen years! Somebody told us "go right," we went right, "turn left," we turned left. I even went to Tepelen.[5] We had to fight against the Italians. The mountains were covered with snow, everywhere."

---

[4]A suburb of Jannina.

[5]Present Albania, north of Gjirokastër.

Panayiotis reminisces about his armed fights.

"Many of my army buddies died. The others are friends. It's important to have friends in this life. But young people today don't know that. When someone goes through hard times, he knows what a friend is."

Frontiers are not only geographic lines. They are people who fix them and afterwards deal with them every day.

"If you seek for me, you'll find me here each evening. We have the same mind, both of us, I think."

2.3.4 *1993–2006: Winter and Spring Between (FY)ROM, Greece, and Bulgaria*

Bulgaria narrowly missed a global crisis in 1996–1997. The situation never got out of hand as it did in Albania, but all necessary conditions were in place. By the end of 1996, a fall of the financial pyramids resulted in a food shortage arousing popular riots. The transition to market-based economy had taken place in a general confusion, as in most CEI countries. A predatory finance developed in the early 1990s, between private sector, state activities, and informal economy. Over the ashes of an already-corrupted communist system, almost all public enterprise assets had come to be plundered. Capital outflows reached alarming levels, and organized mafias gained enough power even to threaten state authorities. All specialists agree in speaking of "unachieved or unmastered transition." In the everyday life, the Bulgarians got used to a mean annual monetary inflation of around 100 % in the 1990s. In 1997, the rate even climbed to 1 000 %. The Bulgarian lev was not already worth buying. In 1996, the GDP falls by 10 %. Even purchasing bread had become problematic (Cellarius 2000). In these difficult and dire circumstances, emigration was envisioned as an obvious solution by most able-bodied workers. Seasonal farm jobs and tourist undeclared services developed into large-scale exercise for most expatriated citizens. Simultaneously, the Western countries grew accustomed to recruiting a skilled workforce at low costs.

As a consequence, Bulgaria was coping with a dramatic decay. Its economic misery was a striking evidence of structural collapse. *Poor* and *old* had become synonymous epithets when speaking about seniors. Without any subsidies, people had acquired the habit of sowing few acres of distributed land in city suburbs. The old "Balkan garden" image associated with Bulgaria over the late Ottoman period erupted once again through the back door. Subsistence farming allowed hundreds of thousands of people to go through bad times. They dug, hoed, and watered plantations by hand: This was surprising in the eyes of foreign observers used to watch big tractors engaged in fieldwork rather than bending human backs. Beside this, land infrastructures, public transport means, and industrial facilities were down. Ruined roads, ageless buses and trains, decaying private houses, devastated collective blocks, and plundered industrial buildings stretched their dreary landscapes everywhere.

I discovered Bulgaria during long stays in Salonika. In the early 1990s, traveling to (FY)ROM was quite risky for a researcher who intended to fulfill his work in Greece. Northern Yugoslavia was won by fire, steel, and tears. Bulgaria was a close and charming destination and, regularly spending months in the Greek Macedonian metropolis, I sometimes needed a change. It gave me the opportunity to put into practice the Bulgarian I was studying at the university. In 1994–1995, visas were still in use. As my trips were decided in a last-minute rush, the entrance permission had to be obtained at the French Consulate in Salonika.

Entering Kulata's gate was an outlandish experience. All cars had to support sanitary disinfection. They went through a basin filled with a disgusting gray liquid, the same juice being thrown simultaneously by hosepipes over windows—the opposite of a car wash, in short. The process had a symbolic force. Later, the border guards checked IDs and attached a paper card to any passport. This document was to be presented to each hotel or residence during the stay. If choosing to use homestay accommodations, local police stations had to stamp the card. The paper was collected back when leaving the country. All foreigners were watched—relics of the old *Balkantourist* system.

Traveling through Bulgaria by bus with a backpack was not difficult in the early 1990s. After arriving at his destination, wherever it was, a challenge awaited the guest: finding suitable accommodation. Nowadays, many private hotels have grown everywhere. At that time of limited human flows, the task was not easy. Cities with 20–50 000 inhabitants often counted no more than one or two hotels. One of them was invariably the old state building. Such venerable institutions applied dual pricing, distinguishing foreign visitors from Bulgarian fellow citizens. The price for a room could reach hundred dollars the night, heating excluded. When possible, I preferred adopting local B&B solutions. It gave me the opportunity to meet chatty old ladies.

Eating was not that easy either. I remember wandering in many streets of small cities in search of a meal. Restaurants existed and kept open doors. But once placed around a table, a long menu in hand, the unexpected client could ask anything. The answer was always the same: "niama"—there isn't. After fierce negotiations, the "cooker" dug up a tomato, a cucumber, and sometimes an onion and named this dish "shopska salata."

On both sides of the Greek–Bulgarian boundary, traffic controls were managed by police forces. Traveling along the borderline was quite difficult. Good maps were unavailable for public use. In Bulgaria, as well as in Greece, low-scale road maps were kept next to high-scale city plans in bookshops. Between both documents, there was hardly anything. Following pathways in remote districts always brought surprises. Many mapped ways did not exist and, on the other hand, many existing roads happened to remain unmapped. All drivers had to plan trips instinctively. Sometimes it worked, sometimes not, and deviations were frequent. Military barracks, observation posts, bunkers, barbed wires, and other defense devices punctuated all roads and spread out in the countryside. The Nestos/Mesta gorges, south of Kulata, were outstanding corridor placed under close surveillance. Shopping in the area was limited to the border point of Kulata itself. The few Greek travelers who dared to cross the boundary were leading to Sofia for business. Buying duty-free cigarettes was their obsession. Across the borderline, Bulgarian shops looked much more appealing to me. Regularly disposed on dusty stalls, "local Bulgarian" products attracted the attention: Sellers offered a great choice of canned food, chocolate, yogurt, famous wines from Melnik, and pieces of pottery at rock-bottom prices.

Things have much changed. How did Bulgaria manage to accomplish such an internal revolution after ten years, celebrating its European integration as soon as 2007? The challenge seemed almost impossible to meet. Athens was doubtlessly the heartland of a Greek *miracle* during the 1950s–1960s. Could Sofia expect to be the primary place of the first Balkan *miracle* of the new century—together with Bucharest?

Between 2005 and 2006, a coordinated research program gives me the opportunity to visit places I had known 10 years before and to discover other pleasant settlements with good fellowship (Darques and Deslondes 2008). Spring instills a breezing atmosphere in the whole countryside. Floodlit landscapes replace rain, snow, and fog. Impressive changes have occurred. The frontier still impulses its rules, albeit with softened effects. North of Drama, a new checkpoint has come into being. Effective normalized relationships now benefit both Bulgarian and Greek people.

Kulata has become the first road entrance to Greece, in front of Gevgelija. Its marketplace is gone. Petrich and Sandanski, two head cities of Pirin Macedonia, have diverted tourist flows. Busloads of Greek travelers are no longer afraid of crossing the boundary to do shopping out of the Euro zone and take advantage of low prices. Chinese goods invade the stalls of open-air markets, beside cheap clothes produced in the surrounding factories. Spas and other *related services* attract many visitors. Enticed by low wages, financial advantages, and other European/national incentives, many foreign assets find here favorable investment conditions. A large number of clothing factories now cover the lower Mesta Basin. The countryside breathes again. One-half of farmlands are still uncultivated and left to wild nature. The majority of young women and men live abroad, and no one knows whether they will return one day, but the European integration undoubtedly brings hope.

Though, past ghosts remain visible here and there. Next to the Iron Curtain, formerly "frozen" settlements are now accessible. They receive modernity reluctantly (Fig. 2.5). Crossed by the fence and its endless barbed wires, the village of Ilinden seems untouched since the Ottoman era (Darques and Deslondes 2008). Ruined Turkish style farmhouses spread everywhere. The most recent public

Fig. 2.5 Traditional farm houses, Openitsa in 1913 (present Macedonia) and Kato Nevrokopi in 2005 (Greece). *Sources* www.lostbulgaria.com, R. Darques, 2005

facilities date back to the 1960s. The mean population age is around seventy, presumably. Somehow through this misery, a handful of people struggle to survive. There is almost no car traffic, just mules wandering through the streets. People around here went through WWII. They remember the first conquering communist period and survived the collapse of the regime. In their eyes, still standing barriers are visible signs of an unfinished war. Are borderlands doomed to remain badlands?

References

Burgel G (2002) Le miracle athénien au XXe siècle. CNRS, Paris

Cellarius BA (2000) You can buy almost anything with potatoes: an examination of barter during economic crisis in Bulgaria. Ethnology 39(1):73–92

Darques R (2004) L'Albanie en transition: mutations démographiques et recomposition territoriale (1989–2001). Espace, populations, sociétés 3:559–575

Darques R, Deslondes O (2008) Bulgarie-Roumanie, les nouveaux horizons de l'Union européenne. Méditerranée 110:136

Egbert H (2006) Cross-border small-scale trading in south-eastern Europe: Do embeddedness and social capital explain enough? Int J Urban Reg Res 30(2):346–361

Fuchs R, Verburg PH, Clevers JGPW, Herold M (2015) The potential of old maps and encyclopaedias for reconstructing historic European land cover/use change. Applied Geography 59:43–55

Gelbman A, Timothy DJ (2011) Border complexity, tourism and international exclaves: a case study. Ann Tourism Res 38(1):110–131

Green SF (2005) Notes from the balkans: locating marginality and ambiguity on the greek-albanian border. Princeton University Press, Princeton

Hart LK (1999) Culture, civilization, and demarcation at the northwest borders of Greece. American Ethnologist 26(1):196–220

Magris C (1999) Microcosms. Harvill, London

Panzac D (1999) Politique sanitaire et fixation des frontières: L'exemple ottoman (XVIIIe–XIXe siècles). Turcica 31:87–108

Power D, Standen N (1999) Frontiers in question: Eurasian borderlands, 700–1700. Palgrave Macmillan

Schubert FN (2011) Hungarian borderlands: from the habsburg empire to the axis alliance, the warsaw pact and the European union. Continuum Publishing Corporation, London, New-York

Sintès P (2010) La raison du mouvement. Territoires et réseaux de migrants albanais en Grèce, Karthala-MMSH-EFA, Paris

Timár G, Székely B, Molnár G, Ferencz C, Kern A, Galambos C, Gercsák G, Zentai L (2008) Combination of historical maps and satellite images of the Banat region—Re-appearance of an old wetland area. Global Planet Change 62(1–2):29–38

Vogel C (1908) Die Balkan-Halbinsel In 4 Blättern: Türkei, Rumänien, Griechenland, Serbien, Montenegro. Bulgarien, Justus Perthes, Gotha

Wachowiak H (ed) (2016) Tourism and borders: contemporary issues. Politics and International Research, Routledge, London-New York

Chapter 3
From Empires to Nations, 1800–2015

Abstract There is a tense debate within the Balkan research community. No one knows the exact location of most past and present political boundaries. After the downfall of the Ottoman and Austrian Empires, the political map of the region never stopped changing. Such instability coupled with acute territorial cross-claims over land has generated border mappings that suffer from a number of defects. Besides, maps are made to express national points of view. They consistently understate the importance of unorthodox "in-between" spaces. The multitude of mistakes in historical cartography suggests that uncertainty is a basic principle of land categorization.

Keywords Balkanization · Land fragmentation · Micro-territorial disputes · Communism · Empires · Collapse · Mapping shortage · Mapping mistakes · Accuracy · Uncertainty

Over the past few decades, the European integration and the establishment of the Schengen area have widely spread the idea that international boundaries have now entered into an era of progressive—yet decisive and maybe not reversible-weakening.[1] *Globalization*[2] has found a privileged experimental field in Eastern Europe and in post-Communist states, more broadly. A few years after the socialist collapse, many countries that had been living for decades behind physical and mental walls opened up to outside influences, as probably never before. For many reasons and in different ways, the recent weakening of most boundaries appears to be a historical singularity whose impact cannot be contested.

Greece has joined the European Union in 1981, Slovenia has taken the same way in 2004, Bulgaria and Romania have followed in 2007, and Croatia in 2013.

---

[1]Many references to *debordering processes* or *blurred borders* are to be found in the relevant literature. See for instance (Newman 2006; O'Dowd 2010; Omae 1990; Paasi 2009; Yeung 1998).

[2]We shall handle this highly successful and controversial concept with extreme care (Yeung 1998). The introduction of such a term from economics to geography induces low significance levels inside this last discipline. As usual relatively to neologisms, language practices shall certainly refine its meaning in the future. For now, we will consider that this notion is quite ineffective in applied geography at local and regional scales.

© Springer International Publishing Switzerland 2017
R. Darques, *Mapping Versatile Boundaries*,
Springer Geography, DOI 10.1007/978-3-319-40925-2_3

The remaining Balkan countries, including Turkey, have been negotiating their accession for years. The process is ongoing. This overall trend would normally drive curious onlookers to conclude that, in the end, history has adopted a converging line for many "little" southeastern European states that have spent their time, power, and strengths establishing sovereign perimeters and securing districts against conflicting neighbors in the past.

Those who went through the Communist era remember the repeated obstacles posed by *blind* government administrations. Daily life was submitted to ordinary embarrassments that are characteristic of excessive bureaucracies. These annoyances were—and somehow still are—visible expressions of practices deeply anchored in history. Of course, the Eastern and Western "Blocs" have brought the supervision of people and goods to its utmost during the Cold War. Despite the Balkan border opening and *globalization* perspectives, many hassling behaviors persist today, as years past legacy: The "open space" is a hard reality to deal with, after two hundred years spent in defining boundaries and improving external security. Moving from a paranoid attitude to its exact opposite is not that simple. The explosion of illegal trafficking over the past two decades worsens the problem, leading to schizophrenic positions and missions of the authorities in charge of the surveillance (Sui 2008).

After the Yugoslav fragmentation of the 1990s, travelers have never crossed as many boundaries as today. If customs procedures had not been lightened, the area would have become a geographic "trap" of international communications, i.e., a "must-avoid" European sector. Building an independent state and securing its boundaries produce contradictory schemes: Laying the foundations for long-term economic growth hardly combines with political and military needs to cut off borderlands from their "background" and to protect large perimeters from outside army intrusions.

3.1 Post-Communist "Open Balkans"

Throughout the Mediterranean Basin, the Balkans could be seen as a specific subregion submitted to recursive instability and great internal tensions. The repeated conflicts in former Yugoslavia—ranging from Croatia to Kosovo through Bosnia–Herzegovina—and many other underlying disputes randomly reported by the Western Press did not deny the reputation of *powder keg* associated with Eastern European countries. Claims of ethnic or confessional minorities, military operations, diplomatic and political coups, and drawn-out quarrels between capitals are here daily routine. At least, this is the image conveyed by popular discourses. Straight fully involved in disputes, the national press is tightly controlled and willingly adopts quarrelsome positions.

Understanding the Balkans through such a prism, without considering divergent viewpoints, is certainly a mistake. Confronted with intricate situations, researchers have to manage complexity. The division of land and people is much more characteristic of the region than any official speech will ever admit. Many political and historical reasons explain why "national homogeneity" is always emphasized as a *doxa* against all evidence. All discourses on national development have grown as far as the Ottoman and Austro-Hungarian Empires have gone deeper in their decline. Despite this, land fragmentation has survived the downfall of the latter, allowing an insidious threat to weigh on each country's fate. Classical analyses present the Jacobinism promoted by most Balkan capitals as a delayed answer to the social–spatial *dilution* of the Empires. Other studies point that the Western powers who had precisely introduced and supported the nation-state model against the Empires revealed unable to control its further application.

These high-scale viewpoints raise a problem: The Balkan area is all the more complex given our limited knowledge. Many practical reasons explain the fundamental lack of understanding: The confinement of Albania, Bulgaria, Yugoslavia, and Romania beyond the Iron Curtain over half a century has adversely affected mutual knowledge. The studies, mappings, and expertise developed by many Western states in the first half of the twentieth century never found later continuation. Without any overlook since then, most changes that have taken place behind the *fence* remain unclear. The question then arises as to whether this knowledge gap could be made up. Nothing is less certain.

Whole worlds are gone, leaving hordes of mute and tired witnesses. The Communist collapse and the end of the Ottoman and Austrian Empires are not only abrupt upheavals. These are historical episodes akin to total breakdowns. Time is measured in reference to these events. Fundamentally, Balkan peoples know what the word *revolution* means. From the Ottoman world to the first nations, then to the Communist regimes, and finally through the post-Communist era, each period disowns the previous one and decides on new standards of living, generating global territorial restructurings. By taking these elements into consideration, scientists usually adopt a schizophrenic attitude: To find an answer to their questions, they have to investigate touchy subjects. To complete their mission, they will rely neither on dated official historiographies, nor on bibliographical or statistical tools bequeathed by the authorities of each corresponding period. New worlds come into being through a process that breaks with past models.

Resorbing such research deficit is a difficult task, especially in geography, even more so because the Balkan states have been facing accelerated changes after they entered the contested *transition* period[3] (Bunce 1995; Karl and Schmitter 1995). Outside observers feel intimidated when confronted with such dramatic adjustments deeply impacting the daily lives of people. Who could guess in the middle of the 1990s that Bulgaria and Romania would join the European Union a decade after? These countries were suffering severe underdevelopment: destroyed industrial plants, out of order communication means, massive emigration, *frozen* state apparatus, crunch of the agricultural production, and global economic

[3]This debated term illustrates the shift from a centrally planned economy to a market economy. Both capitalist and communist systems are not as much bipolar as they pretended to be.

impoverishment. The socioeconomic regression was obvious to everyone. Quite clearly, it would take decades to catch up with the West. Misleading memories.

The present state of Bulgaria, Albania, Romania, and other ex-Yugoslavian countries sends back the late Communist collapse to times that are no more. The remaining old-aged testifiers of these years are now stuck to their social marginality and left behind with a creeping guilty. The landscapes themselves keep fewer traces of that time: From the main cities to the most remote settlements, there have been considerable spatial changes meanwhile.

In line with local or national scenarios, the so-called *transitions* have created social and economic inequality, hence renewed spatial discrimination. Beside the highest dwelling *blocks* and the Gypsies' entrenched neighborhoods, dispersed groups of luxury villas have come into being in the outskirts of the main cities and capitals. The *nouveaux riches* show off material comfort without complexes any more. Many districts undergo a revival by proxy, thanks to money transfers from migrants working in wealthy nations. For now, the post-Communist *middle class* faces a challenging situation. Its emergence is not effective and its future unclear, given the fact that production facilities have not achieved full transformation. Eastern societies maintain unbalanced systems where a privileged minority challenges a multitude of impoverished households that cannot forecast their future without programming long periods of expatriation. Addressing such momentous events and understanding sequential changes entails adopting a "stratigraphic" approach.

With the opening-up of borders, a new chapter of the Balkan demographic history has been opened (Sintès 2010; Sardon 2004; Sikimić et al. 2012). The number of economic migrants, whose movements were closely controlled or forbidden until the late 1980s, has increased to such an extent that Western countries feel the need for raising the alarm and limit their exposure to uncontrolled flows of people. Surprisingly in the eyes of Westerners, the Balkans have found a way out of their "dark zone" status thanks to these spontaneous migrations. New multilateral flows have succeeded to the past one-sided movements from Albania, Greece, Yugoslavia, and Turkey to Western Europe (mainly Germany), periodic journeys and stays of Bulgarians, Albanians, Romanians, and Serbians in Greece; ceaseless travels between Russia, Turkey, and Greece. Furthermore, the Balkans have become a destination for new migrants originating from Asia, Africa, and the Middle East. Today, everybody knows what happens in the refugee camps of Patras, or how custom officers deal with hundreds of weekly illegal border crossings along the lower Evros, in Western Thrace, not to mention the dramatic situation of Kos.

While thousands of Bulgarians have left their homeland for France, the UK, Greece, or Spain, Bulgaria in its turn learns to manage the lasting presence of immigrants within its boundaries. It is clear to everyone that the problem of aging populations claimed solutions. However, the situation has something profoundly unusual. Crossing a black Senegalese in the streets of Dubrovnik, Sofia, or Athens was an exceptional affair until recently. Cities and countryside thus discover a new form of *ethnic heterogeneity*, far from standard *clichés*.

3.2 Balkanization: A Structural Issue

Balkanization initially qualifies geopolitical situations that become unstable because of intricate and mutually unsatisfied territorial claims of several neighboring states. The word is commonly applied in a metaphorical sense to any land submitted to spatial division, mainly drawn forth by ethnic and confessional communities fighting for their *independence*, whatever form this latter would take. Is the concept still valid today in its native region? It is far from obvious. In the shade of the European Union and following the principles of *territorial economy*, many partnership programs have been financed these last years. Large amounts of money promote the economic development of the Balkan borderlands. Most programs are applied seamlessly, entirely ignoring the geographic and historical backgrounds of the areas they are supposed to promote. The exercise is difficult, indeed. Erecting bridges across boundaries settled by bloody wars is no easy task. Cooperation, partnership, combined development, army disengagement, growing number of custom stations, easy goods' transit, funding for transport and communication means, and cultural exchanges are among the objectives pursued: Borders and boundaries are expected to fade away stage by stage, through a series of incentives (Gropas 2004; Bechev 2004).

For the European Union and free market, balkanization is a geographic and historical nonsense to overcome, not to solve. The Communist past of many countries has opened highways to all forms of Western influence, mainly in the cultural and economic fields. It is a striking fact that the socialist breakdown has first aroused from the inside. Some mechanisms just have to come into being. However, resolving political and military disputes is much more challenging. For years now, as opposed to other "model pupils," a number of "badly performing" countries display a great resistance to the EU integration.

The best example is provided by the Yugoslav crisis and Kosovo–Metohija trap. These are—or were—sticky cases able to thwart the plans of foreign investors seeking new markets. The process is known for long. At the end of the nineteenth century, the Western companies that had made significant investments in the Austro-Hungarian or Ottoman Empires had already faced serious management problems issuing from the unpredictable future reserved to their assets. In the Balkans, risks are hard to assess. Social disorders and political uprisings are known threats to long-term investment plans. For greater safety, economic agents should consider the *irremovable* geographic context. Who still remembers that each present boundary is a spatial *compromise*[4] achieved between fiercely competing appetites?

The term *Balkans* itself creates difficulties. In the 1990s, a growing body of literature has hold serious debates on the Balkan "tragic" issue, explaining the necessary Western intervention, imagining ways to solve conflicts, disclosing the ethnic complexity, and finding a *normalized* end to an ongoing "self-destructive"

---

[4]The word is here deliberately understated.

process. To explain events beyond understanding, a permanent exceptional status has been granted to the area. Yugoslavia was often assimilated to the whole Peninsula through a broadening pattern hard to conceive for geographers (Todorova 2009; Sivignon 2009; Garde 1999).

The word *Balkans* supposes the existence of a geographic unity behind the obvious fragmentation of the area. Many authors have spent time spatially defining the region and its limits. The subject will not be addressed in these terms. In light of history, the Balkans might be seen as a contact zone between former geopolitical entities: Habsburg Monarchy, Ottoman Empire, Russian Empire, and other Western Kingdoms. The geographic dimension of marginality is not an artifact. Unlike other more central lands, the Balkans as an interfacing area have supported constant boundary changes that reveal the fluctuating balance of power between Empires and nations. In the end, the *frontier*—as a conceptual category—explains much of local geography.

In semantics, marginality and division account for limited attempts to map the Peninsula as a whole—hence, the impossibility to define geographic limits. Do we have to include Romania, Slovenia, the northern Danube districts, the European part of Turkey and Eastern Anatolia, maybe Cyprus? The question is ineffective, as we shall see further on, for a simple reason: It would be nonsense to fix limits to "mobile territories,"[5] once attached to a suzerainty, another time removed and placed under rival authority.

Better questions could rather be: How are the Balkan lands structured, considering their chronic instability? Are nation-states proper prisms for acknowledging the strong attachment of people to local *homelands*? Is the regional development model promoted by the EU a framework adapted to countries whose dimensions are so limited? Our experience of the area testifies to the feeling that the Balkans are not apprehended through right scale analyses (Sheppard and McMaster 2004; Mamadouh et al. 2004). Why should we plaster the official *Euroregions* to back-countries that obviously live on much smaller spatial units? Sometimes, crossing a mountain range makes you move from one world to another.

The present work intentionally adopts a structural approach of geography, to get past *clichés* spread out here and there by unscrupulous observers. The sharing of culinary knowledge and other analyses aiming to foster supposedly common Balkan traits shall be viewed as irrelevant. Less "volatile"[6] geographic events are cause of concern. The exercise requires the application of two complementary methods—field explorations and quantitative analyses—that strengthen one another constantly. Both arc much time-consuming: Most GIS databases are not immediately available as it happens in other "more developed" countries and need to be entirely built, starting from zero.

[5]The expression is used in a metaphoric sense. It is now common to treat places and spaces like subjects/actors and no more like a support for human activities. The stylistic device is hyperbolic and often misplaced.

[6]There is no negative connotation. Although interesting, the issue stands out of our fieldwork.

Before coming to the GIS project, a brief review of the context within which constant border restructurings take place shall be useful. Few key elements provide a comprehensive framework for the study. These may be seen as premises to further reasoning:

- Since the early nineteenth century, the Ottoman withdrawal has been countered by the reverse growth of the nation-states. The enlargement process is, at first, proportional to the Ottoman drawback; then, it runs up against the expansionism of competing neighbors (Dogo 2003; Dimitropoulos 2008). This mechanism is supposed to be completed when borders achieve stabilization and spatial confines are eventually shared by a number of newly established sovereign actors.
- Moved boundaries determine new territorial centralities. External reshaping and internal restructuring are always associated.
- The degree of centralization of the new states depends on the way they develop: When annexations ("liberations") are abrupt and involve large areas, local selfgovernment will likely persist; on the contrary, when development is based on a process of gradual expansion (staggered integration of small districts), the advent of a highly centralized state apparatus subsequently becomes probable.
- Some regions deviate from the scheme thus defined: the ones that hardly find reliable authority to hang on. They are torn between different states that claim conflicting interests over them. These "blind spots" operate with some autonomy, which is often compensated for by an advanced underdevelopment driving to an almost complete human and economic disengagement.

Of course, local situations are always complex. Only serious investigations can help properly qualify individual cases.

For two centuries, continuously roaming boundaries have sparked a structural instability with major impacts on local development and governance policies. Such mutability profoundly affects civil societies. The "geometric diktat" imposed by official mapmakers is lived by the concerned populations as constant comings and goings. In the end, a kind of schizophrenic perception of their spatial rooting seizes people through a combination of ardent nationalist feelings and a relative disinterest for the ethnic-confessional—eventually political issues. In most cases, this last attitude leads to a temporary or definitive exodus from contested places.

Forced deportations always accompany border changes. These are often preceded by massive spontaneous migrations anticipating the progress of the armed forces (McCarthy 1995). Settlements and cities just next to the boundaries may survive these relocations, paying the price of heavy demographic changes. A great number of places are damaged; others experience slower decays, both events leading to the same end: dereliction of fields and buildings. Border fluctuations often imply agricultural withdrawal and forest advance.

Understanding many issues arising from this approach requires the development of a mapping project whose methodology is now to be explained. Before all, readers are encouraged to keep in mind that any forthcoming map must be interpreted with an "Eastern eye": Urbanization is quite a recent process in the Balkans. Most rural districts have retained high population densities and large workforce, even after WWII. History is not a *blank page*.

3.3 A Border Obsession... Without Mapping Support

Borders have long been ruling the mind of geographers and political scientists. Generations of researchers have developed critical analyses on this topic in a variety of ways. The bibliographical registry is overwhelming, in such a way that the *Balkan studies* seem entirely committed to addressing border issues.

Major hindrances to these works are often their subjection to national perspectives, the ideological background they embrace, and the underestimated spatial implications of debordering/rebordering processes (Wood 2010; Foucher 2011). Balkan lands compile the characteristics of highly fragmented spaces—this is usually the way they are identified from the outside—but this *balkanization* process finds an equivalent inside the scientific production itself (Hoffmann 2008; Todorova 1994, 2002). The scale of analysis is also problematic: Some observers straighten their attention to narrow perimeters; others limit their investigations to "blind" macroscale overviews. In both cases, scientific debates lose some of their effectiveness, given that comparative works are made impossible.

To overcome some of the above-mentioned traps, this work introduces a GIS database entirely devoted to studying border and boundary issues. Although the time period covers two centuries, the approach adopted is the one of a geographer, not a historian. All borders/boundaries that have come into being between 1800 and today have been referenced (Adler et al. 2001; Blake 1995, 1999).

Building the GIS project was a considerable challenge. Surprisingly, no such mapping initiative has ever been pursued up to now—as far as we know. Of course, some atlases highlight a select few historical sequences (Sellier and Sellier 1995; Magocsi 2002), but the absence of any systematic database evidences the complexity of the mapping method to develop. This also means that political instability has reached such a high level it has put everybody off mapping the events.

There are many reasons to the mapping shortage, among which:

1. Policies of the *national military institutes*

These institutions are traditionally engaged in developing topographic coverage derived from acquisition campaigns of aerial imagery. The effort is not costless for small countries. Lacking both financial resources and professional skills, many of them delegated the task to outer companies, or borrowed techniques and materials to foreign institutes. This is how things worked almost everywhere, especially before WWII: Greece has required the assistance of the British Air Army; Albania has been helped by the Italian armed forces; and all Communist states have solved their incapacity in appealing to the Russian Soviet military agency during the Cold War.

Therefore, the heterogeneity of collected data grows to epic proportions. Anyone wishing to create a seamless topographic map of the entire Balkan area would not see the project come to an end. Among the available map sheets, scales differ, acquisition dates do not match, and data quality varies greatly among countries. In other words, national data are incompatible. This makes sense as they were thought to escape such compatibility, on purpose (Monmonier 1991).

2. Diverging mapping systems

The geodetic systems to be used have nothing in common. This defect induces further difficulties in mapping cross-border areas. The variety of national mapping systems is in line with the age of the documents, their scale, and the proper geophysical framework of each country (Robinson et al. 1995). Table 3.1 only mentions little basic information. Each Balkan state has adopted its own geodetic system and mapping tools. Over the years, changing geopolitical conditions and new technical resources has allowed for the development of successive updates.

Defining national reference systems entails decisions that have crucial implications. Adjacent maps covering neighboring areas will never match on a geometrical basis. Any transboundary region is *intrinsically* impossible to map, because no correspondence between *coordinate reference systems* (CRS) can be established. A parallel mapping *vs.* geographic self-isolation of each country is organized.

By way of a specific example, the Hatt system adopted by Greece is certainly one of the most remarkable. The *Hatt equidistant azimuthal projection* works with a reference plane surface tangent to one point on the ellipsoid. This point is called the *Map Sheet Center*. Each map sheet uses a unique center. To put it otherwise, each of the 130 maps covering Greece has its own spheroid. Consequently, distance and shape distortions are of limited impact, but two adjacent sheets cannot be "joined" unless being fully reprojected: The operation may be performed through heavy GIS or photogrammetric softwares. GIS professionals are a bit nervous when processing such topographic maps that escape any geometric transformation referenced by the most notorious softwares. Against this background, these map sheets are ultimately of little interest for users who need to cover more than the 30' × 30' quadrangles (1/100.000). (Fig. 3.1)

Such CRS incompatibility also applies to by-products, i.e., all derived public mappings, including cadastral plans. The huge amount of documents to process makes any update critical and extremely costly. This explains why the "old" reference systems are still in use everywhere, although they were developed before WWII. It will take years to produce new national covers adapted to the current geodetic standards. Many vector graphics worldly disseminated by commercial offices inherit their poor geometric quality and accuracy from such unsolved problems. Misregistered maps may lead to location errors varying from several decimeters to hundreds of meters.

Table 3.1 Official national coordinate reference systems—CRS (position)

Country	Pre-WWII system			Post-WWII system			Latest system		
	Ellipsoid	Map projection	Central point/datum	Ellipsoid	Map projection	Central point/datum	Ellipsoid	Map projection	Central point/datum
Albania	Bessel 1841	Bonne	Lapraka, Tirana	Krassovsky	Gauß-Krüger zone 4	Kamëz, Tirana	Krassovsky	Transverse Mercator	ALB 8 7
Bulgaria	Hayford 1924	Gauß-Krüger	Hermannskögel, Vienna	Krassovsky	Lambert conformal conic (1 ref. parallel)	4 different central points	GRS80	Lambert conformal conic (2 ref. parallels)	–
Greece	Bessel 1841	Hatt	Observatory, Athens	Hayford 1924	Transverse Mercator	ED50	GRS80	Transverse Mercator	GGRS87
Romania	Clarke 1880	Lambert-Cholesky	Observatory, Bucharest	Krassovsky	Gauß-Krüger	Pulkovo, Saint Petersburg	GRS80	Transverse Mercator	Pulkovo, Saint Petersburg
Turkey	Hayford 1924	Gauß-Krüger	Aya Sofia, Istanbul	Hayford 1924	Lambert conform al conic	ED50	GRS80	Gauß-Krüger	TUREF
Former Yugoslavia									
Bosnia–Herzegovina	Bessel 1841	Gauß-Krüger	Hermannskögel, Vienna	Bessel 1841	Gauß-Krüger	Hermannskögel, Vienna	–	–	–
Croatia	Bessel 1841	Gauß-Krüger	Hermannskögel, Vienna	Bessel 1841	Gauß-Krüger	Hermannskögel, Vienna	GRS80	Transverse Mercator	HDKS
Macedonia (FYROM)	Bessel 1841	Gauß-Krüger	Hermannskögel, Vienna	Bessel 1841	Gauß-Krüger	Hermannskögel, Vienna	–	–	–
Montenegro	Bessel 1841	Gauß-Krüger	Hermannskögel, Vienna	Bessel 1841	Gauß-Krüger	Hermannskögel, Vienna	–	–	–
Serbia	Bessel 1841	Gauß-Krüger	Hermannskögel, Vienna	Bessel 1841	Gauß-Krüger	Hermannskögel, Vienna	GRS80	UTM	SREF98
Slovenia	Bessel 1841	Gauß-Krüger	Hermannskögel, Vienna	Bessel 1841	Gauß-Krüger	Hermannskögel, Vienna	Bessel 1841	Transverse Mercator	D48

Source ASPRS, National Geographic & Military Institutes

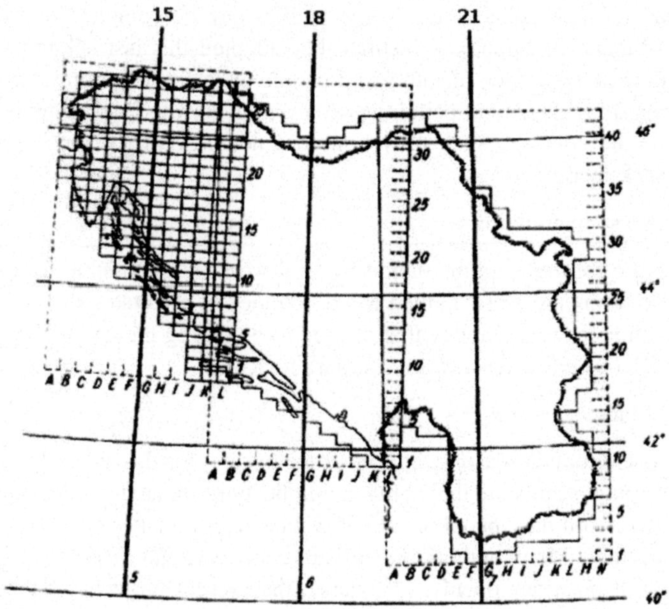

Fig 3.1 Meridians of coordinate reference systems in former Yugoslavia. *Source* GIJNA (Geographic Institute of the Yugoslav People's Army)

3. Limited or forbidden data dissemination

Until lately, CRS characteristics have been kept secret. Some of them remain so: The information presented in Table 3.1 is sometimes doubtful, given that most military officials refuse disclosure on the subject. Bulgaria and Turkey reach high "protection" levels. In Bulgaria, topographic maps are still considered confidential materials, although one could replace them by the Russian Military maps at 1/50.000 sold worldwide. The post-WWII Bulgarian CRS consists of four Lambert Conformal Conic projections based on the 1940 Krassovsky ellipsoid. Each zone is defined by criteria of its own: latitude of origin, scale factor, and azimuth of the central meridian (for rotation purpose). Detailed parameters were never disclosed to the public.[7] As a result, no similar points might be found among documents, simply leaving these sheets for *non-earth* image processing when available.

Thus, high-quality maps are reserved for military use in most Balkan countries. Impoverished paper sheets happened to be delivered to civil administrations and certified companies ("civilian" version). Such a ban on disclosure seems outdated today, as global positioning devices (GPS) have come into common usage. Any VHR (very high resolution) satellite imagery gives now access to much more information than contained in any conventional topographic map.

[7]Similarly, the Pulkovo 1942 datum parameters created by the Soviets are still kept secret.

The ban on delivering sheets to the public may be complete or partial. The Hellenic Military Geographical Institute has adopted this last solution. In accordance with applicable law, unauthorized people or institutions were denied access to any geographic document pertaining to a *yellow* corridor[8] covering the northern confines of the country. Such provisions did not prevent the development of a specific black market.

4. Land access interdiction

Compensating for the mapping deficit by field works is impossible. Until recently, access restrictions are were applied to all confines of the states (Sect. 3.6.1). The phenomenon is less acute today than it used to be during the Cold War. However, many boundaries under tension are still hard to access.

5. Risky topics

Another "background" reason—no longer technical—for this virtually nonexistent border mapping stands in the political implications related to the subject. The self-restraint of ordinary people is obvious. In science, analyses are also biased by political leanings. Many authors in geography, history, and markedly ethnography adopt partisan positions openly. Of course, the present project is not immune to criticism. *Neutrality* is often expected from a research work. Is it even within reach? Possibly. When cautiously used, data processing may increase the convincing power of any investigation and bring the weight of evidence.

Clearly put, the process leading to the establishment of national frontiers has left deep aftereffects in minds and landscapes. Should we not just say that any research initiative on such a matter is irresponsible because it runs the risk of reviving sleeping tensions? Blame can rightly be attributed although encompassing strong countereffects behind good intentions. Keeping frontiers *out of the reach* entails serious disadvantages:

- Large areas of the Balkans would be doomed to remain unobserved for a long time to come.
- The *vacuum* thus created impedes the process of developing an effective knowledge on contiguous territories. The question arises as to whether any "internal" geographic study is relevant.
- As will be explained later, leaving aside border issues would lead to a stalemate because most Balkan provinces have experienced the effects of a border situation over the past two centuries.

At last, the circumstances which made it possible to bring this work to its end must not be left unmentioned. Had the post-2000 interstate relationships not been quite appeased, any research would have remained unachieved. Submitted to latent or open pressures, geography is pushed into a dead-end. A straightforward exercise

[8]In the Athenian IGME office, this color is the one adopted for displaying the forbidden areas (1/50.000 scale).

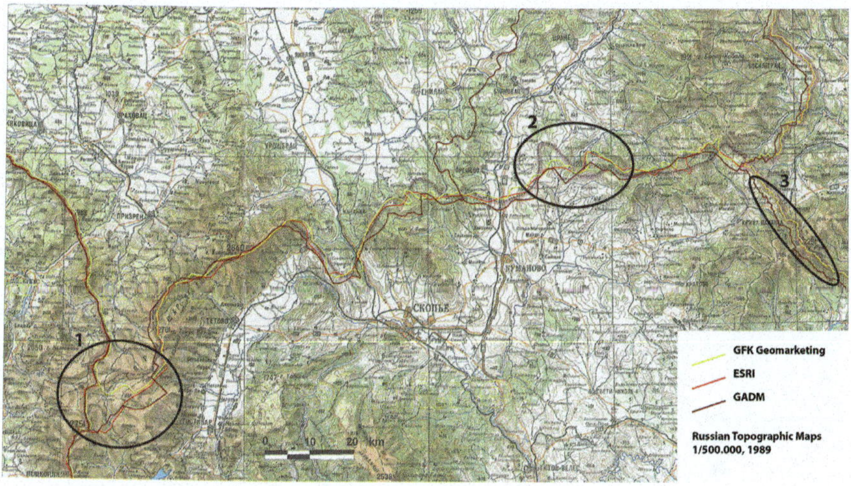

Fig. 3.2 Present position of the international boundaries between (FY) Macedonia, Kosovo, and Serbia (raster and vector graphics). *Source* GFK Geomarketing, ESRI, GADM databases

illustrates the problem: Let us try to locate the current international boundaries on a map—accurately, not roughly as usual. The task is not that stupid.

GFK Geomarketing and ESRI are two worldwide known providers of GIS databases and softwares. Through their powerful distribution network, they reach tens of thousands of users who make the most of their data into self-developed applications. GADM is a free GIS database on worldwide administrative boundaries developed through global collaborative efforts. Altogether, these three organisms are among the most notorious institutions providing GIS data. Figures 3.2 and 3.3 also introduce Russian topographic sheets at scale 1/500.000 as background maps, thus adding another fourth reference.

Surprisingly, the four competitors present nonconformal boundaries in geometry and position. This is a problem only for mainland: Shorelines are rightfully disposed. The statement is all the more amazing that location errors reach unbelievable distances. The area n°1 reveals an offset of more than 10 km (!) between (FY) Macedonia, Albania, and Kosovo. N°2 shows a longitudinal drift between Russian coverage and other vector features of almost identical distance. One could consider that such inaccuracy originates from the nature of the former Yugoslavian administrative limits lately turned into international boundaries. Ellipse n°3 in Fig. 3.2 denies this assumption. The error applies to the old Yugoslavian–Bulgarian border—as opposed to the Yugoslavian–Albanian one. The Dalmatian demarcation, between Croatia and Bosnia–Herzegovina (Fig. 3.3), is another straightforward example: Once again, its shape does not match with Russian maps, leaving undetermined the fate of sizeable areas. It is also noticeable that a single GIS database mentions the new and disputed boundary of Kosovo.

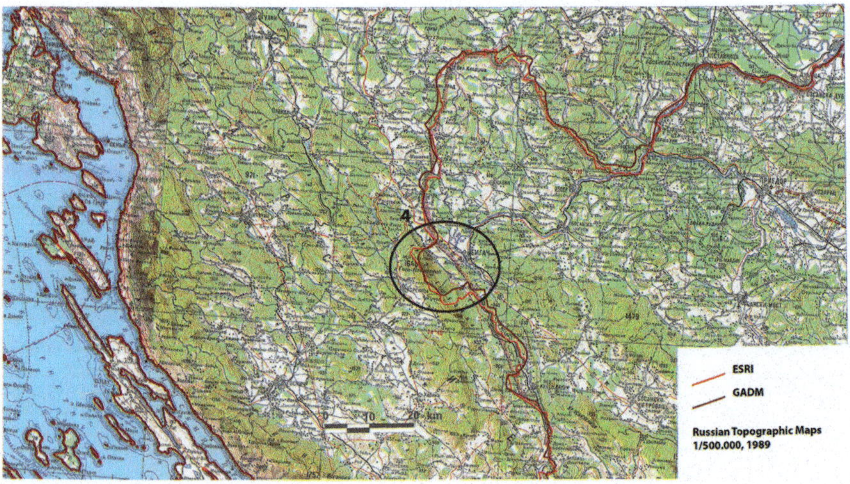

Fig. 3.3 Present position of the international boundaries between Croatia and Bosnia–Herzegovina (raster and vector graphics). *Source* ESRI, GADM databases

Although appeased, the Balkan theater is obviously submitted to a certain mapping confusion, two decades after the disintegration of the Yugoslav Federation and even though some of its successor states have joined the European Union. A number of hypotheses may be advanced to explain these systematic spatial variations, among which:

- GIS features inherit their geometrical irregularities from the topographic maps used during the digitizing process. Distorted and mislocated vector features demonstrate the absence of any reliable information about coordinate reference systems. In view of this difficulty, the commercial companies seated in Western Europe and the USA adopt *homemade* workflows. None of them has agreed to disclose the details of their production process, despite our requests.
- The varying scale of the sheets may explain some shape distortions. The use of the USGS 1/200.000 tiles is likely, given that no systematic coverage of the Balkan area exists apart from those prepared during the Cold War by the US and USSR authorities. This may account for varied levels of generalization of vector features.
- Most data are certainly outdated. The original topographic sheets do not integrate the outcome of the work done by the multilateral border commissions established after the implosion of Yugoslavia.

What is striking is that tens of thousands of people daily use false data. However, the *diagnosis* has to be softened. Low accuracy is the hallmark of most geo-databases used commercially (Thapa and Bossler 1992). On the other hand, it does not mean that ground delineations conform to such inaccuracy. Field surveys confirm this last statement. There is a large gap between maps and ground-based demarcations.

3.4 Some Latent Microterritorial Disputes

These cartographic wanderings show that, although supported by full legal provisions, delineating frontiers is not that simple (Prescott 2014). Borders are not static and everlasting phenomena. Their fixation always follows lively discussions involving nation-states and local communities. The epistemological problem is widely known in geosciences: Land and human communities that form *territories* are not motionless objects. Physical landscapes are constantly changing, human pressures vary, land uses are reallocated, counties come into being or disappear, and transboundary planning projects are implemented despite all these. Engaging in dialogue can be difficult for military institutions, governmental administrations, and local authorities.

By way of illustration, we shall now briefly introduce a few Balkan microterritorial disputes, more or less dormant, that burden the current relations between countries, leaving aside maritime cross-claims.[9] These disputes may help us make the distinction between mapping mistakes and lack of agreement between neighboring states. The difference is convenient, although most often fictitious. A bad conflation of georeferenced outlines is a sure sign of underlying disagreements. The ambition is not to provide a comprehensive list of conflicts, but to focus on the prerequisites that have to be met before fixing boundaries. This will settle grounds for further developments on the issue. The works of B. Lory (Lory 2002) and C. Schofield-M. Klemenčić (Schofield and Klemenčić 1996) are good complementary readings.

3.4.1 Bosnian–Croatian Litigation

Keeping with the above Dalmatian example, it must be stressed that several agreements signed after the breakup of Yugoslavia have resulted in clear demarcations (Klemenčić 1999; Nicod 2004). The boundary between Croatia and Bosnia–Herzegovina has been redefined in 1999 on the basis of the developmental work carried out by a mixed international commission. The new outline has been drawn over 1/25.000 maps and fixed infield, taking into account three different inputs:

- border cadastres,
- limits between census settlements as defined in 1991,
- position of the dividing line between Bosnia–Herzegovina and Croatia at a time when these were republics within Yugoslavia.

---

[9]To maintain the consistency of this work maritime boundaries shall be considered not relevant to this discussion. Their study supposes the introduction of different investigation methods and concepts. The GIS database would have been thought in a completely different way (Prescott and Schofield 1986; Schofield et al. 2013).

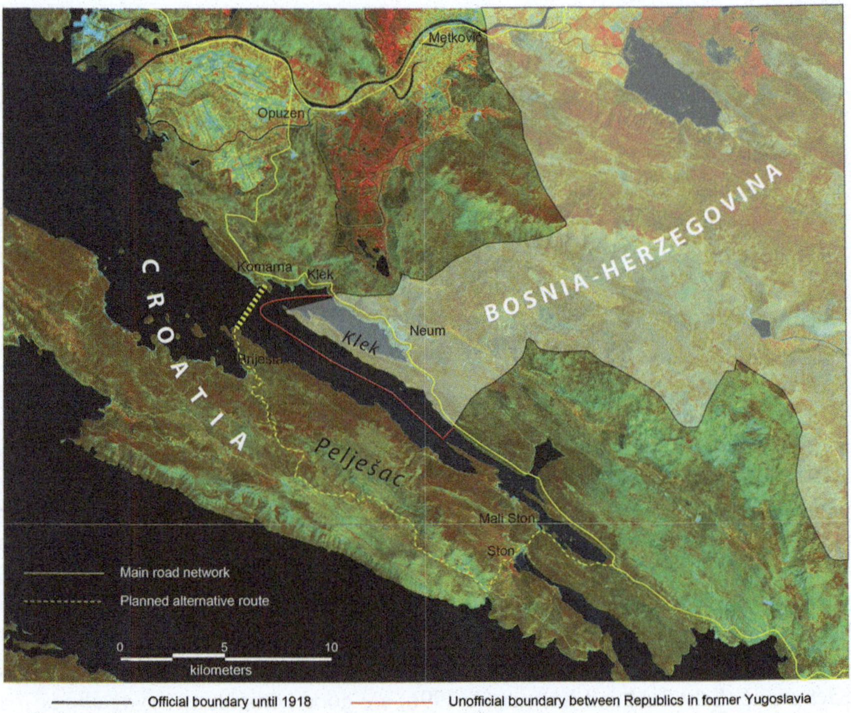

Fig. 3.4 Klek Peninsula: the new borderline between Croatia and Bosnia–Herzegovina. *Source* USGS, Landsat ETM image, 188/30, 14/06/2002, WGS84/NUTM34, NDVI over Pan, ESRI vector features, after (Schofield and Klemenčić 1996)

Details of the outline have been challenged around Bihać, along the Una River (a claimed island by Republika Srpska), and across the Klek Peninsula (Fig. 3.4). The example of Klek highlights the problem of Croatia's land continuity and its need to address the problem of the Bosnian strip of Neum, thanks to a bridge project that would join Komarna to Pelješac. The 1999 agreement has opened onto demarcation field works. The borderline has to be drawn on 1/5.000 map sheets. This last disposition supposes to solve key mapping issues: At this scale, only few border sections had been located with modern equipment. Other segments have to be redrawn on the basis of past demarcations—those fixed by the Austro-Hungarian administration in the nineteenth century.

3.4.2 Croatian–Slovenian Litigation

Other delineations are much more problematic. In 2009, for instance, Croatia and Slovenia had found no solution to the lasting disputes raised by their respective independence. Both countries were to ask the arbitration of the International Court of Justice with the goal of clarifying many issues: Maritime outlines in the Bay of Piran, land properties along the lower Dragonja River, military installations on top of the Žumberak/Gorjanci Mountain, and pieces of land north of the Medimurje area were among the questions to discuss (Ballinger 2003; Klemenčić and Schofield 1995). This last case is interesting because it awakens old border disputes that had not been negotiated through agreements over decades. The demarcation line between Medimurje (Croatia) and Prekmurje (Slovenia) is an old Austro-Hungarian internal frontier—probably defined in the seventeenth century—that follows the course of the Mura River. Unfortunately, the Mura has changed position, as it happens to all rivers, without any recorded revision of the borderline. Recently, Slovenia has proposed that the demarcation line should follow the current thalweg. Croatia has rejected the offer. The dispute centres on the fate of few agricultural estates. The litigation has been used to stop or postpone Croatia's accession to the European Union. In November 2009, Slovenia accepts to lift its blockade in return for new negotiations to be opened. Both sides refer to their good neighboring relations and mention the creation of a special condominium status for the contested places to put an end to the quarrel.

3.4.3 Kosovan–Macedonian Litigation

The latest moot point concerns Kosovo. A few months after the Kosovan declaration of independence (February 2008), a single Macedonian commission starts to fix landmarks along its 85-km boundary. Works are achieved in a few weeks, without reported incidents. The main problem is the determination of the borderline along the Šar Planina, a southern Kosovan mountain range. The settlement, however, prompts critical reactions in the outskirts of Skopje (Kodra Pura, Dëbëllldeh, and Tanuševci).

The operation reveals far political implications. As with Slovenia toward Croatia's European integration (FY), Macedonia has put into the balance the recognition of the borderline as a preliminary to the official acknowledgment of Kosovo's independence. In 2009, bilateral agreements had been reached leading to a transfer of a few land parcels from Kosovo to (FY) Macedonia in the county of Dëbëllldeh. Many security problems have been reported so far, especially in Tanuševci's area. Cross-border traffics are going well. Individuals wear weapons without being troubled in any way and armed groups operate undisguised. Media outlets report clashes close to the *Tanuševci Republic*. These jagged landscapes are not as quiet as they seem to be. The fate of the Macedonian–Serbian frontier should not be tackled without stabilizing at first the Serbian–Kosovan relationships. Belgrade has denied the validity of the 2009 agreement and accuses the "self-proclaimed government of Priština of" granting part of Serbia's territory to Macedonia.

3.4.4 Kosovan–Serbian Litigation

The situation is worse along the Kosovan–Serbian "boundary" where no agreement has been found. The northern and eastern borders of Kosovo are still officially managed by the *uti possidetis* principle adopted by the Badinter Commission in 1991 for delimiting newly emerged states. *Uti possidetis* as a part of international law establishes the inviolability of borders as a means for limiting cross-territorial claims when new states gain independence against former colonial powers (Vidmar 2013). When applied in Yugoslavia, this principle implies that previously existing internal boundaries become international demarcations through an "upgrading" process. The major paradox is that past administrative boundaries are not designed to become international borders. *Uti possidetis* thus freezes potential land reclamations and transforms the arbitrariness of administrative delineations into a permanent form of land legality at the expense of local identities and communities.

In that respect, it is clear that the "weight" of any new demarcation results from an inherited internal legitimacy and consistency. Kosovo received limited administrative political recognition within the government apparatus of Yugoslavia. The area did not have the status of Yugoslavian republic, only wearing an autonomous province grade. A further reason casting doubts on the legitimacy of the process is that the Kosovo status process initiated in 2005 as a result of the UN Security Council Resolution 1244 is not fully resolved. The independence of Kosovo declared in 2008 has been recognized by most "Western" UN member states, but it is still contested by Serbia, Russia, and many other nonaligned countries. Without any consensus reached at an international level, Kosovo remains an *uncertain* area placed under UN control (Capussela 2015). The 2013 Brussels deal is the latest demonstration of a desire to pave way for appeased bilateral relationships between Kosovan–Albanian and Serbian authorities. This agreement aims to integrate North Kosovo municipalities into the Kosovo legal system, thus putting an end to the dual de facto government of Albanian/Serb-majority communities. As always, Serbia's EU candidacy is put into the balance of negotiations. To put it clearly, stepwise bottom-up agreements are negotiated on a regular basis in the hope of finding compromise solutions to an intractable issue—the independence of Kosovo is considered to be unconstitutional by the Serbian law. The European institutional crisis, the political–military conflict in Ukraine, and the recent refugee crisis are events that might unexpectedly change future developments on the issue.

Against this background, Belgrade still considers today the borderline with Kosovo as an administrative boundary line. Bilateral demarcation works are made simply impossible because it would imply the constitution of a joint commission. With Montenegro, the problem is almost similar although the demarcation process has some advance. The Kosovo State Commission for Marking and Maintenance of the State Border has recently issued a report in relation to the detailed delimitation line with Montenegro. In both cases, all necessary precautions must be taken against even referring to the demarcation process. Defining accurate boundaries could inevitably lead local communities to confrontation. The 2004, 2008, and

2011 civilian unrest in Kosovo and the difficult situation of Mitrovica remain engraved in memories. The International Security Force (KFOR) controls a 25-km air safety zone and a 5-km ground perimeter. Protests, clashes, and frictions are ongoing in parallel with intergovernmental negotiations.

The dilemma of Northern Kosovo is instructive because this area illustrates a complex post-conflict situation that has given rise to unofficial land divisions (Cattaruzza and Dérens 2015). When the legal status of land is not safe and international relations are like a dialogue of the deaf, whole districts live under a special regime commonly referred to as "autonomy." Over the past 20 years, this word has been used wherever unsolved Yugoslavian issues were identified. This is true for the *Autonomous Province of Western Bosnia* in the 1990s, the Brčko district, and now North Kosovo—among other possible examples.

For historians, this is no surprise. "Autonomy" is a generic label given to local communities forced to live by own means because of the limited institutional presence of sovereign states that are supposed to afford them protection. Within the Ottoman Empire, many Balkan lands had been awarded such practical *semi-independence*. Local potentates occupy key positions. Political and military disputes, administrative withdrawal, and economic collapse make a typical *Ottoman cocktail*.

The above examples are not exhaustive. Litigations have emerged despite the decision of the Badinter Commission, taken between 1991 and 1993, that the international recognition of any independent state would happen on the exclusive condition that the Yugoslavian internal frontiers remain unchanged. At the same time, it has become increasingly evident that disputes are not the monopoly of former Yugoslavia. In fact, the whole Peninsula is riddled with such microterritorial conflicts unearthed from remote pasts. Some disagreements are just smoldering. Others come up under the pressure of public discontent. All else being equal, apart from Yugoslavia, most Balkan quarrels have been kept deliberately dormant these past decades not to wake up old demons (Hassner 2006).

Let us take the example of a virtually uncontested area today, although crossed by a major Balkan transport axis: the Turkish–Bulgarian boundary. Returning to mapping issues, the frontier alignment between these two countries has been legally defined for the last time by the Treaty of Lausanne (1923). With the exception of the Greek *Trigon*, its demarcation goes back to the Treaty of Neuilly (1919). This boundary had been established for the first time at the end of the second Balkan War, in September 1913. Since then, no change has occurred. Although inaccurate as regards current standards, the delineation seems to satisfy both neighboring states so far, as it happens in other places. That is to say, uncertain diplomatic relations do not necessarily imply repeated border revisions or transboundary cross-claims. The *Cold War* does not only apply to East–West interrelationships over the post-WWII era; the concept may extend beyond its usual limits to include other geographic frameworks. A great number of territorial claims undergo long-term *freezing* in the Balkans.

3.5 Pitfalls of the Academic Border Mappings

As already mentioned, the existing border mappings paradoxically suffer from a lack of consistency, geometric inaccuracies, and outdated data. Everyone wants to investigate the *question*, but few people are actually able to offer a comprehensive approach that would move beyond instituted *national perspectives.*[10] The current project aims at covering the gap. To do so, new mapping tools must be engaged that break with "traditional" representations (Sellier and Sellier 1995; Cattaruzza and Sintès 2012; Magocsi 2002; Hupchick and Cox 2001).

Most academic maps stem from historical and political studies. They usually support a static cartography that has many shortcomings (Biggs 1999):

1. These mappings are *snapshots*, instantaneous pictures that bring to the forefront historic "moments" and ignore the "in-between" periods that might contribute to a better understanding of the latter.
2. Depicted as geometric lines, shared boundaries become non-spatial objects, as if deprived of any environmental impact (Mizrahi 2009).
3. Mapped data only focus on *conventional* political boundaries. Subsequently, other momentous "internal" frontiers are ignored, although they sometimes weigh more than their official *twins* (Labrianidis 2001).
4. Borders are seen as one single category, with no awareness of the different statutes, forms, conventions, delineations, controls, etc., that represent their many achievements.
5. In semiology, using lines for drawing interstate frontiers might be seen as a mistake—especially in ancient times: Fuzzy logics may be more relevant (Wang and Hall 1996).
6. Broadly speaking, the higher the interest in border issues, the lower the intelligence of borderlands. The success of broadscale maps that display state boundaries is never greater than in war times, i.e., when the affected areas are no longer open to civilians. Throughout the Balkans, such maps—although of poor quality and blemished by partisan perspectives—have always been circulating: Peace times are scarce.
7. A great number of mistakes reduce the relevance of such "rough" mappings. Scales are often very broad—an option limiting the impact of any graphic approximation. In the end, mapmakers usually feel unsatisfied because local spaces remain outside the scope and escape possible analyses.

In summary, one can easily *guess* what shall be evidenced as a basic characteristic of the area: After the Ottoman withdrawal, the past international boundaries —often changed into internal frontiers,[11] then into simple administrative limits— have left traces of their existence everywhere (Yérasimos 1991; Hajdarpašić 2008;

[10]For this reason, most manuals in geography simply gather information arising from national backgrounds without promoting *bridging* works.

[11]The reverse process may also apply.

Brown 1996). However, in the absence of dedicated mapping tools, no one can define them precisely.

3.6 Some Illustrated Mistakes

As demonstrated in Chap. 3.3, *uncertainty* is a key concept potentially applicable to any border mapping, even the most recent ones. When dealing with recent maps, many mistakes may result from different technical reasons or from intentional impoverishments (Congalton and Green 1999; Lunetta 2004; Foody and Atkinson 2002; Foody 2007). On the contrary, errors are standard inputs of historical maps. The quality of old documents is all the more difficult to assess given that no alternative information exists. Past political maps are irreplaceable sources. They are always used as *references,* although burdened with obvious wrongdoings. The Saints might be faulty: what happens when God himself is failing? It is enough to make you forget your Latin. In any case, errors have to be identified before being fixed.

Mapping mistakes are countless. In order to avoid a scattering of efforts, the present work will address two areas of interest: the Lower Danube basin and Greek Islands.

3.6.1 The Indefinite Wildlands of the Lower Danube

Concluded in September 1829 after the Russo-Turkish War of 1828–1829, the Treaty of Edirne strengthens the Russian presence in Eastern Europe at the expense of the Ottoman Empire. Beside other territorial extensions, Russia takes this opportunity to control the mouths of the Danube. Few mapmakers of the time take this significant border adjustment into account. (Fig. 3.5)

Both upper maps have been published after the war, although Lapie's work must have been prepared before. The most surprising point is that the demarcation remains unchanged, still positioned along the Sulina canal, right in the middle of the delta. Other historical maps of the 1800–1820s confirm the statement, as if no treaty had been signed. Lapie's map mentions a "neutral territory" which is missing in other past treaties. Furthermore, between the principality of Wallachia and the Ottoman Empire, the outline position varies among mainstream and oxbows, which represents a 5–15-km shift along the Danube (Fig. 3.6).

In 1855, Flemming finally hands out the Danube delta to the Russian Empire. Surprisingly, a simultaneous map of the author brings back the boundary to the Sulina canal. Considering its greatest distance, the *shift*—or *uncertainty*—reaches about 50 km! South of Braila, the boundary adopts a specific itinerary, different from those of Lapie and Radefeld, once following the western branch of the Danube, another time passing through more eastern meanders (Fig. 3.7)

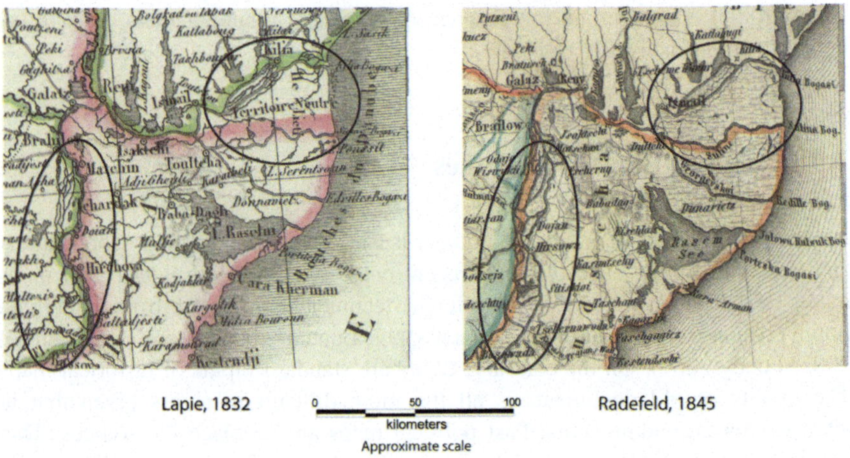

Fig. 3.5 The lower Danube after the Treaty of Edirne (1829). *Source* David Rumsey Map Collection (Lapie 1832; Radefeld 1845)

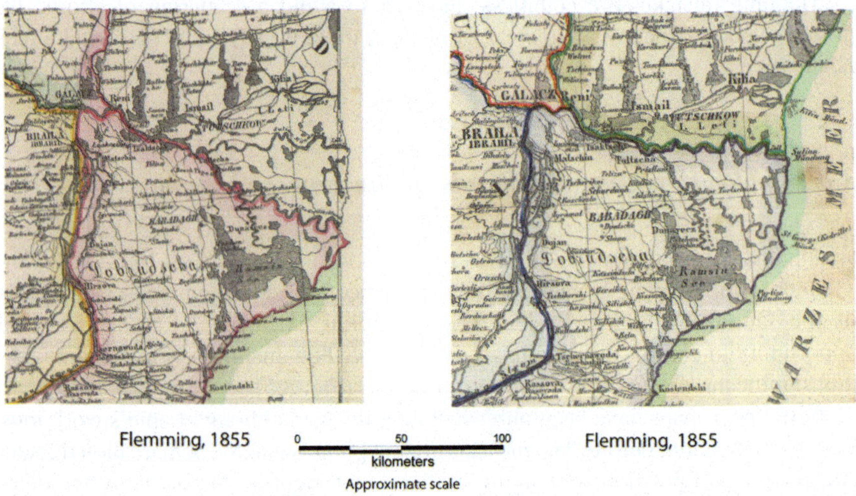

Fig. 3.6 The lower Danube in 1855. *Source* David Rumsey Map Collection (Berghaus et al. 1855a, b)

Following the Crimean War (1853–1856), the Treaty of Paris once again troubles the game. Russia's influence is decreasing in the Black Sea. The Empire loses the mouths of the Danube it had gained 27 years before. Historical maps, however, remain confused about the right position of the boundary. In 1860, Grassl reports once again a "neutral zone." It was not until the 1870s that success arrives: At last, and without any new in-between event to reshuffle spatial attributions, Stieler proposes a demarcation closely similar to the one we know today.

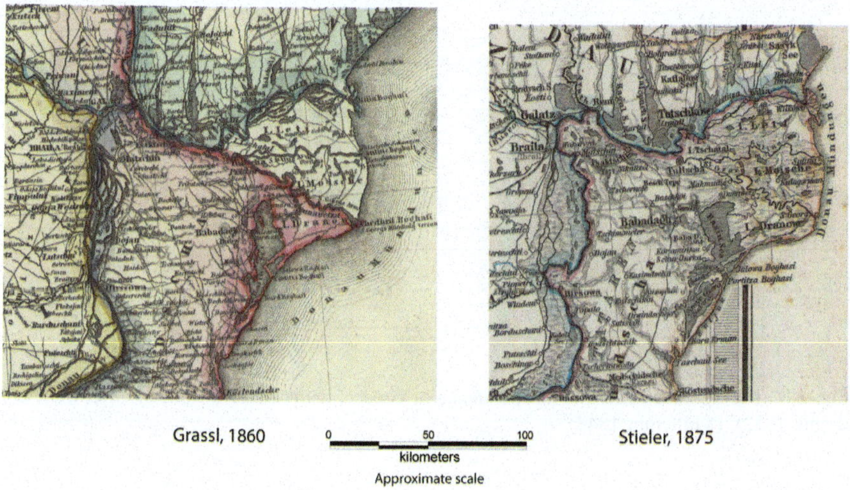

Fig. 3.7 The lower Danube after the Treaty of Paris (1856). *Source* David Rumsey Map Collection (Grassl 1860; Stieler 1875)

This exercise suggests that, until lately, no one knew the exact location of the boundary. Frontiers vary in space and confuse minds altogether. Borderlands are fuzzy zones that cover wide areas. The *no man's land* is first a *nowhere land*. The fact remains that such documents were often used in support of international treaties signed between conflicting parties. Most cartographers are of Western origin and their field knowledge may be limited. Entirely inspired by German mappings and supported by its delegated technicians, the Ottoman cartography inherits the defects of its origins and does not prove greater reliability (Özdemir 2008).

3.6.2 Ambiguous Sovereignty in the Ionian and Aegean Islands

Not surprisingly, the hundreds of Ionian, Adriatic, and Aegean Islands are poorly mapped—without raising the issue of maritime boundaries. The islands in general are spaces that are randomly attached to various sovereign states. The comparison method will not be chronological as before, but spatial.

A rough examination of the historical background is necessary before moving on to the documents. Key dates are as follows:

- Treaty of Paris (1815): The *United States of the Ionian Islands* are created and placed under British protection.
- London Protocol (1830): The official independence of Greece is proclaimed.

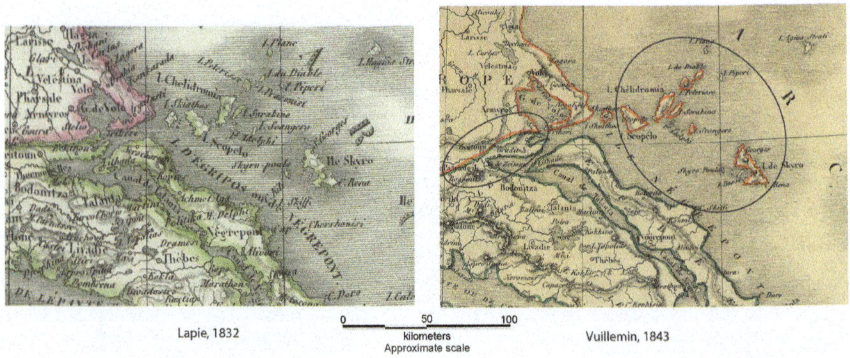

Fig. 3.8 The Sporades after the independence of Greece. *Source* David Rumsey Map Collection (Lapie 1832), (Vuillemin 1843)

- Treaty of London (1832): Samos is an autonomous principality within the Ottoman Empire.
- Treaty of London (1864): The Ionian Islands unite with Greece.
- Treaty of Bucharest (1913): Samos is attributed to Greece with other Aegean Islands.

Unfortunately, these events are mapped with wide allowances. Greece's independence is doubtfully managed by official mappings—in both maritime and mainland areas (Fig. 3.8). The Sporades Islands, for instance, still belong to the Ottoman Empire in 1843 according to Vuillemin, even though they should be separated. The fate of many minor oriental islands is also undetermined. West of Lamia's Gulf, the position of the boundary is uncertain.

The islands positioned on the external face of the Peloponnese are another example of geographic indeterminacy. The British protectorate of the Heptanese, established at the beginning of the nineteenth century, has remained separated from the Kingdom of Greece until 1864. Defining such a pelagic country spatially is not a trivial issue, because of its scattered islands. Few historical maps agree to find similar borders of the state. In 1832, for instance (Fig. 3.9), Lapie and Arrowsmith rightly mention Cythera as one of the seven main constitutive islands of the Heptanese. But the fate of Anticythera, Sapienza, and Schiza is something of a fluke, once attached to Greece, the next time to the Ionian Islands. Other islets inherit worse destiny: They are seen as "nowhere lands." Kiepert, whose skills and knowledge have been widely used for the preparation of various international treaties, heightens the uncertainty to an upper level: He successively grants Kalamos and Kastos to Greece or to the Ionian Islands, with other microislands off Acarnania. Last but not least, the specific case of the island of Sazan (Saseno), 80 km north of Corfu, off Vlorë, must be emphasized. Such *confetti* has probably never been mapped properly. Many documents simply "forget" this Ionian possession transferred to Greece in 1864, abandoned to Italy in 1913, and handed over to Albania in 1920. (Fig. 3.10)

Fig. 3.9 The Ionian Islands after the independence of Greece. *Source* David Rumsey Map Collection (Lapie 1832; Kiepert and Weiland 1855; Kiepert 1853; Arrowsmith 1832)

In the Aegean Sea, the situation of Samos is also precarious. At the Treaty of London (1832), the island gains a status of autonomous principality placed under the guardianship of the Ottoman Empire. This position is close to the one granted to Wallachia and Moldavia, but few maps mention the specific case of Samos, even though they all picture the trans-Danubian principalities as "semi-independent." The error persists despite significant advances registered by the existing mapping means at the beginning of the twentieth century. Vogel's map describes the island as simply belonging to the late Ottoman Empire. This example underscores the fact that the legal status of land is a serious issue to address in traditional mapping. When the status is "ambiguous"—which is frequently the case for areas placed de facto under the guidance of foreign nations and Empires, maps are likely to adopt basic rather than complex political categorizations.

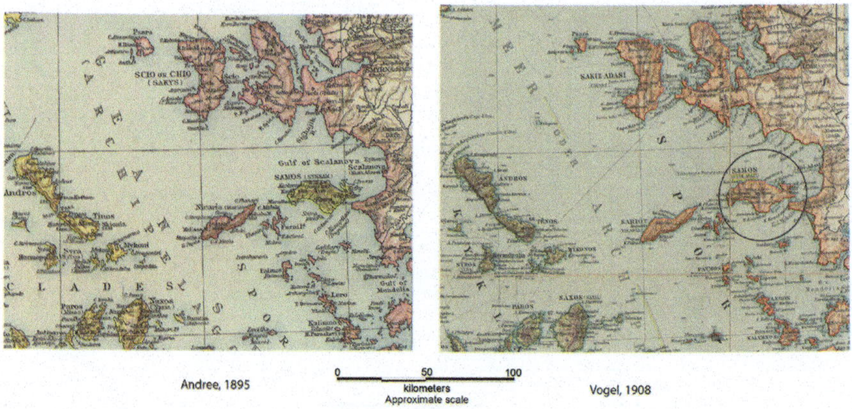

Fig. 3.10 The principality of Samos at the end of the nineteenth century. *Source* David Rumsey Map Collection (Andree 1895), Harvard Geospatial Library (Vogel 1908)

There are many such mapping mistakes. It is interesting to note that the inaccuracies even involve famous cartographers of the nineteenth century. Heinrich and Richard Kiepert (father and son), for instance, remain at the forefront of the advancement of Balkan mapping until the end of the Ottoman Empire. They respond positively to the request for assistance made by the Ottoman army and develop high-scale topographic maps of Rumelia and Anatolia that will be used during the Balkan Wars and WWI. On the eve of the twentieth century, a page of the Empires' cartography is being turned. The first aerial photographs and the advances of photogrammetry allow the creation of new "industrialized" military institutes that will replace the old school mapmakers and triangulation techniques. Large numbers of engineers solve mapping uncertainties through elaborated means and transfer the confusing potential in the political field. With the emergence of the nation-states, topographic maps gain a new confidential status proportional to their *efficiency*. Mistakes might be public. Accurate information shall be kept secret.

Such a statement gives credence to the arguments developed by Denis Wood who identifies the parallel emergence of modern state and cartography. Historically, mapping never aimed at reproducing some *spatial reality*. The entire process of mapmaking is dedicated to building the *State* rather than measuring land. In other words, mapping is a pure expression of an institutional power *against* political opposing forces. Not surprisingly, maps promote land integrity against any spatial fragmentation. Therefore, unorthodox political subsets are easily "erased" from pictures, the purpose of which is to tell to the world how much each state applies its sovereignty up to the most remote districts.

"It was the maps that conjured up borders where none had existed; the maps that summoned unity from chaos; the maps that enrobed the shapeless, that is, the maps that endowed with form what from the beginning had been no more than a dream" (Wood 2010), p. 33.

References

Adler RK, Furness S, Schofield CH (2001) Geographical information in delimitation, demarcation, and management of international land boundaries, vol 3. International Boundaries Research Unit, University of Durham, Durham, Boundary and Security Briefing

Andree R (1895) Greece, with Crete or Candia. The Times, London

Arrowsmith J (1832) Turkey in Europe. J. Arrowsmith, London

Ballinger P (2003) History in exile: memory and identity at the borders of the Balkans. Princeton University Press, Princeton

Bechev D (2004) Contested borders, contested identity: the case of regionalism in southeast Europe. J SE Eur Black Sea Stud 4(1):77–95

Berghaus H, Flemming C, Handtke F, Sohr K (1855a) General-Karte der Europäischen Turkei und des Vladikats Montenegro. Flemming, Carl, Glogau-Leipzig

Berghaus H, Flemming C, Handtke F, Sohr K (1855b) Karte von der Moldau. Walachei und Siebenbürgen, Flemming, Carl, Glogau-Leipzig

Biggs M (1999) Putting the state on the map: cartography, territory and European state formation. Comp Stud Soc Hist: Int Quart. 41(2):374–405

Blake G (1995) The depiction of international boundaries on topographic maps. Bound Secur Bull Bound Secur Bull 3(1):44–51

Blake G (1999) Geographers and international boundaries. Bound Secur Bull Bound Secur Bull 7 (4):55–62

Brown LC (1996) Imperial legacy: the Ottoman imprint on the Balkans and the middle east. Columbia University Press, New York

Bunce V (1995) Paper curtains and paper tigers. Slav Rev 54(4):979–987

Capussela AL (2015) State-building in Kosovo: democracy. Corruption and the EU in the Balkans. I. B, Tauris, London-New York

Cattaruzza A, Dérens J-A (2015) Créer une frontière dans le post-conflit: le cas du Nord-Kosovo et de Mitrovica. Hérodote 3:58–75

Cattaruzza A, Sintès P (2012) Atlas géopolitique des Balkans. Autrement, Paris

Congalton RG, Green K (1999) Assessing the accuracy of remotely sensed data: principles and practices. Lewis, Boca Raton

Dimitropoulos D (2008) Limites intérieures dans l'espace grec au temps de la domination ottomane. L'aspect géographique des administrations politiques et ecclésiastiques. Hist Rev-La Revue Hist 5:239–253

Dogo M (2003) Holding the empire together. The Ottoman decline and frontier provinces in the Balkans. Riv Storica Ital 115(2):516–542

Foody GM (2007) Map comparison in GIS. Prog Phys Geogr 31(4):439–445

Foody GM, Atkinson PM (2002) Uncertainty in remote sensing and GIS. John Wiley, Chichester

Foucher M (2011) La bataille des cartes. Bourin, Paris

Garde P (1999) Vie et mort de la Yougoslavie. Fayard, Paris

Grassl J (1860) Neueste Special & Kriegs-Karte der Europäischen Türkey in IV Blättern. Bibliographischen Instituts Hildburghausen, Hildburghausen

Gropas R (2004) Functional Borders, Sustainable Security and EU-Balkan Relations. SE Eur Black Sea Stud 4(1):49–76

Hajdarpašić E (2008) Out of the ruins of the Ottoman empire: reflections on the Ottoman legacy in south-eastern Europe. Middle East Stud 44(5)

Hassner RE (2006) The path to intractability: time and the entrenchment of territorial disputes. Int Secur 31(3):107–138

Hoffmann C (2008) The Balkanization of Ottoman rule: premodern origins of the modern international system in southeastern Europe. Coop Confl 43(4):373–396

Hupchick DP, Cox HE (2001) The Palgrave concise historical atlas of the Balkans. Palgrave, New-York

Karl TL, Schmitter PC (1995) From an iron-curtain to a paper curtain: grounding transitologists or students of postcommunism? Slav Rev 54(4):965–978

Kiepert H (1853) Das Königreich Hellas oder Griechenland und die Republik der Sieben Ionischen Inseln. Geographisches Institut, Weimar

Kiepert H, Weiland C (1855) Das Osmanische Europa oder die Europäische Türkey nebst dem Königreiche Griechenland und den Ionischen Inseln. Geographisches Institut, Weimar

Klemenčić M (1999) The border agreement between Croatia and Bosnia-Herzegovina: the first but not the last. Bound Secur Bull 7(4):96–101

Klemenčić M, Schofield C (1995) Croatia and Slovenia: the "four hamlets" case. Boun Secur Bull 2(4):65–77

Labrianidis L (2001) 'Internal frontiers' as a hindrance to development. Eur Plan Stud 9(1):85–103

Lapie M (1832) Carte de la Turquie d'Europe et de la Grèce moderne. Eymery, Fruger et Cie, Paris

Lory B (ed) (2002) Contentieux micro-territoriaux dans les Balkans, XIXe–XXIe siècles, vol VI. Balkanologie, vol 1–2, Paris

Lunetta RS (2004) Remote sensing and GIS accuracy assessment. CRC Press, Boca Raton

Magocsi PR (2002) Historical atlas of central Europe. University of Washington Press, Seattle

Mamadouh V, Kramsch O, Van Der Velde M (2004) Articulating local and global scales. TESG Tijdschrift voor economische en sociale geografie 95(5):455–466

McCarthy J (1995) Death and exile: the ethnic cleansing of Ottoman muslims, 1821–1922. Darwin Press, Princeton

Mizrahi JD (2009) De la région frontière à la ligne frontière. Les confins méridionaux de la Syrie de la fin de l'Empire ottoman au début des mandats. Vingtième Siècle 3:77–90

Monmonier M (1991) How to lie with maps. University of Chicago Press, Chicago

Newman D (2006) The lines that continue to separate us: borders in our 'borderless' world. Prog Hum Geogr 30(2):143–161

Nicod J (2004) Les nouvelles frontières dans l'espace dinarique: résurgences historiques et ethniques, adaptations locales au relief et quelques limites aberrantes. Méditerranée 103(3–4):5–20

O'Dowd L (2010) From a 'borderless world' to a 'world of borders': 'bringing history back in'. Environ Plan D: Soc Space 28(6):1031–1050

Ki Omae (1990) The borderless world: power and strategy in the interlinked economy. HarperBusiness, New York

Özdemir K (2008) Osmanlı haritaları. Avea, Istanbul

Paasi A (2009) Bounded spaces in a 'borderless world': border studies, power and the anatomy of territory. J Power 2(2):213–234

Prescott JRV (2014) Political frontiers and boundaries. Routledge, London, New York

Prescott JRV, Schofield C (1986) Maritime political boundaries of the world. Nijhoff, Leiden, Boston

Radefeld CCF (1845) Europäische Türkey. Bibliographischen Instituts, Hildburghausen

Robinson AH, Morrison JL, Muehrcke PC, Kimerling AJ, Guptill SC (1995) Elements of cartography, 6th edn. John Wiley & Sons, New York, N.Y

Sardon J-P (2004) Les populations des Balkans. Espace Popul Soc 3:459–717

Schofield C, Klemenčić M (eds) (1996) The maritime boundaries of the Adriatic Sea, vol 1. Maritime Briefing, vol 8. University of Durham

Schofield CH, Lee S, Kwon M-S (eds) (2013) The limits of maritime jurisdiction. Martinus Nijhoff, Brill, Leiden; Boston

Sellier A, Sellier J (1995) Atlas des peuples d'Europe centrale. La Découverte, Paris

Sheppard ES, McMaster RB (eds) (2004) Scale and geographic inquiry: nature, society, and method. Blackwell, Malden, MA

Sikimić B, Hristov P, Golubović B (2012) Labour migrations in the Balkans. Verlag Otto Sagner, München-Berlin

Sintès P (2010) La raison du mouvement. Territoires et réseaux de migrants albanais en Grèce, Karthala-MMSH-EFA, Paris

Sivignon M (2009) Les Balkans. Une géopolitique de la violence, Belin, Paris

Stieler A (1875) Die Europäische Türkei. Justus Perthes Gotha

Sui DZ (ed) (2008) Geospatial technologies and homeland security: research frontiers and future challenges. Springer, Dordrecht, London

Thapa K, Bossler J (1992) Accuracy of spatial data used in geographic information systems. Photogram Eng Remote Sens 58(6):835–841

Todorova M (1994) The Balkans: From discovery to invention. Slav Rev Interdisc Q Russ Eurasian East Eur Stud 53(2):453

Todorova M (2002) The Balkans as category of analysis: borders, space, time. Geschichte Und Gesellschaft 28(3):470–492

Todorova M (2009) Imagining the Balkans. Oxford University Press, New-York

Vidmar J (2013) Democratic statehood in international law: the emergence of new states in post-cold war practice. Hart Publishing, Oxford-Portland

Vogel C (1908) Die Balkan-Halbinsel in 4 Blättern: Türkei, Rumänien, Griechenland, Serbien, Montenegro. Bulgarien, Justus Perthes, Gotha

Vuillemin AA (1843) Etat de la Grèce. Langlumé, Paris

Wang F, Hall GB (1996) Fuzzy representation of geographical boundaries in GIS. Int J Geogr Inf Syst 10(5):573–590

Wood D (2010) Rethinking the power of maps. Guilford Press, New-York

Yérasimos S (1991) Balkans: frontières d'aujourd'hui, d'hier et de demain? Hérodote 63:80–98

Yeung HWC (1998) Capital, state and space: contesting the borderless world. Trans Inst Br Geogr 23(3):291–309

Chapter 4
Balkan Border Mapping: Building the GIS Project

Abstract This section presents the methodology. Thanks to a corpus of approximately 400 maps, a GIS database was built with the intent of identifying all official and semi official boundaries from 1800 to present day. The geocoding workflow allows the development of graphic representations blurring conventional images of the Balkans, far beyond official mappings. Border areas leave their "anonymity" and gain a new status that allows unveiling the "hidden" geographic features of the Peninsula.

Keywords Data processing · Methodology · Image rectification · Censorship · Border status · Independence · Autonomy · Border mobility · Spatial modeling

Providing new insights into the issue of border mobility supposes to map past demarcations through an integrated GIS tool combining geographic and statistical databases. The initial work program should have been conventional (Bailey and Schick 2009; Atack 2009). Unexpectedly, however, this undertaking came with many challenges. The processing of data turned out to be highly complex and time-consuming, for different reasons (Gregory and Ell 2008; Hiller and Knowles 2008; Von Lünen and Travis 2013; Ott and Swiaczny 2001; Caluwe et al. 2004).

This project aims to provide a full georeferenced database addressing the border issue over the past two centuries, i.e., since the decline of the Austro-Hungarian and Ottoman Empires and the emergence of nation-states. The ambition is not limited to that point. The database shall be able to cover high-scale accuracy needs (Donaldson 2008; Jones 1945; Mamadouh et al. 2004). It must provide access to confined areas severely affected by the goings and comings of a great number of versatile boundaries. It shall also include different other limits/frontiers whose status is not strictly "international," and whose importance is often greater than the "official" ones. To achieve these goals, a comprehensive strategy has been developed.

© Springer International Publishing Switzerland 2017
R. Darques, *Mapping Versatile Boundaries*,
Springer Geography, DOI 10.1007/978-3-319-40925-2_4

4.1 Chronological and Spatial Limits of the Mapping Work

4.1.1 *Defining the Balkan Area on a Spatial Basis*

The initial project limited its scope to an area defined by the former Ottoman Empire, i.e., to provinces that had been part of *Rumelia* at least once. But soon it was found that the goal could not be reached while excluding many outlying parts of the Austro-Hungarian, Russian, French, and British Empires neighboring the Ottoman area. The primary perimeter had to grow to keep records of the spatial variability of the Empires. The same operating principle has been adopted for states that have succeeded the Empires: Each present country had to be studied in light of its successive spatial configurations. Given that the contours and limits of the states vary in time, the GIS database came to reference any land *legally* owned at least *once* by the concerned countries, even when such annexations ran for a limited period.

At the beginning of the nineteenth century, for instance, Wallachia and Moldavia used to be vassal Principalities of the Ottoman Empire. As such, they had to be included in the database. The same applies for states that have achieved independence later. For this reason, Slovenia and Croatia—two constitutive republics of a former Yugoslav Federation whose center of gravity was located in Belgrade— have been maintained within the study area, although their Balkan identity is much discussed.

As we will define it further on, the Peninsula has a distinctive, unlikely shape (Kavalski 2006). The adopted spatial limits roughly correspond to an extended classical definition (Fig. 4.1). This includes present Greece, Albania, Bulgaria, Romania, and former Yugoslavia. Intentionally, Cyprus and Turkey are left aside. The final surface gain is significant: +26 %. From the "narrow" present pattern to the "extended" *diachronic* perimeter, the area receives 200 000 km^2—a surface almost equal to present Romania (Table 4.1).

Figure 4.1 illustrates the *full extent* of the Balkan area, as a bridge between Western states, Central Europe, Asia Minor, and the Commonwealth of Independent States. Far from being disconnected with the rest of Europe, the region is not isolated, but porous to the influence of its neighbors. Romania and Greece encroach on more eastern lands and testify to the futility of many political debates on where to draw the line between *Europe* and *Asia*. The same statement applies to the supposed division between western and eastern Europe in Slovenia. In other words, the Balkan area trespasses its frontiers.

4.1.2 *Temporal Limits*

The mapping work covers a period beginning with the establishment of the sovereign nation-states. The census starts in the 1810s with the first steps of the Serbian autonomy and Greek revolution and continues down to the present day,

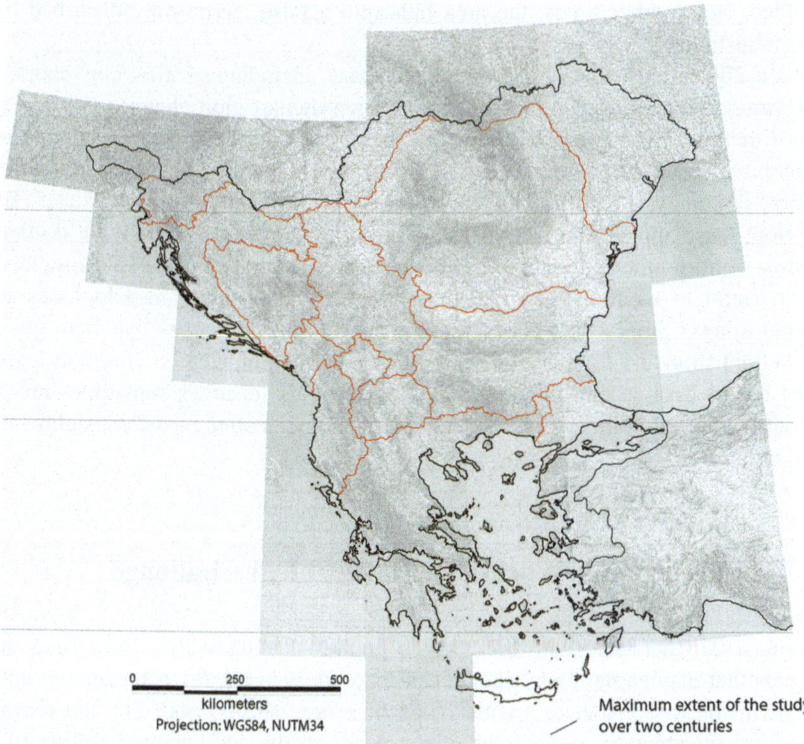

Fig. 4.1 Balkan extents: a regional versus national approach. *Source* Russian Military Topographic Maps 1/500.000, author's own GIS database, ESRI

Table 4.1 Compared conventional versus extended Balkan perimeters and geographic units

	Present land areas	Extended Balken perimeter[b]	Gain	%
Surface km^2	763 843	962 952	199 109	26
Land surface "units"[a]	15	431	416	2773

[a]For now, eleven countries share space, including Kosovo. The amount reaches fifteen when adding other various "autonomous areas" introduced in the database. The *unit* is defined as the smallest polygon left undivided by any boundary, regardless of the period
[b]Synthetic and non-simultaneous data over 2 centuries
Source ESRI, author's own GIS database

including the Yugoslav crisis and latest spatial developments. In that purpose, producing a comprehensive chronology of all wars and political events leading to any frontier change was a prerequisite. Taking back a necessary distance, the Peninsula reveals endless border fluctuations randomly interrupted by peace periods. Sometimes, obsolete spatial orders break into pieces. All cards are then reshuffled. Some other times, the area falls into a false inertia only disturbed by minor adjustments.

About **200 dated events** form the geodatabase, including treaties, conventions, open wars, and uprisings that lead to any known demarcation change. This series raises difficulties: The events' flow is not regular; changes can be insignificant or considerable, leading to overall upsets. This means comparing uneven episodes. For instance, the analysis testifies to the existence of a series of short-lived states. By definition, these appear and fade away within a few days or weeks. What do they weigh in front of other political *mastodons*, such as several centuries-old Empires? Was it useful to mention their existence? Was it not a better idea to focus on essential things? Similarly, who matters the fate of minor *counties* that were given some administrative autonomy in front of Empires submitted to a common legislation? Let us assume that main political and territorial changes sometimes reveal their nature through ordinary events. Limited phenomena often provide insights into major issues.

4.2 The Geocoding Workflow: A Technical Challenge

Building a GIS database with these goals implies dealing with serious mapping problems that may explain why such a census work escaped any achievement until present. In the first instance, a wide range of source sheets, both old and recent, suffer from information weaknesses, which leads to the cumbersome feeling of a general *uncertainty*. Modern GIS-based mapping techniques may even enhance this feeling: The raster to vector transform preceding the process of any geodatabase development causes further impoverishment of *original* maps. This is "double penalty" (Longley et al. 2010).

Today, most GIS databases derive from digital topographic sources endowed with exact geometric characteristics. Such sources usually originate from governmental and/or military programs. It is well known that these organizations adopt conventional approaches. Heavy financial resources enable them to develop complex techniques that deliver, in the end, spatial information with measurable quality assessment. With such level of investment and for security reasons, the authorities often limit the dissemination of data through legal regulatory frameworks that decide how geographic information might be released.

In mapmaking, one can identify three historical periods over the years 1800s–2010s[1]:

1. Until the early twentieth century, lasting ground investigations and triangulation techniques produce "classical" maps (Gartner and Ortag 2010). The collected data are by definition inaccurate. The limited advances in geodetic knowledge

[1]The information is here reduced to a strict minimum. Hundreds of books provide serious reflection on this subject.

do not ensure an equal geometric accuracy to all delivered sheets. *Countervailing measures* based on extrapolation techniques need to be introduced. The overall information shortage is thus compensated by the introduction of many graphic artifacts. Any map is valuable, although each sheet inherits varying qualities and defaults. Scales randomly exceed 1/300.000.

2. Aerial photography opens up a new period for further mapping developments (1900–1980). National topographic covers become a "must-have" tool any modern country is supposed to earn. Effective systems enabling methodical data gathering determine the emergence of reliable information. The task of developing a production line is entrusted to powerful military organizations responsible for implementing firm controls on data dissemination. The scarcity of *published* data becomes a rule—this is planned on purpose. Civil map sheets are impoverished and include deliberate mistakes when available (Monmonier 1991). They may reach high scales (randomly more than 1/25.000) (Fig. 4.2).
3. The industrial development of satellite imagery—especially through very high-resolution products—paves the way for a large-scale dissemination of information (from the 1980s onward). Image accuracy reaches peak levels never imagined before (absolute error positions below one meter or decimeter). This situation implies that national military institutes lose their monopoly. By the way of different types of digital media, geographic information becomes a product of "general commodity." The former publication shortage turns into an overabundance. However, the development of complex algorithms in the GIS and remote sensing fields is hampered by a rise in misuses. The vast amount of digital data and the free circulation of software tools introduces a new era of qualitative variability related to the skills of each operator and to the relevance of his methodological choices.

The project designed to process such weighty cartographic databases has proved to be challenging, because reference sheets suffer from an extreme qualitative variability. Significant hurdles have also impeded data collection: Sometimes, gathering information was an easy task; some other times, simple access requests faced a wall of rejection. Furthermore, any data correspondence between countries was impossible to achieve. States have developed very personal visions of their spatial environment, to such an extent that they virtually ignore any physical attachment to neighboring *companions*: Beyond linear boundaries, a yawning void usually spreads out on the outer face of national territories. The official mappings also *forget* minor historical "accidents"—temporary spatial expansion or contraction of *homeland*—considered as worth not depicting. Beside these one-sided positions, systematic transnational mappings find random expressions (Branch 2011; Ruggie 1993). Most historical atlases conventionally cover wider regions, but few of them specifically address the Balkan area. They also lack topographic accuracy.[2]

[2]This is an understatement. Many of them widely use "rough" linear Bezier curves.

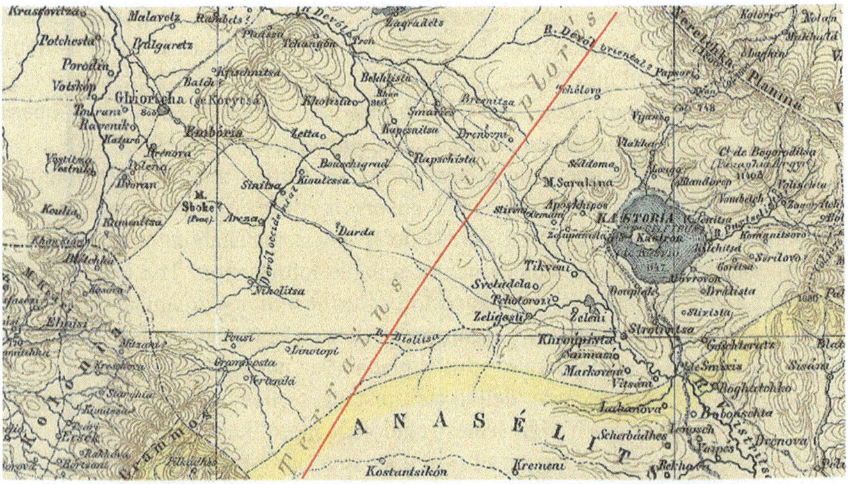
Malavotz
Potchesta
Zagradets
R. Devol
Papnori
Planina
Smardes
Brenitsa
Emboria
Zetta
Rapschista
Kisutessa
Bouschigrad
Sinitsa
Darda
Sbohe
Arena
M. Saraluna
Aposkhipos
KASTORIA
Kastron
Tikveni
Sveladela
Zelegosti
Zeleni
Nikolitsa
Stroltsa
Khroupista
Marhowsa
Vitsam
Linotopi
Bistitsa
Kostantsikon
Kremeni
A N A S É L I T
Lubanova
Scherbadhes
Leousch
Dresnova
Bekhaba
Grammos
Drakista
Gochtertz
Bobonschta

22339

◀ **Fig. 4.2** From mapping shortage to censorship on geographic information. *Sources* Carte de l'Epire et de la Thessalie (Kiepert 1880), censured aerial photograph of Salonika (western suburbs), close to its historic center. Hellenic Military Geographical Institute, 1965, estimated scale 1/16.000. *The first high scale (1/500.000) and rather late sheet of H. Kiepert mentions the existence of "uncharted territories", but the observation is missing on lower scale maps of the author. The area covers the current Greek–Albanian confines, still Ottoman at that time. On the bottom aerial photo, Black areas are erased from all circulated documents delivered on both films and printed paper. The prohibition applies to military and other "strategic" fields, although Salonika can hardly be counted as a border city*

In view of this, it was decided to build a geodatabase from scratch and to ignore second-hand sources studiously. The *Harvard Geospatial Library*[3] and *David Rumsey Historical Map Collection*[4] provided us some "primary" raster files. These online libraries are rich, although incomplete. Other maps originate from different national collections disseminated in Europe, the USA, and Russia. The *Hellenic National Map Library* keeps records of valuable sheets in particular. The book series published by the *Cambridge Archive Editions* (Wagstaff 2002; Jarman 1997; Destani 2001; Destani 1999; Burdett 1995) or the body of texts collected by E. Hertslet (Hertslet 1875–1891) also proved to be useful. Many other maps have been gathered during stays in the Balkans, some of them being edited through large sheets or reproduced within finely illustrated books and atlases.

The result is an eclectic database made up with mixed documents. To achieve our goal, we have processed about **400** historical maps of different qualities, purposes, and origins. Among this corpus, about **120** were selected and geocoded. During the geocoding workflow, a few sheets revealed such a poor geometric conformity that they were rejected. The remaining maps went through a process of digitization with the intent of producing vector graphics.

The geometric processing supposed to identify a basic document that would become a *reference* for correcting all other maps. The accuracy and quality of this universal "background" would affect the outcome of the whole project. The absence of any homogeneous modern topographic coverage punched its weight. One step toward resolution was to use the Russian sheets dating from the end of the 1980s, whose scale ranges from 1/100.000 to 1/500.000. Insofar as some regions, especially in former Yugoslavia, are not covered, we also fell back on Landsat satellite images and other by-products like *Corine Land Cover*. These reference raster files are geometrically qualified as ortho-images, which is of prime importance when exploring mountain areas.

The geocoding workflow has been completed with the ERDAS-ErMapper remote sensing program. Considering that no *direct model* of geometric rectification[5] could be used because of the diversity of map sources and in the absence of any information (metadata) about older sheets, it was decided to apply a systematic polynomial rectification with varying degrees granted to the polynomial, in light of

[3]http://hgl.harvard.edu.

[4]http://www.davidrumsey.com.

[5]Otherwise called "physical" model.

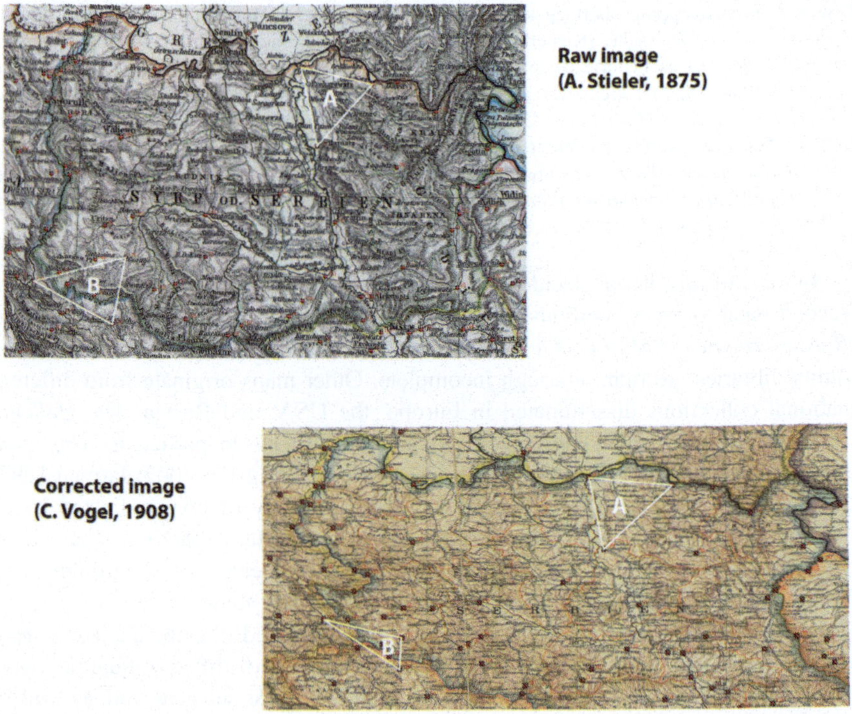

Fig. 4.3 Ground control points and geometric conformity. Uneven map distortions and absolute position accuracy. *Source* David Rumsey Map Collection (Stieler 1875), Harvard Geospatial Library (Vogel 1908)

each image quality. Dozens of *pair points* (between **50 and 150** pairs) were necessary for geocoding each sheet. Remote sensing experts perfectly know that the polynomial method is a last recourse's solution as it does not ensure that the rectified final documents receive the same equivalent qualities. Therefore, many geometric distortions have not been solved, especially with the oldest sheets where coordinate delta errors widely vary within the maps themselves—hence, the pressing needs to capture a maximum number of homologous points. Several months were needed to achieve this especially time-consuming task (Fig. 4.3).

Why take the risk of launching such a large project? Is the maximum scale precision a legitimate aim? Determining the exact place of countless boundaries—considering the technical limits of each historical period—was, in fact, a *sine qua non* condition for bringing this exercise to its achievement. There is a straightforward reason for this: A position uncertainty applied to hundreds of borderline segments would have deeply reduced the benefits of our GIS database. In addition, a shift of several 100 m may have no importance when applied to uninhabited places. In our case, the fate of many hamlets, villages, and cities was put in jeopardy. Here again, a location error applied to hundreds of places over hundreds of border sections would have invalidated numerous results.

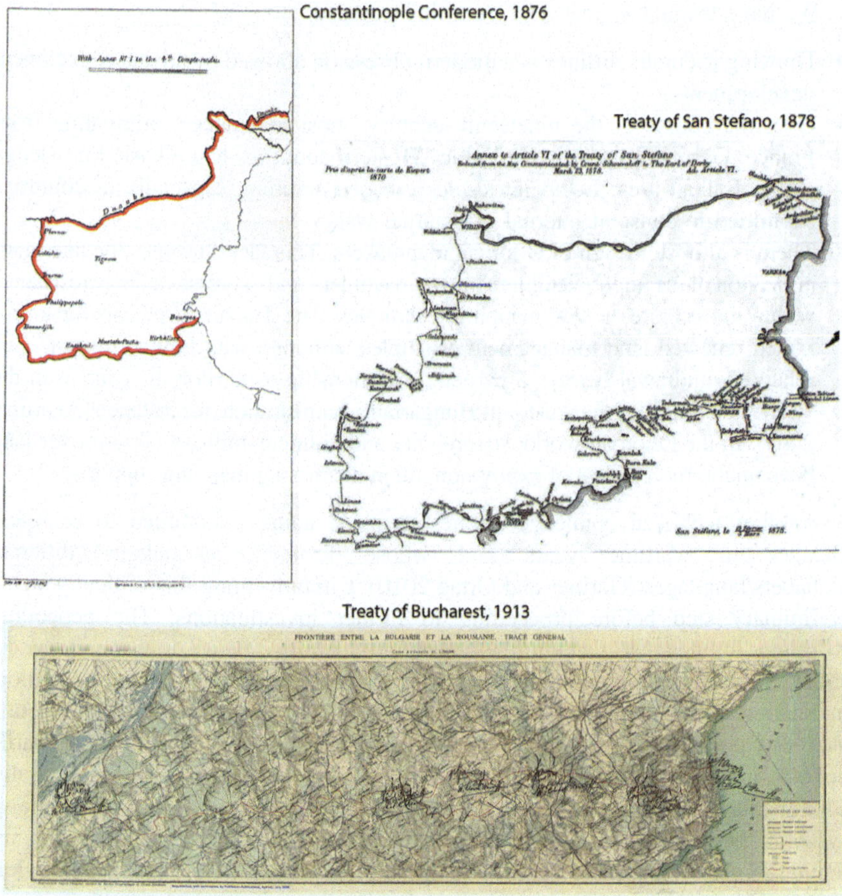

Fig. 4.4 Some maps of Bulgaria endorsed to official treaties. *Sources* Foreign Office of Great Britain, eastern Vilayet of the Constantinople Conference and Great Bulgaria of San Stefano (Foreign Office of Great Britain 1878a, b), Ministère des Affaires Etrangères, southern Dobrudja section (Ministère des Affaires Etrangères 1913)

To keep its relevance, the entire workflow has been thought for precision measurement. Even though, rectifying the geometry of 120 old maps will inevitably produce uneven outcomes. As a result, the final accuracy of vector graphics shall be appreciated on a case-by-case basis (Gartner and Ortag 2010; Thapa and Bossler 1992). These partially inherit characteristics of their originals, despite rectifications (Fig. 4.4).

No digital processing can solve scale discrepancies between original documents that suffer either from *too* accurate or *too* rough demarcations. The respective positions of all borderlines recorded in the final geodatabase may thus be interpreted as *best estimates*. When old map sheets could be compared, various mapping errors were detected and solved.

We have learned a great deal from this exercise:

- Drawing accurate outlines is a modern obsession allowed by advanced technical developments.
- Over the course of the nineteenth century, most mapmakers originating from France, Germany, Austria, and other Western countries had a basic knowledge of the Balkan area. Local intricacies escaped scrutiny, especially in countries submitted to constant internal or external wars.
- There is almost no sign of Ottoman mapmakers. This fact illustrates the late modernization of the imperial administrative machinery and expresses a fierce resistance within the Empire to new mapping techniques. The Tanzimat reforms have promoted renewed land management principles, although with limited impact. This relative withdrawal betrays a particular relationship to territoriality, far from the "Cartesian mind" of the Austrian–Hungarian administration, for instance (Akerman 2009). In the Ottoman world, sovereignty and political–military control over land have uncommon means of expression, often dismissing mapping supports.

Another technical challenge applies to place names, submitted to ceaseless changes in wartime years and whose formalization adopts different alphabets/languages (Gartner and Ortag 2010). Circumventing this difficulty was a preliminary step before proceeding to further investigations. The geocoding workflow implies that matching landmarks are identified between past and present sheets. The main problem comes from the various forms each toponym may adopt in relation to the nationality of mapmakers. Most names have been frequently altered: Upon leaving their Ottoman or Austrian–Hungarian form, they primarily uphold national patterns (i.e., alphabets/languages); afterward, the socialist administrations manage to leave their mark by revising old-fashioned national systems; the latest wave of change takes place after the collapse of the communist regimes. On average, these successive decisions represent thousands of toponymic changes. Fortunately, the experience gained from previous works has contributed to dodging many traps. When applied to familiar places, the issue could be solved without too much difficulty. Conversely, identifying ground control points pertaining to hamlets ruined decades ago and referenced under various toponymic forms could sometimes be a daunting experience.

The accuracy thus varies from one borderline to the other. This inconvenience leaves the GIS database coherence untouched. It just tells how it should be interpreted. Although each boundary position has been validated by cross-comparison, some misgiving degree persists. As already said, uncertainty is a key concept to put forward when dealing with Balkan borders and boundaries. *Uncertainty* is also a way of qualifying the difference between *borderlines* and *borderlands*. This concept allows us to shift from geometric outlines—a technical artifact indeed—to surface areas (Shields 2006). Frontiers should rather be viewed as contacting areas, extensive regions. In map processing, this standpoint introduces *fuzzy logics* in the research arena. All else being equal, the most advanced mapping techniques unexpectedly meet the "old" border management system established by the Ottomans.

4.3 Legal Independences and Semiofficial Autonomies: Toward a Border Classification

The main goal of this work is not only to help identify accurate boundary positions but to observe these objects through the lens of multiple categorizations. Infield experiences early taught us that studying political–national boundaries is a real challenge. Many geographic limits are much more "violent" or simply obvious to landscape observers than international boundaries "peacefully" gained by lush vegetation. These internal limits deserve much less attention from scientists despite their prevailing incidence (Péchoux 1977; Smith and Varzi 2000). Thus, interstate boundaries revealed too poor a category to satisfy our curiosity.

In view of this, the survey shall be completed with nonstandard frontier units. In the Balkans, all nation-states are unsure about where to fix their outer limits. In addition, the issue of human mixing is always critical. This leads us to believe that borders and boundaries adopt many nuanced forms. In other terms, traditional mappings artificially establish *consistent* countries that, in reality, are neither uniform nor homogeneous. This bias should be partially solved by the introduction of levels of regional and local autonomy.

Two categories of "boundaries," more or less conventional, have been defined:

1. The first group is made of international boundaries, **legally** settled through formal treaties and recognized by the international law, regardless of their existence period and without considering the conditions of their being developed.
2. The "semiofficial" group is a motley collection of five subcategories:
   - *short-lived Republics* emerging here and there in the Balkans across time.
   - *war annexations* not recognized by later treaties, although officially approved by few belligerent states. This category does not include the occupation zones, whose existence is by definition temporary and non-official.
   - *contested*, although effective, *boundaries* not fully recognized by the international community.
   - *self-proclaimed states* of the Yugoslav Wars.
   - *autonomous provinces* and *special status regions* within the Austrian–Hungarian and Ottoman Empires. They often hide an official and negotiated local "semi-independence"—just think about the trusteeship of the Ottoman "Principalities," the Saxon settlements in Transylvania, the *Pashalık* of Ali, or Mount Athos. This category might be assimilated to "internal frontiers" or boundaries "to be."

This second large and diverse group has been designed to highlight the critical importance of local territories against fully accredited "countries." Many boundaries officially supported by very official treaties and conventions remain without application in practice. They are smoky concepts once drawn over a map in the secret of padded conferences. In contrast, very few mapmakers address the fate of truly autonomous regions—seeds for future independence. The gap between official and "semiofficial" groups should be reduced, at least partially, with the integration of other *demarcations* into the GIS databases. There is no pretension to be exhaustive: The "semiofficial" list is not complete and is thought to remain open.

It is no coincidence that any national claim will remain outside the scope of the analysis. Serious attention needs to be paid to the essential difference between *effective* and *dreamed* borders. For what we are concerned, *fantasized* perspectives on land will not be discussed.

4.4 The Structure of GIS Databases

At this point, few other elements are worth mentioning:

1. The digitizing process has produced two vector databases:
   - The first one is made of POLYGONS defining primary surface "units" surrounded by current or past boundaries. These narrow areas might be viewed as territories—often microterritories—that have escaped any spatial division over the past two centuries. They are *supposed* to keep some geographic cohesion depending, of course, on local conditions in the long-term history.
   - The second one is made of POLYLINES and derives from the latter. These are linear segments that begin and end when boundaries overlap—disregarding any time correspondence.
2. Two statistical databases will complete the set:
   - The first statistical database refers to the POLYGONS and defines the *identity card* of each geographic unit, its affiliation in terms of sovereignty, and its category (official, semiofficial). Of course, any status change is reported.
   - The second statistical database also derives from the latter and applies to the POLYLINES. It systematically combines the trusteeship with the category of both tangent polygons, thus defining the attributes of each border section and its changes across time. The external Balkan boundaries are kept "undefined" because their outer neighborhoods remain unidentified.

A table can help evaluate the amount of energy devoted to these works (Tables 4.2 and 4.3).

Deriving polylines from polygons is a complex but necessary operation. In the first instance, the process is useful for cross-checking data and detecting errors. This may be extremely time-consuming. For instance, the statistical database which refers to polylines results from the application of about 40 sequential algorithms. At each stage, any error detection could validate the entire processing operation or disclose improper calculations. Three times it brought us to repeat the whole work chain

Table 4.2 Graphic databases' structure

	Unit numberb	Total surface and length	Min. area and distance	Max. area and distance	Mean	Mean std deviation
Polygons	431[a]	962 605 km^{2c}	0.4 km^2	74 196 km^2	2249 km^2	2894
Polylines	1207[b]	47 306 km	0.2 km	746 km	39.2 km	33.39

[a]A unit may include several polygons, for instance, when gathering land and/or island areas
[b]1138 land sections (27 external, 1165 internal), 65 sea sections, 85 coastline segments and 1579 islands are excluded from the count
[c]Sea surfaces excluded. Internal water surfaces included
Source Author's own GIS databases

Table 4.3 Statistical databases' structure (MS Access and Excel platforms)

	ID number (rows)	Number of primary variables (columns)
Polygons	431	226[a]
Polylines	1207	73

[a]113 dates × 2 (ID + category)
Source Author's own GIS databases

from the beginning: A two-month full-time performance when applied to datasheets composed of thousands of individuals and variables! These considerations only refer to primary variables. Of course, additional second variables were derived to complete the panel. All this gives substance to the oncoming analysis.

4.5 Border Mobility, Border Uncertainty

The diachronic analysis produces sobering figures. With 1 200 line segments and a cumulated length of 50 000 km, borders and boundaries throw all their weight behind Balkan geography. These data apply to an "extended" area that covers one million square kilometers. Such information makes sense when compared to the present 30 000 km of land borders crisscrossing the entire European Union, whose surface area has stabilized at approximately four million square kilometers. Of course, the census process is cumulative. Boundaries of all kinds are aggregated over two centuries. Even though, the obtained results are meaningful: Land fragmentation has eventually come to be measured.

Through cross-referenced official and semiofficial frontiers, a new way of viewing the Peninsula arises (Fig. 4.5). In general, people are unable to remember the location of a single Balkan country. We add fuel to the fire. The Balkans are almost entirely covered by a spaghetti-like network that deeply structures the area. The picture is so impressive that any post-Yugoslavian map with its unlikely microstates might look like a child drawing comparatively.

Panoptic and diachronic illustrations reveal a fragmentation that has few equivalents in the world. *Balkanization* is not an emphatic word used to depict

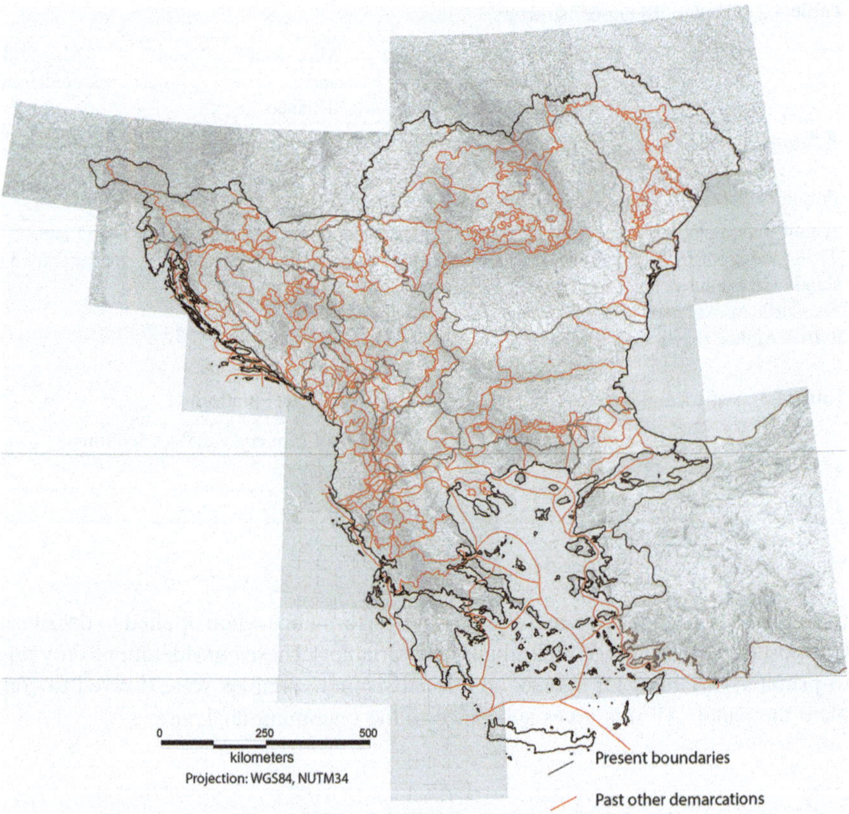

Fig. 4.5 Balkan official and semiofficial boundaries, 1800–2015. *Source* Russian Military Topographic Maps 1/500.000, author's own GIS database, ESRI

a complicated political and territorial situation. The term qualifies a smashed up geographic *body*. As a result, it becomes obvious that each current country results from the spatial combination of various surface elements, sometimes added or some other times withdrawn. The process finally produces what will be recognized as *national unity*. A simple glance at Fig. 4.5 is sufficient to persuade us that nations are eclectic creations. This cartography emphasizes the Balkan complexity much better than many typical comments pertaining to ethnic and confessional mixing—the icing on the cake of most political debates.

Boundaries are—or were—almost everywhere.[6] Notwithstanding some parts of the Danube and Maritsa basins, plus southern Greece, the Peninsula presents an entirely marbled/rippled face, with a significant difference between its western and

[6]This assertion has to be understood in a narrow sense. E. Balibar (Balibar 1998) and other authors use the same words as an opposite metaphor to express the dilution of border territoriality in the post-communist era.

eastern parts. Each "central area" has been—or may still be, albeit to a limited degree—a border region for a while. Alternatively, the eccentricity of many edge places varies strongly across time. Neither centrality nor marginality are adapted concepts for qualifying any territory in the long term.

Surprisingly, this graphic representation helps understand the European expansion and solves an enigma: the integration of Bulgaria and Romania in 2007. Greece, Bulgaria, and Romania appear here as fairly "stable" lands, keen on preserving significant central regions from any dissection effect. Their spatial cohesion seems higher than that of the western Balkan areas. A reversed geography shows up: *Western* countries are more *balkanized* than the "oriental" lands bordering the Black Sea. The "Barbarians" are not where they ought to be. Of course, this a *posteriori* assessment is of limited weight and ignores other elements.

Beside such "static" perspective, a dynamic approach should bring a new angle to the long-standing frontier issue. After two centuries of roaming boundaries and political upsets, border mobility cannot be viewed as a technical hitch, but rather as a phenomenon deeply embedded in space. To take an ordinary image, the situation might be seen as tidal movements arriving from different incident angles and opposing their energy until a temporary balance of forces is reached.

Boundary and territory are two sides of the same coin, as if neither concepts could be distinguished. At different degrees, they shape the Balkan landscapes. Considering between two and four generations, nearly all families have experienced a border change and its attendant calamities: open wars, latent armed conflicts, ethnic-confessional oppositions, forced displacements of populations, and property losses. Far from being fully achieved by a material existence, the "border effect" is a "way of life."

It does not seem far-fetched to argue that the European "open space" is an external injunction disrupting centuries-old processes. The border stability implied by the European enlargement deviates from the ordinary local scheme.

To the 9400 km of present international boundaries, the GIS database adds 32 000 extra kilometers. Data categorizations do not allow overall calculations because of redundant measured distances. This point deserves explanation. Usually, a border is defined by a single attribute: international or administrative, for instance. The adopted methodology is not compatible with this simple categorization, which supposes both neighboring units to stand in a same "equal" position. Introducing varied degrees of *efficiency* requires defining each boundary by two different references at least. In some cases, four simultaneous levels of attachment were used. For instance, in the first half of the nineteenth century, the Saxon settlements of Transylvania were "autonomous regions" sometimes included in the Military Frontier zone, itself part of the Grand Principality of Transylvania, which belonged to the Austrian Empire. Such combinations describe the conditions that prevail in the database production system. Each border may be referenced in different categories simultaneously.

The phenomenon becomes even more complex when such interlocking positions are observed across time, i.e., when all data will have to change constantly.

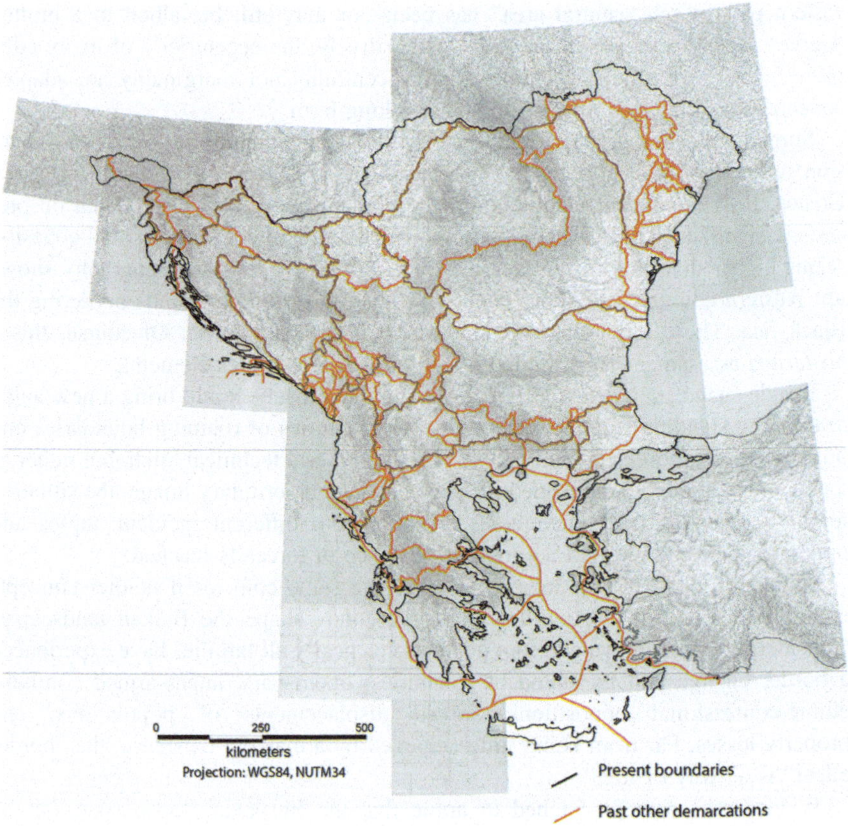

Fig. 4.6 Balkan international official boundaries, 1800–2015. *Source* Russian Military Topographic Maps 1/500.000, author's own GIS database, ESRI

Therefore, the following *synthetic* tables and maps possess a significant degree of redundancy Figs. 4.6, 4.7, 4.8 and 4.9

The decision not to limit the analysis to the official boundaries generates vast quantities of relevant research results. The often forgotten **short-lived Republics** reach more than 11 000 km of border distance. First of all, the category has emerged as a way of identifying underlying forms of local *autonomy* that could eventually face the new centralities created by the advent of modern nation states. Some of these *republics* might be seen as lasting legacies of the autonomous provinces that had thrived within the Empires, although with limited life expectancy. The short-lived Republics include various cases:

- Regions under uprising that unilaterally declare their independence, sometimes called "mountain republics." Most of them are issued from socialist armed groups, in both the nineteenth and twentieth centuries. They emerge in troubled

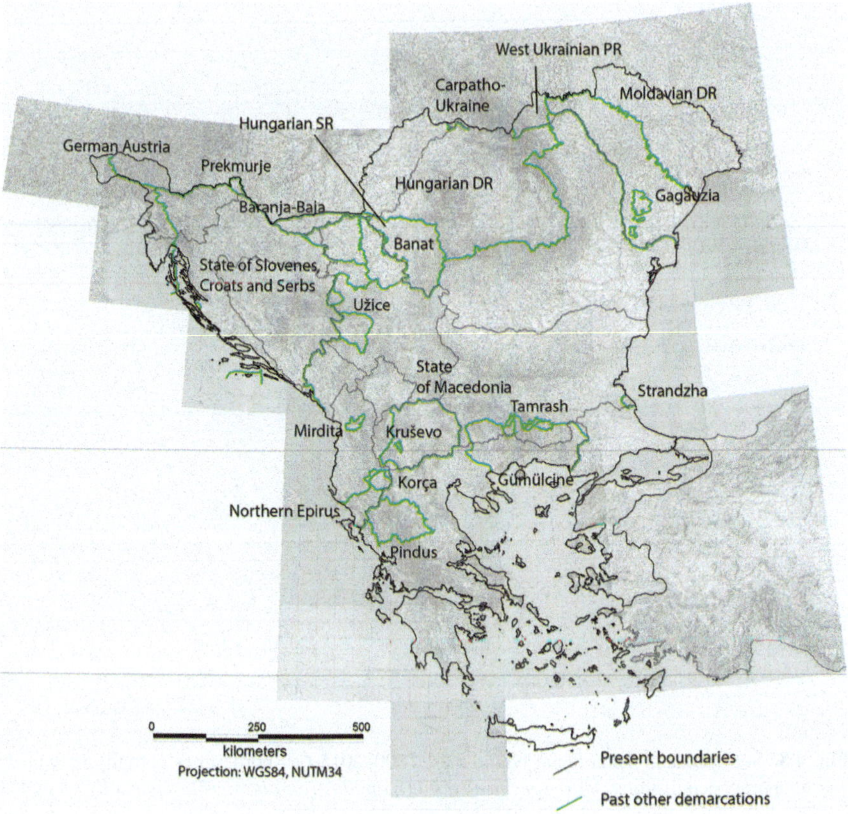

Fig. 4.7 Some Balkan semiofficial boundaries, 1800–2015 (1): Short-lived Republics. *Source* Russian Military Topographic Maps 1/500.000, author's own GIS database, ESRI

times, when central governments are so weak that applicant local organizations look forward to playing leading roles.

- Fake states imposed by foreign occupying forces during wartime.
- Buffer states that emerge as a parenthesis during periods of political crisis: Unlike previously, this mainly applies to chief nation-states.

These political bodies have two main characteristics. Their life expectancy is extremely limited, ranging from a few hours to a few weeks. Their territory is barely defined, which reduces proportionately the significance of digitized outlines. They are in-between formations that usually yield in front of strong opponents able to muster regular military forces. Despite or perhaps even because of their intrinsic weakness, they reveal the existence of alternative spatial-political models able to frustrate the logic of nation-states. Many of them claimed their attachment to a multiethnic identity model.

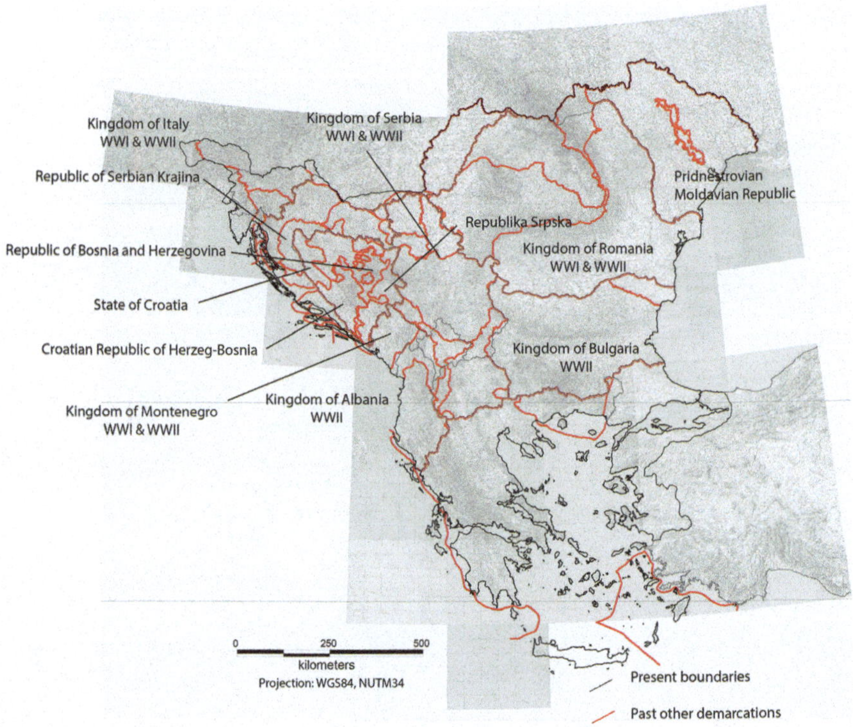

Fig. 4.8 Some Balkan semiofficial boundaries, 1800–2015 (2): borderlines partially recognized, war temporary acquisitions. *More specifically, this map illustrates borderlines partially recognized by the international community, planned war annexations not recognized by later treaties, postwar de facto autonomous communities (Yugoslav Wars, mainly). Occupation zones and other stabilized fronts are not figured. Non-exhaustive list. Source* Russian Military Topographic Maps 1/500.000, author's own GIS database, ESRI

The **autonomous regions** of the Empires could be depicted as early forms of such entities. Their position is not far from the one of other autonomous provinces belonging to nation-states or from the one of former Yugoslavian republics. They are often seeds to further independence gained through wars and ruthless negotiations. With 26 000 km in length, the fourth category of Table 4.4 comes first before the international boundaries (category 1). The group is abounding and shows, by the way, that late nation-based "monopolistic" maps rely on varied substrates. As usual, the cartographic material covers distinct entities:

- Present or past regional autonomous communities within nation-states or federations (autonomous provinces and republics).
- Regional divisions granted by diplomatic negotiations (Vilayets of western and eastern Bulgaria).

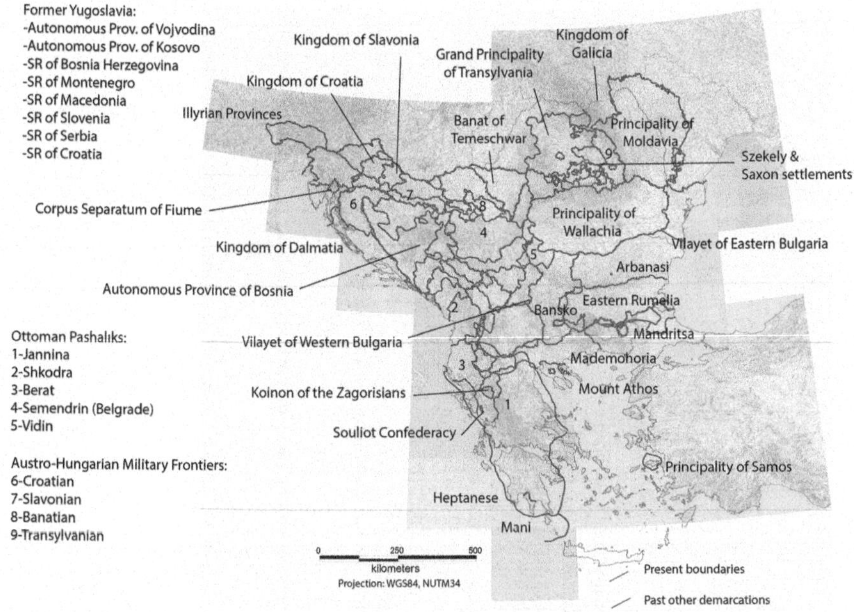

Fig. 4.9 Some Balkan semiofficial boundaries, 1800–2015 (3): autonomous "provinces," special status regions. *This category groups varied political bodies*: *Autonomous "provinces,"* imperial dependencies, *special status regions within the Ottoman, Austro-Hungarian, Russian, and French Empires, autonomous regions or republics within nation-states or federations. Source* Russian Military Topographic Maps 1/500.000, author's own GIS database, ESRI

- Quite the opposite, centuries-old autonomous communities based on elaborated legal statutes (Mount Athos, and Székely and Saxon settlements).
- Local semi-independences validated for instance by tax-exempt status, commercial privileges, transfer of special police powers (Mademohoria, Koinon of the Zagorisians, Corpus Separatum of Fiume, Arbanasi, and Bansko).
- Imperial autonomous provinces whose governors benefit from extended military and executive powers. These organizations face one another: The *Military Frontier* (Militärgrenze) is conceived as a protection for Austria against the Ottoman external threat; the extended powers of Pashas are meant to ensure land control over the imperial confines and prevent internal rebellions or external attacks unless the Pashas themselves instigate such uprisings.

This mapping portrays a fundamental pattern. First of all, distancing ourselves with the issue, nation-states are not among the most stable territories. Prime long-lasting bodies are local formations/communities that *resist* upper governing entities (nations, Empires) while avoiding direct confrontations. The Saxon settlements are established in the twelfth century, and their rights are abolished in 1867

Table 4.4 Land border length, 1800–2015[a]

	Total	Boundaries in 2015[b]	Other demarcations 1800–2015	Historical international official boundaries[c]	Historical semiofficial borders[d]		
					Short-lived Republics	Borderlines partially recognized, war temporary acquisitions	Atonomous "provinces," special status regions
				1	2	3	4
Length (km)	**41 875**	**9372**	**32 503**	20 998	11 307	18 165	28 021
%	**100**	**22.4**	**77.6**	50.1	27.0	43.4	62.1

[a]Not including land and sea coastlines. Mainland coastlines: 11 526 km (extended Balkans). Insular coastlines: 14 451 km (extended Balkans)

[b]Kosovo included, Turkey excluded

[c]Including the current boundaries

[d]Redundant data across time and space between the four categories. Each boundary polyline is defined by two simultaneous attachments for a single date. The sum has no meaning

Source Author's own GIS database, ESRI

with the *Ausgleich*. The status of Mount Athos is also of medieval versus Byzantine descent (tenth century). Originally granted by the emperor himself, it has survived countless foreign invasions, four centuries of Ottoman rule, and is still legally recognized by the Hellenic Constitution today.

Mapping such territories may turn out to be either easy or hazardous. For instance, the boundary of Mount Athos has been clearly defined centuries ago and has undergone no modification since then. On the other hand, older maps ignore the early Ottoman Vilayets and Pashalıks. Moreover, these *dominions* are spatially flexible. Their areas vary with the ambitions of their leaders. The Pashalık of Ali of Tepelen, otherwise called *Pashalık of Jannina*, is typical of such inconstancy. Beginning from the confines of southern Epirus, its surface never ceased to increase until the 1820s. At this date, it includes half of today's Albania, the whole Thessaly, Central Greece, and the Peloponnese. No mapping can encompass such instability. Few rare documents have kept records of these changes.

For the purpose of clarification, it should also be pointed out that many lands are totally deprived of any known mapping. This absence may have various causes, but uncharted territories exist beside—or despite—official mappings. These are embodied by human communities—village groupings—that live under a common economic, political, and social regime. They share a "common fate." After carefully reflecting on the matter, it was decided not to include the self-contained communities—like the Mirditë clans—in the GIS database. Although undefined by any kind of established "borders," these local autonomous *systems* play a major role. As far as we know, the Mirditë valleys have never been ruled by any official written status. In spite of this, all *foreign* governmental bodies endowed with full sovereign capabilities on these mountains proved to be unable to impose their authority. The "strongest" borders may neither be mapped nor managed by legal contingencies.

The picture would not be complete without examining the **international boundaries**. The first and third categories (Figs. 4.6 and 4.8) meet each other. They describe the same reality. The distinction is formal, and both categories could have been grouped in a single document. The difference lies in the way these boundaries are recognized at an international level: Some may be fully acknowledged by the Great Powers; others are accepted by a limited number of partner states. Considering their everyday existence, the distinction may be of no importance. Furthermore, an international recognition is no warranty that cross-territorial claims might be abandoned.

The GIS databases include no occupation zone. The illegal boundaries unrecognized by international law had to be kept aside. The decision has mapping grounds: War fronts are moving ceaselessly. From the perspective of geographers, this choice may be seen as a mistake because the outcomes of wars are often more lasting than any planning project managed in time of peace. Revolts, fights, and uprisings are unremitting events of the Balkan history. However, mapping such erratic phenomena is a challenge that seemed impossible to achieve in the absence of reliable map sources.

Category 3 is made of two main inputs:

- "Legal" state limits established through war plans between allied countries. This relates primarily to WWI (London Pact, 1915, Treaty of Bucharest, 1916) and WWII (Vienna Award and Treaty of Craiova, 1940, partition of Yugoslavia, 1941 and 1943).[7] Leaving aside their practical application, these multilateral agreements assume legal forms that had to be considered seriously. If these treaties were to be confirmed by lasting military successes, they would have probably become permanent boundaries. With regard to their potential life span, these semi-legal frontiers were usually defined for a term of few years. This is enough time to leave deep traces in the landscapes and in people's mind. These demarcations often exceed the impact of short-lived Republics.
- *De facto* states unilaterally proclaimed and recognized by a limited number of Great Powers or neighboring countries. This class relates to the self-established states of the Yugoslav Wars and Pridnestrovian Moldavian Republic, this last case being an example of breakaway territory unrecognized for many years.

Because most of these political areas went into being during wartimes, their graphic transcription is once again marred by a relatively low significance. They are "potential" control zones regularly submitted to spatial adjustments decided by the armed forces' advances or withdrawals. The limit between occupation zone and legal suzerainty is extremely thin. This is based on a *pretension* to legality.

As a result, the third category cumulates a global border distance almost equivalent to the official boundaries (category 1). This gives an idea of the spatial ambitions of the states in the twentieth century. It is a hallmark of the area: Border disputes do not only apply to confines; they embrace the whole space. Our mapping project thus validates the idea of a domino game that comes into being as soon as any border change is validated.

As a final note, these reflections demonstrate the extreme complexity of the border issue in the Balkans. The adopted methodology will have disruptive effects on minds that are used to deal with "isochronous" and "iso-level" boundaries. Overcoming this "flat" picture was a goal we hope to have reached.

Until present, the implications of such multilevel territorial fragmentation on the administrative networks have not been discussed (Anderson and O'Dowd 1999; Anderson 1996). The question of the existence of "regional"—i.e., infra-national—geographic units is here clearly raised. Most nation-states have put emphasis on clearing regional singularities directly or indirectly, thanks to active land centralization policies. This process is a clear response to the "failing" Ottoman and Austro-Hungarian planning policies that had led to land dislocation. The districts whose ethnic-confessional situation has not been solved by the international law are exceptions to such territorial leveling.

---

[7]Some other war time borders duly accredited by legal treaties have been kept in the first international group: for instance, the boundaries defined by the Treaty of London or by the Treaty of Bucharest in 1913. Of course, such classification is open to debate.

The question is posed much more starkly when recalling that *regional identities*, traditions, and cultures have been almost entirely reinvented afterward by the nation-states that, in a first time, had put a great deal of energy into obliterating past local forms of local "autonomy". This is the price to be paid for fully asserting national suzerainty and ensuring effective land control. Previously considered as "unsafe" science, ethnography is asked today to build seeming transnational bridges. Musical folklore, rural museums, and cross-national costume-based reconciliations duly financed by international institutions prosper and are intended to provide some cultural consistency to a series of abandoned districts strictly controlled by military troops for decades; this time lapse is just enough for the angry generations who witnessed the wars to pass away. As a result of land division, territorial leveling and centralization also explain why the implementation of wide-scale administrative reforms is easier than in Western countries.[8]

All in all, this work provides elements of a response to many questions about the Peninsula. People find here key tools for interpreting local specificities that would have remained otherwise unseen. This comes down to the question of whether the *boundary system* alters the usual relationships between geography and territory. Is the standard center–periphery model still valid in the Balkans? Can we still speak of border *mobility* or *ubiquity*? Obviously, studying the Peninsula implies to develop new approaches, far from the conventional views based on monolithic nations.

4.6 From the Balkan to the Balkans: Graphically Modeling the Peninsula

Such quantitative works allow further graphic interpretations. The assignment of a digital terrain model (DTM) to Fig. 4.5 shall improve its readability. The NASA SRTM mission supplies a high-resolution imagery (90 m) to be combined with vector layouts. This illustration shows a host of interweaving frontiers stretching between mountain ranges. Their ground settlement hardly finds any human or physical object to hang on. In contrast, a limited number of borders firmly anchored in space stretch out across the Rumanian and Bulgarian plains. The imbalance is obvious, although the spatial distribution needs to be weighted cautiously. We have to distinguish the spatial instability attached to the western fronts from the *frequency* of changes applied to stable frontiers in the eastern Balkans. Many boundaries may be highly disputed and the surrounding regions submitted to regular runoffs while land demarcations remain accepted by antagonistic parties. Spatial distribution and conceptual categorization determine two distinct Balkan areas.

[8]The *Kapodistria* regulatory reform of 1998 has abolished about 80 % of all Greek local districts. The *Kallikrati* plan of 2011 has erased 65 % of the first-level territorial units left by the Kapodistria law. Other major administrative reforms have been carried out in former socialist countries as well, resulting in the invention of new territorialities. Late imperial organizational structures had experienced a similar extinction in their time. Most local communities went through sequential spatial redefinitions that mirror the global border instability under discussion.

From the various interpretations that may be given to Fig. 4.5, three spatial models shall be hold.

The low-scale approach exhibits an east–west pattern where boundaries concentrate on the western part of the Balkan area (Fig. 4.10). From Slovenia to northern Greece, west of the Vardar–Morava–Sava axis, about 700 border sections concentrate. They cumulate 20 000 km within a 316 000 km^2 area. This is half of the overall Balkan boundary length for one-third of the area. The discrepancy between western and eastern lands is once again strongly asserted. The picture perfectly matches with the famous *baffling hole* that ruins the official EU cartography with systematic blank areas in Yugoslavia and Albania.

Higher scales provide other interpretation keys. As already noticed, border mobility and inconsistency are typical of mountain regions such as Rhodope, northern Pindus, Sara Planina, the Dinaric Alps, and the Carpathians. Of course there are exceptions to the east-west border pattern. Slight infringements to this rule may exist in certain areas. For instance, central and northern Bulgaria escape the

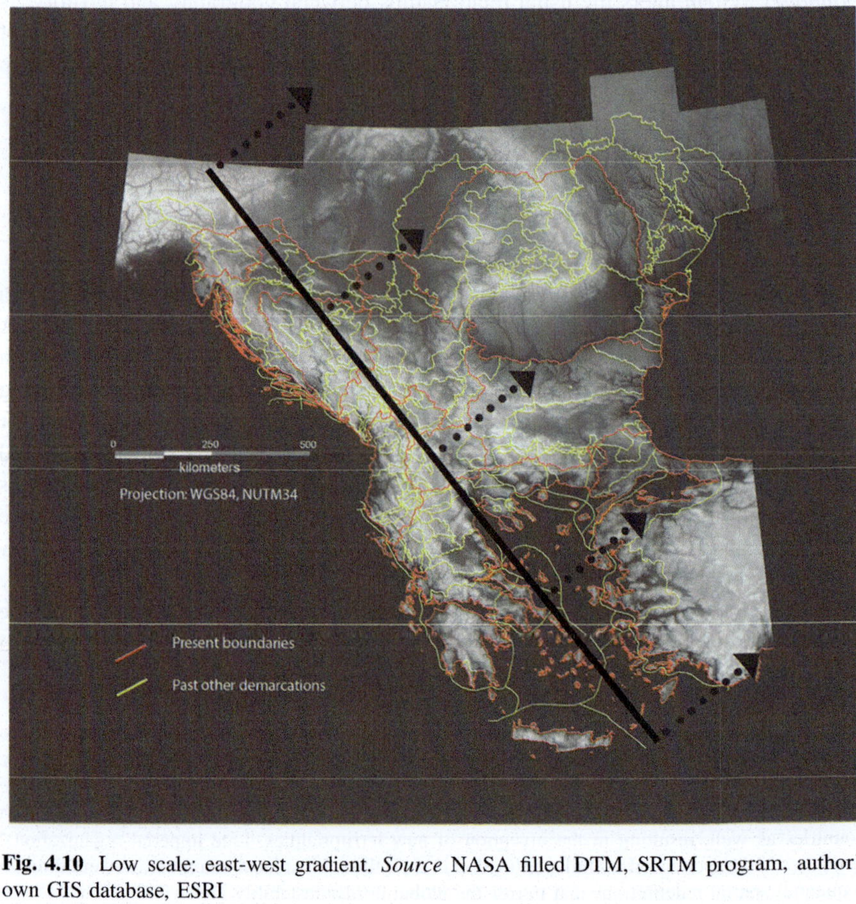

Fig. 4.10 Low scale: east-west gradient. *Source* NASA filled DTM, SRTM program, author's own GIS database, ESRI

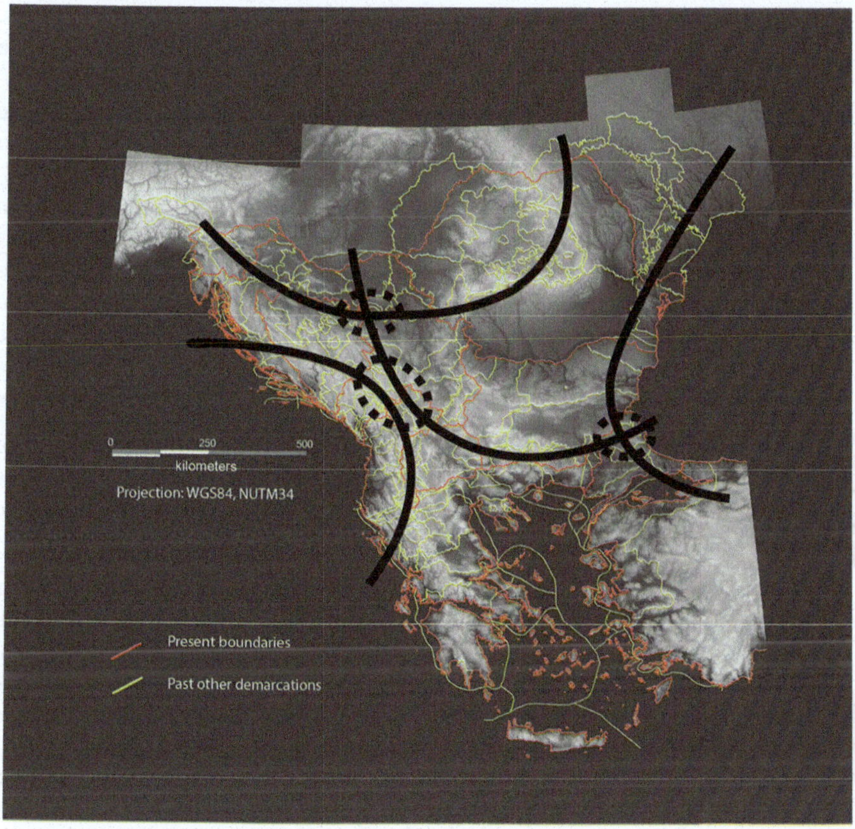

Fig. 4.11 Middle scale: conflicting fronts. *Source* NASA filled DTM, SRTM program, author's own GIS database, ESRI

fuzzy space category depicted by this mapping, despite its topographic profile (Balkan range). The mountain areas of central Greece and Peloponnese are not affected either. By contrast, the lowlands of the Drava Valley offer a dense network of moving borders, as well as the lower Dniester, Prut, and Danube.

Another reading method supplements these remarks (Fig. 4.11). Four linear curves emphasize a series of highly connected vectors:

- A first curve delineates Albania and Montenegro—one of the first countries to have reached independence, and one of the last. Surprisingly, both states are found to be associated. The astonishment effect disappears when we look a little closer at the stunning resilience/resistance developed by these two countries in response to foreign "occupying forces." In the eyes of the Austro-Hungarian and Ottoman Empires, these mountain areas have always been a Balkan "Far West" virtually impossible to control and endlessly afflicted by rebellion.
- The second most noticeable loop takes its origin in the Black Sea (Strandzha Mountains) and follows southern Rhodope. Afterward, it reaches the Kosovo

Polje and Tisza River, north of Belgrade. The unity of this front is less obvious. One might see residual traces of late Ottoman boundaries, still active on the eve of the Balkan Wars.

- The third bow follows the Carpathian ridge, crosses the Danube, and reaches the Sava River basin. This case is clearly opposite to the previous one: It applies to late Austro-Hungarian frontiers.
- The final curve may be more difficult to identify. It covers the Black Sea coast. In this area, density and high border mobility are not inconsistent with one another. *Nomadic sovereignties* align their institutional frameworks on recursive landmarks. This arc relates to the northern and southern limits of Bulgaria in eastern Thrace and Dobrudja. It brings to the fore the Romanian access to the Black Sea and the strategic positions of Moldavia, Bessarabia, and Budjak—these districts form the coastal corridor that would potentially connect Russia to Istanbul.

Lastly, a high-scale picture of the Balkan area (Fig. 4.12) discloses a series of geopolitical "hot points" and identifies the most disputed sites. By "hot points," we mean areas where border changes have reached the highest levels. Three zones are clearly evidenced:

- Torn between Turkey, Greece, and Bulgaria, the Thracian area extends as far as southern Rhodope to the west and the lower Evros valley to the east. Such concentration of demarcations reveals the tensions that weigh on trilateral relations and the difficulties experienced in fixing stable borderlines in the middle of highly populated and mixed areas. The famous *Trigon*[9] and the strategic place of Edirne—the old Ottoman capital forced into border exile—have particularly attracted the attention of competing forces. When compared with other forthcoming *hot points* that share relative mountain remoteness, this district holds an obvious strategic value: It controls the main route between Europe and Anatolia.
- The Prespa Lake area caught between Greece, Macedonia, and Albania also belongs to this group of highly disputed regions. Different vector layouts cross one another and run in all directions. Decoding such a *bazaar* is a challenge in itself. Southern Albania has long been waiting for its limits to be fixed, despite the recognition of the country's independence. The pathway is strategic: It is all about controlling the junction between central Macedonia and the Adriatic.
- Trapped between Serbia, Kosovo, Montenegro, Albania, and Herzegovina, the Sandžak corridor is in a similar situation. The strategic function of this area is well known and derives from its locking position. Serbia and Montenegro have long applied pressure on this corridor, which used to link the late Ottoman Bosnia–Herzegovina with other southern imperial possessions. Among southern Serbia, eastern Montenegro, and northern Albania, the competition was rude.

Finally, Fig. 4.12 allows for the identification of other secondary challenged lands:

[9] A land triangle formed by the confluence of the Ardas and Maritsa–Evros rivers, west of Edirne.

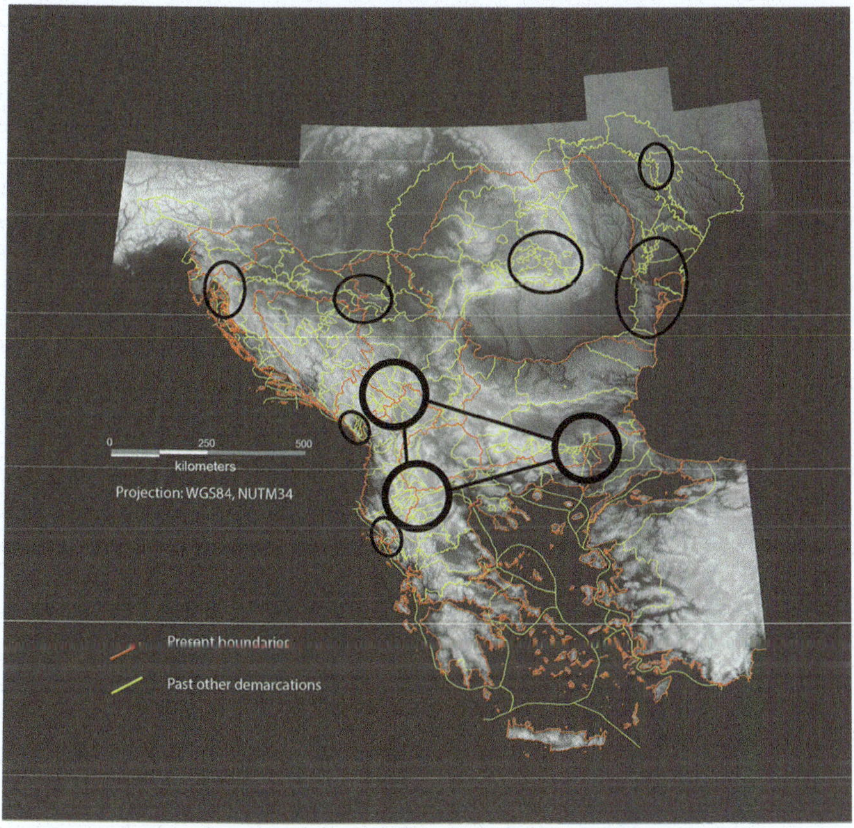

Fig. 4.12 High scale: hot points and most disputed lands. *Source* NASA filled DTM, SRTM program, author's own GIS database, ESRI

- The sector of Shkodër–Kotor, between southern Montenegro and northern Albania.
- The district of "Epirus," caught between Greece and Albania.
- The Slavonian area.
- The Dobrudjan region and its surroundings (Bessarabia).
- Transylvania.
- Moldavia and Transnistria.
- Istria.

The list is not complete and is not meant to remain closed. There are many examples of disputed lands, and this book would not count enough pages to review them all. In concluding on this point, it may be assumed that the three main hot points define a triangle at the center of which *Macedonia* finds its place. This work thus confirms the central position of Macedonia in the Balkans—should it be necessary to recall it once more?

4.7 Border Resilience

Measuring border resilience is a good way to illustrate the overall political and territorial instability. This is possible through our geodatabases. As the polyline database registers individual border changes, the cumulative impact of repeated events can be studied. Results are highly agglomerated. Time interval calculations have been performed through dedicated algorithms. In an attempt to maintain some consistency to the analysis, it has been decided to give up the distinction between official and semiofficial outlines. Accordingly, Figs. 4.13 and 4.14 merge all referenced border classes.

In terms of periodicity, one shall keep in mind that the Cold War determines much of Fig. 4.13. This historical episode has made the mean existence term of

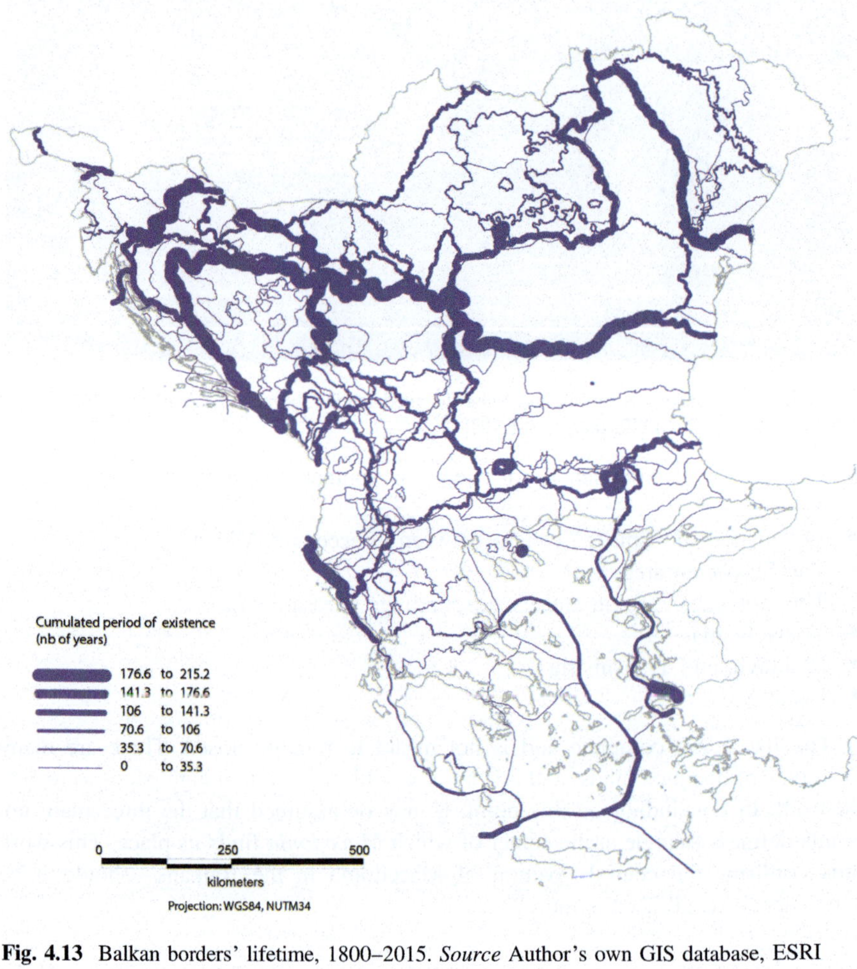

Fig. 4.13 Balkan borders' lifetime, 1800–2015. *Source* Author's own GIS database, ESRI

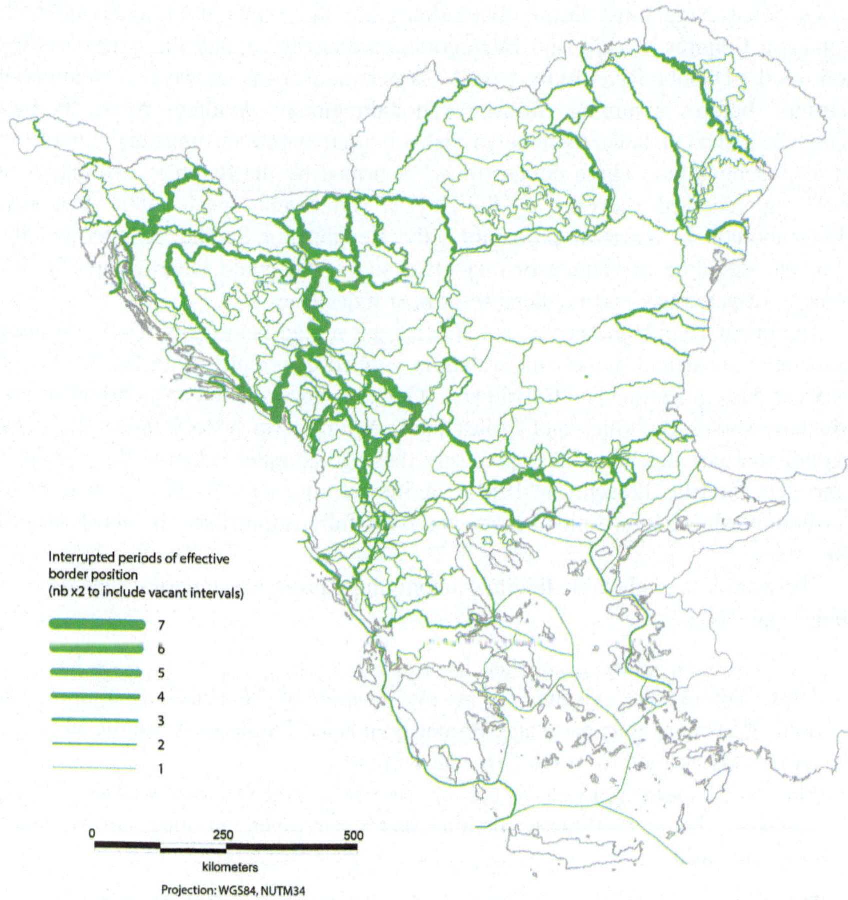

Fig. 4.14 Borders' instability, 1800–2015. *Source* Author's own GIS database, ESRI

many line segments climb to all-time record highs. For half a century, frozen frameworks have compensated for historical land instabilities, leading to a significant growth of the overall border resilience. In the long term, however, this episode might be qualified as exceptional. The Yugoslav Wars have made this statement obvious. The *life span* concept applied to a boundary/frontier serves as a convenient means of assessing geographic impacts, assuming that the longer the borderline life, the deeper the effect on natural and human environments.

The results are startling. Against all expectations, the most stable boundaries are to be found in former Yugoslavia, a country that recent events taught us to consider as eminently unsteady. This has valid reasons to appear. Until the 1990s, Yugoslavia was marked by unachieved national awakening. Wavering between the Habsburg and Ottoman influences up until WWI, the "country" had adopted an institutional framework that sharply contrasts with its neighborhood.

Obviously, the most lasting borderlines are those that used to divide both opposing Empires. Bosnia and Herzegovina are encircled, and the largest outline follows the Danube River to the east. Most perennial segments were cross-imperial barriers before assuming further national/regional dividing roles. Former Yugoslavia thus included extremely reliable borders within its limits under the form of federal republics. These delineations, confirmed by the Badinter Arbitration in 1991 and renewed afterward by the Dayton Agreements, persist quite logically. When thought in terms of probability, the Republic of Serbian Krajina and the Croatian Republic of Herzeg-Bosnia—two states that could have eventually disrupted old patterns—had predictable *ex post* trajectories.

The upper Danubian and Savian streams are rooted landmarks. They are often associated, in mental models more than in applied geography, with the distinction between eastern Europe and Balkan area (Germans versus Ottomans, Orthodox, and Muslims versus Catholics and Protestants). A long-term history meets the latest geopolitical contingencies. It is precisely these geographic balances that Greece's entry into the EEC breaches in 1981. The integration *by the south door* was a true revolution whose implications have not been fully appreciated by most outside observers.

There are less predictable findings among these lasting boundaries, regardless of their legal status:

- The Slovenian southern limit has become an international demarcation lately, in 1991, although its existence as internal frontier of the Habsburg Empire had come into being at an early age. Primarily, it is used to divide Carniola and other crown possessions from the Croatian Kingdom.
- The same pattern applies to the eastern Slavonian and Syrmian borders, the only difference being that these sections had been given an international status 80 years earlier.

These two examples underline that the internal frontiers/limits often play a more essential structuring role than many outside boundaries submitted to constant changes. The judgment is all the more so grounded that microterritories granted with special statutes are more likely to circumvent political disputes and spatial conflicts. Some of them are clearly delimited; others inherit fuzzy spatial definitions. Mount Athos is the model that brings territorial resilience to its utmost. The Mirditë communities are a contrasting case of long-lasting human organization defying all political regimes without legal definition. Territorial identities/coherence does not systematically rely on strict conceptual or spatial definitions.

To conclude on these long-established borders, two more points are of particular interest:

- The eastern contact zone between the Empires is less visible than its western equivalent, although its existence can be detected along the Prut and lower Danube.
- Dalmatia is a distinct case. Beginning with the Velebit range south of Rijeka, then following part of the Una thalweg and the Dinara mountains through the

enclave of Kotor, a line could be drawn until the Ionian Islands. The western coastal lands and islands are rather cut off from the rest of the Balkan hinterland —remnants of an early geographic feature. In the eyes of senior travelers, the diverging fate of the two main insular groups is abundantly clear: Dalmatian and Ionian Islands live apart from the Balkan mainland, whereas the Aegean Islands are strictly attached to the *continent* they neighbor, with few exceptions.

At the other extreme of the age scale, short-lived boundaries are legion. The overgrowth of insignificant borderlines is typical of the southern Balkans, in contrast to more solidly structured provinces up north. Between these two "worlds," the dividing line extends from Shkodër to Varna. The ex-Ottoman Balkan area cumulates three characteristics:

1. Strong presence of microscale long-lasting units.
2. Sizeable number of roaming borderlines that hardly settle for more than a few years.
3. Late post-WWII strengthening of national borders that, in the end, fail to catch up with their northern "heavy" equivalents.

It is interesting to observe that all "hot points" emphasized in Fig. 4.12 belong to the southern Balkans whose they inherit a low resilience.

As a result, fewer than half of the current borders prove to be "sustainable" in the long term. Over the past two centuries—the *age of nationalism*—the average lifetime of a boundary reaches 20 000 days—i.e. approximately 54 years. Only 200 segments climb to more than 100 years of existence (17 % of 1200 units). What does this tell us? Few borders survive over centuries, and among them, fewer are interstate limits. In the Balkans, political geography is a fantasized science. Politicians and people spend their time and energy imagining new territorial imprints, but they leave untouched an *underground world.*

This life span mapping develops a true portrait of contrasting Balkan lands, easily divided into steady subregions that might be viewed as distinct "systems." The insight here proposed is far from a partitioned state-based cartography: Some recent international boundaries come out with increased weight; others wrongly considered stable are weakened. When applied to present countries, Fig. 4.13 may be used as a gauge of internal heterogeneity (Atzili 2006). Inevitably, this prompts the question of whether this work finally promotes the emergence of a new *regional*, rather than national, geography? What about "uncertainty"—our guiding principle?

The concept of uncertainty mainly arises from in-between cases. Let us look at some examples:

1. The Aegean Islands

Once changed into administrative limits, former international borders continue to shape landscapes. Despite highly centralized nation-states, subnational *levels* outlive *by transparency*. Considering today's Greece and avoiding three exceptions (Mount Athos, Corfu, and Samos)—all with little extent—the oldest Hellenic boundaries are not external, but internal. They are not continental, but maritime. The Heptanese frontier cumulates 117 years of existence, north of Lefkada. The Cycladic–Dodecanesian outline is credited with 115 years, whereas the northern land boundaries with Albania, Bulgaria, Turkey, and (FY)ROM vary between 80 and 100 years only. Similarly, in the eastern Aegean, the present Greek–Turkish boundary is only 10 years older than the Cycladic/Sporadic-eastern Aegean Islands section.

2. Montenegro–Kosovo–Macedonia

Serbia brings together a number of midterm frontiers. The Danube and Drina basins as well as the Stara Planina to the east clearly define a core area. However, and this is the point, this framework does not coincide with the present territorial definition of the country. North of Vojvodina, the borders of Bačka and Banat are established after WWI. These are nearly a century old. On the opposite side, southern Serbia barely settles its edges. Surprisingly, the eastern Montenegrin and Kosovan confine stay ahead of (FY)ROM's demarcation, which is presently far less contested despite its limited 65-year lifetime.

3. Wallachia–Transylvania–Moldavia

The situation is here very similar to that in Greece. Present Romania is determined by borders that have left deep underneath landmarks. Danube and Prut are consistent drawings, despite hesitations along Moldavia. The access to the Black Sea across Dobrudja was denied for almost 80 years over the past 2 centuries. Opposite to the Danube Delta, Transylvania has remained under Austrian government for 117 years, the Carpathian Mountains maintaining the entrenchment of Wallachia and Moldavia. Because of its dimensions, Romania is characterized by a considerable internal heterogeneity—a property that indirectly affects its political and territorial cohesion.

4. Military Frontiers

One last case is worth mentioning. Despite their fragmentation and early withdrawal, the Austrian *Militärgrenze* is clearly visible, stretching from the Croatian shoreline to Slavonia, until Banat and Transylvania. The significance of persisting internal frontiers is once again confirmed when weighted with erratic international boundaries.

Another way to enhance the relevance of this statistical analysis is to produce a reverse mapping. Measuring border *instability* can be achieved by using a straightforward marker: The number of times a boundary is re-established between vacancy episodes (Fig. 4.14). A *dispute indicator* emerges. Using this method, we can observe every change in pressure. In that kind of game, winners are found once again in former Yugoslavia. The federation is brought to an accomplishment: It cumulates the most reliable and most contested frontiers.

The correlation between life span and periods of break is obvious: The longer the period of *border effectiveness*, the higher the likelihood of "suspension." This is a commonplace. However, the argument demonstrates that being among the oldest frontiers make no warranty about a potential collapse. This is especially true in Bosnia and Herzegovina, between Slovenia and Croatia, and in western Bačka. Kosovo and Montenegro are in middle positions: In these remote places, boundaries have always been contested and moved, although some outlines happen to undergo *renewals* at times. Uncertainty has much in common with *vacillation*. The situation of Kosovo is typical: Besieged for ages by Serbia's wrinkling territories and Albania's short-term extensions, the area has its own specific character.

Figure 4.14 brings to the fore five other areas: Banat, Moldavia, eastern Macedonia and Thrace, Vardar Macedonia, and Dobrudja. These cases relate to relatively stable boundaries. Successive stages of spatial extension and withdrawal of neighboring countries determine "in-between" districts that serve as buffer zones. These areas, which were granted with relative spatial integrity when placed under the guidance of imperial powers, are caught in the *cross fire* of confronting national claims. How do they manage to outlive sequential changes? With the idea of reviving old organic relations, transcending claims, and stifling residual conflicts, the EU has recently focused its attention on such contested districts in the frame of a planned oncoming integration. From the outside, these buffer zones are often *hurt lands* occupied by mixed populations. They also are "natural bridges" for the international diplomacy, for the same reasons.

European projects are easy growing in such *unclassified* lands that have been the theater of countless political uprisings aimed at defining independent territories safe from the "rapacity" of powerful neighbors. "Local" claims find common outlets during transitional periods, when upper range states lean to political weakness. Let us remember the Voivodeship of Serbia and Banat of Temeschwar, the short-lived Republic of Banat, the Gümülcine and Tamrash Republics, and all Moldavian government forms. We shall refer back to Fig. 4.7 and, to a lesser extent, Fig. 4.8.

The coincidence of Fig. 4.14 with the EU cross-border mappings is really compelling (Fig. 4.15).

For sure, these cross-border *devices* come to light under peaceful skies. This does not prevent certain no-shows from being suspect: The contact area between Montenegro–Kosovo–Albania–(FY)ROM is remarkably absent. Most toponyms adopted for qualifying Euroregions carefully dodge old geographic names and resort to "neutral" designations. Alongside other river names, the explicit reference to the "Danube" reaches the highest popularity rating, although few terms are less significant. Other slithering semantic combinations cannot be taken seriously. Two eastern Slovenian Euroregions avoid mentioning the term "Prekmurje," the "lower Danube" strictly matches historical Budjak (Popescu 2008), the "Upper Prut" is not Bukovina, the "DKMT" area is not Banat, and so on. Obviously, the EU is afraid of facing skeletons in the closet (Miossec 2013). Strangely enough, water bodies have gained international recognition.

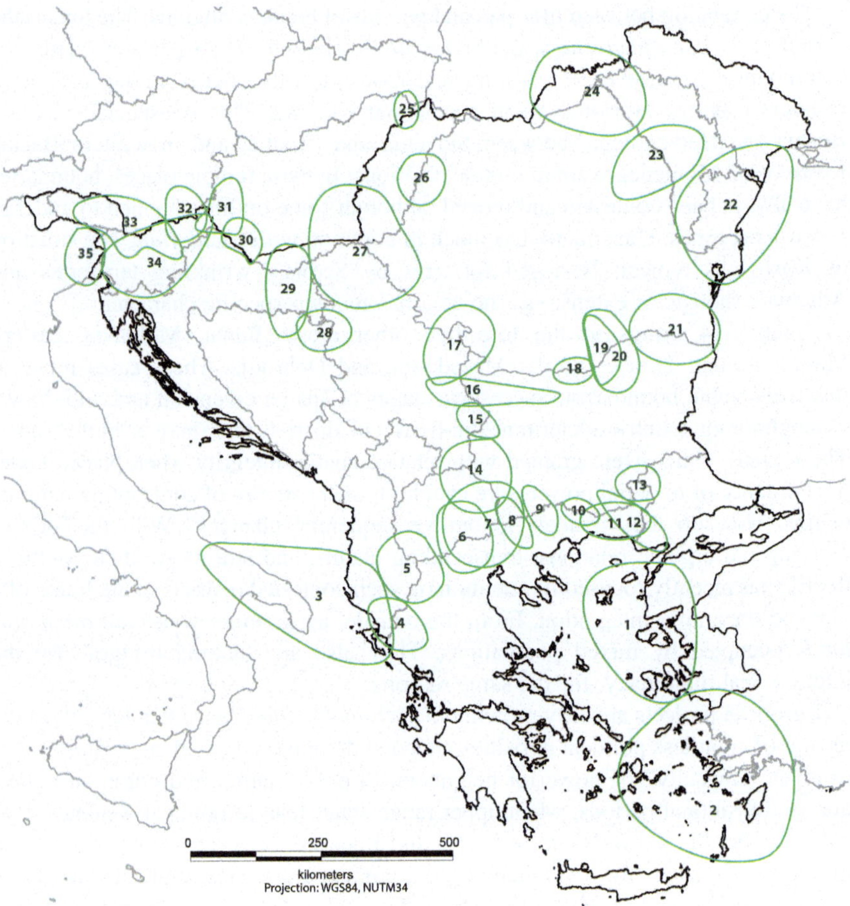

Fig. 4.15 European border and cross-border regions in the Balkans, 2012. *Source* Author's own GIS database, ESRI, www.aebr.eu

Table 4.5 Mean time life of present state boundaries, 1800–2015

Present country	Mean number of years	Legal national independence date	Nb. of years since official indepence
Bosnia Herzegovina	157.5	1995	20
Croatia	153.0	1991	25
Romania	125.0	1878	137
Bulgaria	121.2	1908	107
Greece	107.5	1832	184
Slovenia	106.5	1991	25
Serbia	103.7	1878	137
Albania	99.7	1913	103
Montenegro	94.7	1878	137
(FY)ROM	86.0	1991	24
Kosovo	*75.3*	*2008*	*8*

Source Author's own GIS database, ESRI

It is not a mere paradox to notice that, in the end, geography is called back not to promote war as always, but peace and cooperation,[10] whatever fictional these rising devices might be at present (Harrison and Growe 2014; Turnock 2002). As stated by F. Celata and R. Coletti, such transnational spatial units and joint management authorities are thought to help "overcome some of the limitations of place-based strategies—namely, the risk of policy capture, territorial introversion and communitarian confinement—while incurring in others—institutional isomorphism, technocratic management and democratic deficit" (Celata and Coletti 2014).

A new regionalism is thus called to complement or replace old state scalar organizations. Distinguishing territorial and relational ambitions in these plans is problematic (Harrison and Growe 2014). Not surprisingly, Euroregions turn out to define "soft borderscapes" deprived of definite boundaries—they adopt potato shapes. This is the ultimate expression of an old principle of land management especially valued by the Ottomans: fuzziness. This leads to a paradoxical situation. Can we think of any conditions in which *soft* uncertainty would be now a panacea for those who want to "solve" border problems? Two views of the modern world—places versus flows, government versus governance, sovereignty versus competency transfer—and two discourses about geography are clashing (Jonas 2012; Keating 2004; Kramsch and Hooper 2004; Painter 2008).

In closing the debate on resilience issues, serious attention needs to be paid for a while to national frameworks, expressly avoided up to now, not to be kept locked behind fences. The GIS database may be used to calculate an average lifetime for each country's present border configuration (Table 4.5). This exercise shows how much national frameworks are distorting prisms to explore the significance of borders and boundaries. This is particularly evident when we see that Bosnia

[10]Translated into the EU official language, it is called *cohesion*.

Heezegovina, the last Balkan state to have achieved full independence (1995), inherits the most reliable boundaries in the long run (157 years).

As opposed to that, three countries whose independence dates back to the early 19th century -Greece, Serbia and Montenegro- reach levels as low as 100 years for their mean border lifetime. These figures shake our usual convinctions. The political-territorial stability of nation-states seems to be completely disconnected from their measurable border resilience. No clear law emerges in terms of geostrategic risk assessment. Local contingencies determine more efficiently state stability and territorial evolution than any fantasized "fate."

References

Akerman JR (ed) (2009) The imperial map: cartography and the mastery of empire. University of Chicago Press, Chicago

Anderson J, O'Dowd L (1999) Borders, border regions and territoriality: contradictory meanings, changing significance. Reg Stud 33(7):593–604

Anderson M (1996) Frontiers: territory and state formation in the modern world. Polity Press; Blackwell Publishers, Cambridge; Oxford

Atack J (2009) Historical GIS: technologies, methodologies and scholarship. Econ Hist Rev 62 (3):781–782

Atzili B (2006) When good fences make good neighbors; and when they make bad ones: the international effects of "Border Fixity". Conference Papers—International Studies Association:1–26

Bailey TJ, Schick JBM (2009) Historical GIS: enabling the collision of history and geography (English). Soc Sci Comput Rev 27(3):291–296

Balibar E (1998) The borders of Europe. In: Cheah P, Robbins B (eds) Cosmopolitics: thinking and feeling beyond the nation. University of Minnesota Press, Minneapolis, p 392

Branch J (2011) Mapping the Sovereign State: technology, authority, and systemic change. Int Org 65(1):1–36. doi:10.1017/s0020818310000299

Burdett ALP (ed) (1995) Bosnia, Croatia, Serbia: historical boundaries, 1815–1945. Cambridge Archive Editions, Slough

Caluwe Rd, Tré Gd, Bordogna G (eds) (2004) Spatio-temporal databases: flexible querying and reasoning. Springer, Berlin; Heidelberg; New York

Celata F, Coletti R (2014) Place-based strategies or territorial cooperation? Regional development in transnational perspective in Italy. Local Econ 29(4–5):394–411

Destani B (1999) Albania & Kosovo: political & ethnic boundaries, 1867–1946. Cambridge Archive Editions, Slough

Destani B (2001) Montenegro: political and ethnic boundaries, 1840–1920. Cambridge Archive Editions, Slough

Donaldson JW (2008) Politics and scale in boundary-making: the work of boundary commissions. J Hist Geogr 34(3):393–396

Foreign Office of Great Britain (1878a) Further correspondence respecting the affairs of Turkey, vol Turkey. Harrison, London

Foreign Office of Great Britain (1878b) Maps showing the new boundaries under the preliminary treaty of peace between Russia and Turkey, signed at San Stefano, vol Turkey. Harrison, London

Gartner GF, Ortag F (eds) (2010) Cartography in Central and Eastern Europe. Springer, Berlin

Gregory IN, Ell PS (eds) (2008) Historical GIS: technologies, methodologies, and scholarship. Cambridge University Press, Cambridge

Harrison J, Growe A (2014) From places to flows? Planning for the new 'regional world' in Germany. Eur Urban Reg Stud 21(1):21–41

Hertslet E (1875–1891) The map of Europe by treaty: showing the various political and territorial changes which have taken place since the general peace of 1814. Butterworths-Harrison, London

Hiller A, Knowles AK (eds) (2008) Placing history: how maps, spatial data, and GIS are changing historical scholarship. ESRI Press, Redlands

Jarman RL (1997) Yugoslavia: political diaries, 1918–1965. Cambridge Archive Editions, Slough

Jonas AEG (2012) Region and place: regionalism in question. Prog Hum Geogr 36(2):263–272

Jones SB (1945) Boundary-making: A handbook for statesmen, treaty editors, and boundary commissioners, vol 8. Carnegie Endowment for International Peace, Division of International Law, Washington

Kavalski E (2006) From the western balkans to the greater balkans area: the external conditioning of "Awkward" and "Integrated" States. Mediterr Q 17(3):86–100

Keating M (2004) Regions and regionalism in Europe. Edward Elgar, Cheltenham; Northhampton

Kiepert H (1880) Carte de l'Epire et de la Thessalie. Reimer, Berlin

Kramsch OT, Hooper B (eds) (2004) Cross-border governance in the European Union. Routledge, London-New York

Longley P, Goodchild M, Maguire D, Rhind D (2010) Geographical information systems and science. Wiley, Chichester

Mamadouh V, Kramsch O, Van Der Velde M (2004) Articulating local and global scales. TESG Tijdschrift voor economische en sociale geografie 95(5):455–466

Ministère des Affaires Etrangères (1913) Le traité de paix de Bucarest. Ministère des Affaires Etrangères, Bucarest

Miossec J M (2013) Rivages d'Europe. Personnalité et avenir d'un continent ouvert. L'Harmattan, Paris

Monmonier M (1991) How to lie with maps. University of Chicago Press, Chicago

Ott T, Swiaczny F (2001) Time-integrative geographic information systems: management and analysis of spatio-temporal data. Springer, Berlin, New York

Painter J (2008) Cartographic anxiety and the search for regionality. Environ Plan A 40(2):342–361

Péchoux P-Y (1977) Formation du réseau administratif et intégration du territoire dans la Grèce moderne. In: IEDES (ed) Etat, pouvoir et espace dans le Tiers Monde. Paris, pp 187–212

Popescu G (2008) The conflicting logics of cross-border reterritorialization: geopolitics of euroregions in Eastern Europe. Political Geogr 27(4):418–438

Ruggie JG (1993) Territoriality and beyond: problematizing modernity in international relations. Int Org 47(1):139–174

Shields R (2006) Boundary-thinking in theories of the present: the virtuality of reflexive modernization. Eur J Soc Theory 9(2):223–237

Smith B, Varzi AC (2000) Fiat and bona fide boundaries. Res 60(2):401–420

Stieler A (1875) Die Europäische Türkei. Justus Perthes Gotha

Thapa K, Bossler J (1992) Accuracy of spatial data used in geographic information systems. Photogramm Eng Remote Sens 58(6):835–841

Turnock D (2002) Cross-border cooperation: a major element in regional policy in East Central Europe. Scottish Geogr J 118(1):19–40

Vogel C (1908) Die Balkan-Halbinsel in 4 Blättern: Türkei, Rumänien, Griechenland, Serbien, Montenegro, Bulgarien. Justus Perthes, Gotha

Von Lünen A, Travis C (eds) (2013) History and GIS: epistemologies, considerations and reflections. Springer, Dordrecht; Heidelberg; New York; London

Wagstaff JM (2002) Greece: ethnicity and sovereignty, 1820–1994. Cambridge Archive Editions, Slough

Chapter 5
"Hot Points" with a Hand Lens: The Process of Local Border Sedimentation

> "What does it mean?
> That certain phenomena cannot be observed without scaling up or enlargement; and all this because they would remain unidentified and invisible without this distortion. These phenomena escaping naked eye detection place us before the *following alternative: 'whether overemphasis or knowledge abandonment'*. Microscopy and telescopy are immediate examples of that, insofar as they help us seek for the truth through the process of building images."
> Günther Anders, *The Outdatedness of Human Beings 1. On the Soul in the Era of the Second Industrial Revolution*, 1956

Abstract At the middle of the scale, the research centers on the most sensitive areas. The aim is to observe the local aftereffects caused by constant border changes. The identification of contested lands and blind spots helps illuminate the conflicts between nations and Empires for control of these areas. How do local populations and authorities manage to resist the crushing forces of external ambitions?

Keywords Contested areas · Boundary settlement · Diachronic mapping · Land confines · Mobile interfaces · Historical demarcations · In-between spaces

This work would have only begun to scrape the surface of the issue without further enlargements. Broadscale geopolitics provides a few clues, but they seem to miss the point. Accurate GIS data fortunately enable further field-attached approaches that might explain the genesis of border landscapes (Rumley and Minghi 2015). This chapter focuses attention on geographic "blind spots" cautiously ignored by conventional studies.

Improving the high-scale DTM of the NASA SRTM program was a preliminary step for preparing a new series of maps. The DTM geometry has been improved to match with our vector data, and a specific *lookup table* has been computed. Introducing elevation is no coincidence. Terrain is a fundamental input for those who

© Springer International Publishing Switzerland 2017

R. Darques, *Mapping Versatile Boundaries*,
Springer Geography, DOI 10.1007/978-3-319-40925-2_5

try to understand the process of border fixation in mountain districts. At such fine scales, it is important to keep in mind the uncertainty principle outlined earlier.

Most figures presented hereafter have been reduced in their size to fit to the limited dimensions of a printed work although such imagery could support a much larger visualization (several meters wide) or take advantage of a mapping video animation. The cumulative effect that these maps produce is the natural outcome of an instantaneous display. It could be reduced by a dynamic processing. Thanks to the DTM, a high-scale 3D perspective could even enhance the video animation. All this is possible, but paper is incompatible support to any such display. However, meaningful, printed pictures remain impoverished when compared with their digital counterparts.

Six regions receive attention. These borderlands are complex systems and would need much more than a few words to be explained thoroughly. The aim here is not to cumulate monographs, but just to focus on the specific impact of major territorial changes at a local scale. The synthetic map-based approach shall be illustrated by old map sheets few people have heard about, apart from a few experts. Most of these described territorial frameworks are no more, although they *somehow* continue to weigh on present geography.

5.1 Defining the International Boundary in Rhodope and Thrace

The Thracian lands, Rhodope mountain range, and Aegean Macedonia belong to this group of contested areas enjoying today a quietness only troubled by few border guards trying to contain the "savage attempts" of migrants to enter the EU.[1] Hanging on the Rila and Pirin heights, the Rhodope Mountains are oriented eastward (Fig. 5.1). On their southern edge, they determine a narrow strip of land composed of eastern Macedonia and western Thrace, both areas belonging to Greece. Roughly speaking, two river basins delimit the area under study, east and west: Strymon-Struma and Evros-Maritsa. The Thracian *cornerstone* around Edirne plays a key role in the area through its control of road and railway traffic. Edirne is also the largest Balkan city to be in immediate contact with international boundaries.

The current demarcation is inherited from the Balkan Wars and, after WWI, from the Treaty of Neuilly. Intentionally, the Muslim minority of western Thrace will not be addressed as an issue per se. Neither shall we deal with the Pomak[2] communities (Demetriou 2004), nor shall we study the situation of Slavic-speaking

[1]The elements presented hereunder are preliminary to further investigations (chap. 8). Information is accredited by field experiences acquired during the 1990s and 2000s.

[2]Muslim minority speaking a Slavic-*Bulgarian* language.

Fig. 5.1 Historical borders in the Rhodope region, 1800–2015. *Source* Author's own GIS database, NASA DTM, SRTM Program, ESRI

Fig. 5.2 Political map, Republic of Gümülcine. *Source* Turkish Prime Ministry, General Directorate of State Archives (Turkish Prime Ministry)

people living in Greece. The aim is to focus on the actual mechanics of shifting boundaries.

The premises for further spatial restructuring are laid down in 1877–1885 and 1913–1919. Until the *April Uprising* and the Russo-Turkish War of 1877–1878, the entire region was an "open space" placed under Ottoman supervision. The newly acknowledged Principality of Bulgaria and the autonomous Province of Eastern Rumelia break away from the declining Empire. The new outline adopts a northern position following the contours of the highest peaks of Rila and western Rhodope before wandering across the Thracian Plain without stable landmarks until the Strandzha hills. The delineation was maintained for almost 40 years.

At the end of the 1910s, the next boundary adopts a farther south location. The Rhodope range is no longer divided and changes into a background support separated from the Mediterranean coast. This second demarcation is the one we still know today. All other pictured boundaries are temporary. They express spatial faltering and political hiccups. The transient access phases of Bulgaria to the "White Sea," for instance, are clearly emphasized. Two short-lived Republics highlight the in-between position of this disputed land strip: the *Republic of Tamrash* and the *Republic of Gümülcine* Fig. (5.2). Facing nations and Empire engaged in a constant struggle, they offer dissimilar political patterns: The Tamrash

Fig. 5.3 Edirne's border position in 1908. *Source* Harvard Geospatial Library (Vogel 1908)

Republic affirms the specificity of remote Pomak places; the Gümülcine Republic is an unexpected association of Greek, Muslim, and Jewish populations who stand up against the Bulgarian administration after the Second Balkan War. Both political "accidents" are symptomatic of the serious troubles that accompany the process of border settlement (Chap. 8).

At the end of the nineteenth century, even though Edirne is already a border city, its domination over western and eastern Thrace is uncontested. In particular, emphasis has to be put on the position of Svilengrad (formerly Mustafa Paşa) (Fig. 5.3). In 1878, 4 years after the completion of the construction of a modern railway station in its neighborhood across the line to Istanbul, this small Ottoman city becomes a crossing point close to the newly fixed frontier between the Ottoman Empire and eastern Rumelia. After the Balkan Wars and its annexation by Bulgaria, the place retains its role as a major checkpoint. The borderline moves a few kilometers eastward. The spatial extension of Greece condemns Edirne to remain on the sidelines for long. Consequently, this geographic blind spot has been a major European—or Rumeliot—crossroads for more than 130 years, which is quite exceptional in the Balkans.

Unlike other forthcoming examples, the region has a distinctive character. The boundary is here a rather *late* phenomenon (end of the nineteenth century). Thrace, Rhodope, and eastern Macedonia are areas virtually kept under Ottoman rule until the fall of the Empire. The current boundary was part of the Iron Curtain and, as such, its presence has deeply impacted the surrounding districts. As a corollary to this assumption, the role played by internal Rhodope as an early buffer land between Greek, Bulgarian, and Turkish ambitions shall be kept in mind.

5.2 Incapacitated Lands Surrounding the Lakes of Prespa and Ohrid

Another disputed Balkan *Trigon* is found on the opposite side of the Peninsula (Fig. 5.4) At the intersection of Greece, (FY)Macedonia, and Albania, comings and goings of boundaries are the distinguishing marks of a fierce competition for land control. The region's name is a known problem: "Epirus" is a historical designation claimed by Greece in reference to the area caught between northern Greece and southern Albania (Fig. 5.5). In response, Tirana uses the expression "southern Albania," which denies any local specificity to these remote lands. It must be remembered that the state of war declared in 1940 between Athens and Tirana has never been fully abrogated. Diplomatic ties were restored only in 1971. After the socialist Andreas Papandreou came into power in 1981, bilateral relations have steadily improved. The lifting of the war state happened in 1987. Foreign Minister Karolos Papoulias gave his name to the agreement sanctioning the reconciliation process. These negotiations would pave the way a few years later to the opening of borders for millions of Albanian workers. Since the beginning of the 1990s, the relations between both countries have taken many twists and turns. Most often, the legal situation of the Albanian migrant workers in Greece was used as an adjustment variable for strengthening or relaxing ties between Athens and Tirana. Recently (2013–2015), fresh debates have been held in the public arena. As a background of official discussions about the Albanian application for membership of the EU, Greece has found the opportunity to turn intractable historic-political issues to its advantage. Through its veto power, Athens intends to solve the question of its maritime boundaries with Albania north of Corfu. The issue at stake is the enforcement of the Ionian Sea Border Agreement signed in 2009 and the fate of 225 km^2 of territorial waters where oil and gas reserves are located. The state of war, whose decision has not been officially abolished by the Greek Parliament, is also brought into negotiation. As usual, the Grand Game of international diplomacy collides with small local interests.

Leaving aside issues related to maritime boundaries, the Greek–Albanian mainland border area consists of five sectors, from north to south:

1. **The Lake district**, a traditional mixed zone of Greek–Slavic–Albanian population maintained under military pressure for decades. Although highly fragmented and underdeveloped, this area enjoys today a "special status." Since 2000, a transnational natural park implying tripartite cooperation has come into being. The project aims to attract tourist flows stimulated by breathtaking landscapes that combine fresh air and water bodies. The Lake basins are

Fig. 5.4 Historical borders between Greece and Albania, 1800–2015. *Source* Author's own GIS database, NASA DTM, SRTM Program, ESRI

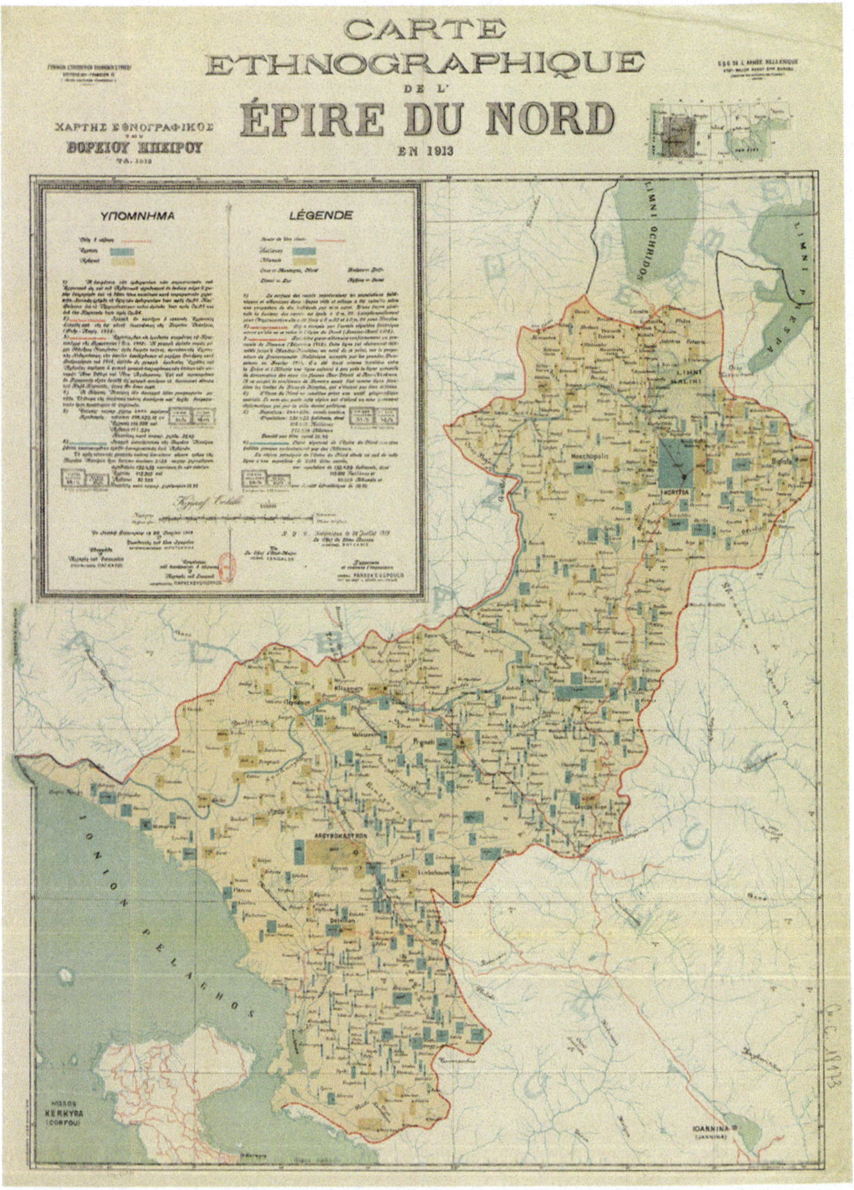

Fig. 5.5 Northern Epirus in 1913. *Source* Greek Army Headquarters (1919)

partitioned, as if antagonist states had preferred a face-to-face meeting across water bodies. Choosing a water demarcation is the right way not to settle boundaries in the countryside. Three large cities (Bitola, Korçë, and Ohrid) and three intermediate (Kastoria, Florina, and Pogradec) occupy the Lake district.

The area is thus more populated than other perimeters overhanging the Ionian Sea. For sure, controlling mountain passes was of the utmost importance until the fall of the Ottoman Empire, especially between the Aegean and Ionian seas. Since the beginning of the twentieth century, although suffering from uncomfortable border situations, these towns have managed to maintain or increase their population, thanks to demographic outflows from their countryside and direct state support.

2. The "fight against physics" continues southward. Against consecutive NW–SE mountain ranges, the Greek–Albanian boundary has to find its way until the Ionian Sea: a challenging task, indeed! **The basin of Korçë** is today an isolated Albanian area. Around it, the heights of Morava, Voskopojë, Valamara, Thatë, and Grammos build a theater. The Grammos-Morava-Thatë line would have been a convenient *natural* fence demarcating the Albanian and Greek sides. But governments have agreed otherwise. The Morava is bypassed by the east. Such topographical oddities are ordinary.
   Once again, interstate boundaries are settled lately. Two events illustrate the local indeterminacy at the end of the nineteenth century: The outline of San Stefano (1878) integrates Korçë's area in a *Great Bulgaria* that includes most of present (FY)ROM; the Balkan Wars and WWI witness the emergence of two short-lived Republics: the *Republic of Korçë*, sustained by the French armed forces and the *Republic of Northern Epirus* supported by Greece, with Gjirokastër as the center of gravity (Fig. 5.4).
   The Republic of Northern Epirus is an avatar—i.e., political outcome—of border fluctuations. From the Ionian Sea to the Prespa and Ohrid lakes, from Chameria[3] (Çamëria-Τσαμουριά) to Korçë, uncommonly mixed ethnic and confessional populations occupied this piece of land that no census ever managed to count adequately. Here, as elsewhere, national partitions have imposed cold-steel dissection to Ottoman lands managed for centuries by imperial politics that had established the ethnic diversity as a governance system.

3. Covering the high Drino, Vjosa, and Kalamas valleys, and the polje of Jannina, the third sector holds an intermediate position. These lands were historically acknowledged to the **Pashalıks of Jannina and Berat**. The shadow memories of Ali Pasha still haunt places and minds. On the western side of the Pindus, the corridor was ruled by the head city of Jannina, which used to interconnect both northern and southern Ionian territories. Afterward, it suffered the outcomes of the partition most severely (Fig. 5.6)
4. Next to it, **former Chameria** occupies the Ionian coastal area. This land originally populated by mixed Albanian and Greek communities has undergone radical changes after the Balkan Wars. Its annexation by Greece opened an era for cross-border "spontaneous" population exchanges, but the Muslim

[3]Old Turkish name applied to present *Thesprotia*. See glossary: *Çam*.

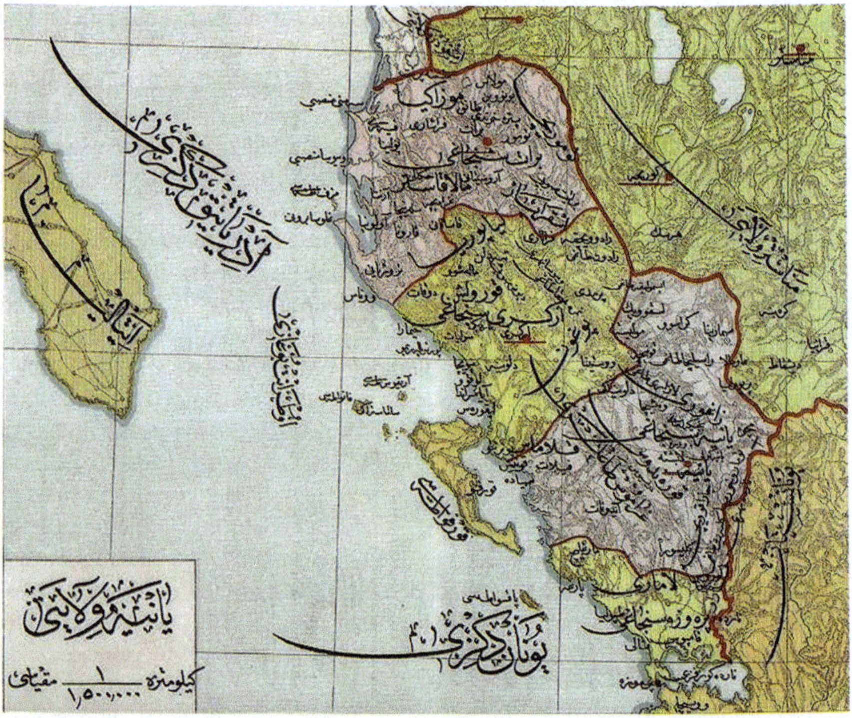

Fig. 5.6 The Vilayet of Jannina at the end of the nineteenth century. *Source* Memalik-i mahruse-i şahaneye mahsus mükemmel ve mufassal atlas (Nasrullah et al. 1907). *The shape of the Vilayet has continually changed after the Tanzimat reforms. Here is presented a "minor" territorial imprint*

Albanian-speaking communities have been "expelled" permanently in 1944 following the advance of Greek resistance fighters (Manta 2009). No territorial cross-claims exist today between Greece and Albania although residual tensions persist especially about the unsettled issue of houses and other land properties left behind. Apart from the animated tourist seaside spots scattered between Igoumenitsa and Preveza, the countryside is rather depressed today. The irrigation systems of lower Kalamas and Acherontas maintain an agricultural activity in otherwise blighted areas. The development of Igoumenitsa as an entry port to Greece dates back to the 1950s–1960s with the Iron Curtain imposing strict controls on the movement of goods and people toward Turkey.

5. Geographically isolated from mainland concerns, **the Ionian insular domain** forms the lowest step of the ladder.

The remoteness of this *Far West* of the Ottoman Empire went from bad to worse during the nineteenth century. With the independence of Greece and the *liberation* of Thessaly, the already advanced backwardness of *Epirus* went from bad to worse. After 1881, the area forms a *swallow hole* caught between two Greek regions, the

Fig. 5.7 Epirus as a funnel, 1881–1912. *Source* David Rumsey Map Collection (Andree 1895)

Islands and Thessaly (Fig. 5.7). The Ottoman administrative and urban system, although historically weak, is condemned to a predictable collapse (Dimitropoulos 2008).

The situation of Preveza is stunning. The city is an Ottoman *end of the world.* In 1864, all surrounding islands become Greek. In 1881, the annexation of Arta divides the upper alluvial plain of the Ambracian Gulf into two parts—despite its narrow dimensions. The semi-blockade lasts for almost 40 years. The late Ottoman dismembering creates many such "dead ends." Heading its Vilayet, Jannina, for instance, faces two successive geographic crises: The first one cuts the city off from its eastern dependencies (from 1881 to 1912). Once reconnected to main Greece after the annexation of Epirus in 1912, the city of Ali Pasha finds itself sidelined by the new Greek–Albanian border (from 1912 onward). The penalty is doubled. The increased number of frontiers necessarily leads to the emergence of such "disabled territories." In any case, the splendor of Ali's palaces that had brought the influence of Jannina far beyond the Pindus range belongs to history textbooks.

Concluding with these observations on the Greek–Albanian fringe lands, we can return to the *uncertainty* principle. The early spatial definition of Albania was not meant to be mapped. Legal treaties cautiously avoid establishing the Albanian independence through fixed borderlines, leaving this task to future accomplishments. Unlike most other examples, the process of settling boundaries follows here a bottom-up course, which partially escapes the cycle of repeated international conferences. A state legally established without accurate territorial definition for 7 years (from 1913 to 1921) is obviously a case study. Therefore, any outline prior to 1921 should be considered almost fictive. When firearms make the law, map-makers have to be patient (Guy 2005, 2008; Milenkoski and Talevski 2001).

5.3 Border Uncertainty in Montenegro and Southern Serbia

The third "hot point" is one of the most challenging. Previous *trigons* give way to an even more complex territory, a *tetragon* involving Montenegro, Kosovo, Albania, and Serbia—with the addition of (FY)ROM, this would produce a *pentagon*. The initial aim was to examine the spatial fluctuations of Montenegro. Extremely versatile boundaries have led us to extend the study area to include Bulgarian confines.

Figure 5.8 should seem bewildering to inexperienced people. This map brings back the Epirus and Rhodope cases to the level of a child's play. The challenge is to make it clearer. From this point on, national versus imperial interests are no longer issues in the spotlight. Nations fight with one another to impose their own territorial order. Unlike previously, the real problem lies in early autonomy/independence. Brought back to the beginning of the nineteenth century, the process of *sedimentation* is thicker. It covers a period twice longer.

The overall area could be viewed as a broad "Serbian South" engaging contact with Bulgarian, Montenegrin, Albanian, and Macedonian lands. The edges of the Republic of Ragusa, the Illyrian Provinces of the French Empire, and Austro-Hungarian Dalmatia are visible on the left side of the map. It is almost like all three attempts to reach Albania had failed to grab the last kilometers, the city of Bar being a final frontier. All around Kotor and until Bar, a very stable border is thus settled: This *end of the world* defines the spatial limits of coastal and insular districts attached to the West. The Orjen and Lovćen Mounts reach 1600–1700 m height and form a serious obstacle for the Ottoman territorial ambitions. This physical protection finds some continuity along the Dinara range. It helps maintain the Central Balkans apart from the Adriatic.

Behind this partition line, the Ottoman *Far West* struggles for its independence. The Ottoman confines suffer from invalidating landlocked positions throughout the nineteenth century. The early geographic profile of Montenegro, for instance, hardly meets its current reputation of picturesque haven for tourists. A forced relinquishment impairs its development potential. Surrounded by "hard" imperial frontiers, the Ottoman Province of Bosnia can be seen as a spatial protrusion. Confronted to the Austrian military marchlands and to Moldavia's autonomy, the Serbian Vilayets experience a similar remoteness. The Serbian, Montenegrin, and Bosniak upheavals of the early nineteenth century may thus be interpreted as attempts to break painful encirclements.

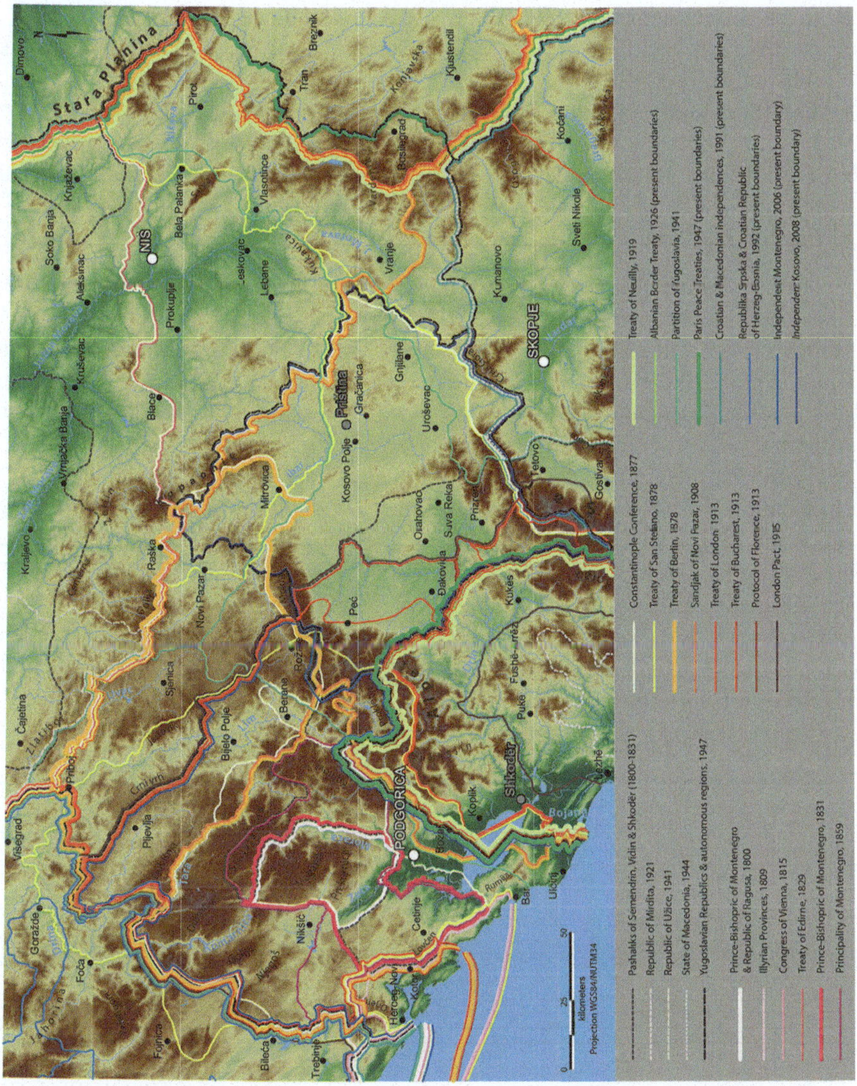

Fig. 5.8 Historical borders between Montenegro, Albania, Kosovo, and South Serbia, 1800–2015. *Source* Author's own GIS database, NASA DTM, SRTM Program, ESRI

It takes a lot of time to bring to its term the process of achieving full national independence. This is not the least of paradoxes to note that dreams for *open land* meet national ambitions. Stage by stage, feudal and ethnic local leaderships gain force—they never ceased to exist in the Ottoman state. The Porte gradually recognizes the Principality of Serbia through a series of decisions that follow ceaseless insurrections and progressive spatial broadening.

- In 1804, the Serbian revolution leads to a first unofficial recognition.
- In 1815, at the end of the second Serbian uprising, the independence is self-proclaimed.
- The formal recognition of Serbia's statehood is decided in 1817.
- In 1830, another *de jure* recognition validates the autonomy already granted.
- The full official independence is gained in 1878, together with the international recognition.
- Finally, the Principality becomes Kingdom in 1882.

The same gradual transition to independence applies to other Ottoman borderlands, with self-determined rhythms. The Prince-Bishopric of Montenegro, for example, is an old institution dating back to the seventeenth century. Always at fight to enforce their sovereignty against the Ottomans, the *Vladikas* of Montenegro gain a formal recognition by the Porte in 1789, thanks to an increasing Russian support.

Figure 5.9 presents the country in its "native" shape at the beginning of the nineteenth century. Surrounded on all sides by Ottoman dependencies, among which the semi-autonomous Pashalık of Shkodër placed under the governance of the Bushati family, this piece of land changes later into a secular Principality (1852) and Kingdom (1910).

Geographic isolation is a common fate at that time, when little chieftains make the law and "maintain order" in confined valleys and mountains. The initial Montenegrin surface reaches 1 300 km^2—only half of present Luxembourg—

Fig. 5.9 Montenegro as defined in 1805. *Source* Digital National Library of Serbia (Traux de and Berken 1805)

covered almost entirely by peaks reaching 2 000 m overhanging the Zeta Valley 1 950 m down. How did this microstate manage to increase its acreage by ten and reach its current 14 000 km^2?

The spatial growth runs in all directions. Access to the sea is gained in 1878. The new maritime corridor stretches south of Kotor until the lower Bojana River. But the real outward opening of Montenegro on the Adriatic Sea comes much later. The "revolution" takes place after WWII: As a constituent republic of the Yugoslavian Federation, Montenegro finally includes the enclave of Kotor, which doubles its coastal area.

Looking landward, the situation is more intricate. Throughout the nineteenth century, Montenegro's northern expansion is enacted simultaneously with Serbia's southward extent. Thus, a geographic landlock is created that leaves Bosnia and Herzegovina apart from the other Rumelian districts of the Ottoman Empire. The Sandžak of Novi Pazar is a corridor that divides both Orthodox nations until the Treaty of Berlin (1878). This date is a milestone. Bosnia and Herzegovina come under Habsburg domination, whereas Austrian troops occupy the Sandžak until 1908. The intermediate corridor disappears after the first Balkan War: Montenegro and Serbia find a common border in 1913 with the carve-up of the Sandžak (Fig. 5.10).

The territorial gains of Montenegro are impressive. The upper limit abuts on the Uvac River and, at the end of the second Balkan War, the country includes part of Metohija (present district of Peć, Kosovo) and part of the northern Albanian Alps. In 1918, Yugoslavia absorbs Montenegro. Fighting for centuries to enforce its sovereignty and finally approving a self-dissolution with the aim of managing a successful political union with the Kingdom of Serbia is a strange destiny. If we bypass the parenthesis of WWII and the temporary revival of Montenegro as an Italian puppet state the actual *resurrection* of the country happens in 2006.

All this is well known. Returning to the core issue implies mentioning that the absence of any spatial definition of Albania just after its declared independence has identical outcomes around Shkodër and in Prokletije as in Epirus. Processing Montenegro's historical maps for further GIS integration turned into something of a nightmare. Advanced image rectification softwares cannot even prevent from significant location errors. Many nineteenth century maps suffered from inaccurate triangulation in measuring distances. Geometric qualities vary widely among and within map products. A further challenge was to identify local places, taking as references old Ottoman names that had fallen into disuse. All in all, remodeling early Montenegrin maps proved to be terribly complicated (Fig. 5.11).

The relative stability of northwestern Montenegro is noticeable. The opposite happens at the north ends of Albania, where an *unspecified zone* is mentioned. This latter area covers southern Serbia—around Novi Pazar—and, further east, includes eastern Kosovo. The instability of these confines is real. Regardless of the late Yugoslav demarcations, the boundaries of Montenegro *stricto sensu* never survived more than 30 years without change. Opposite dynamics preside over the destiny of Montenegro and Bosnia–Herzegovina: Until the 1990s, Bosnia was an untouched haven compared with its nearby eastern regions.

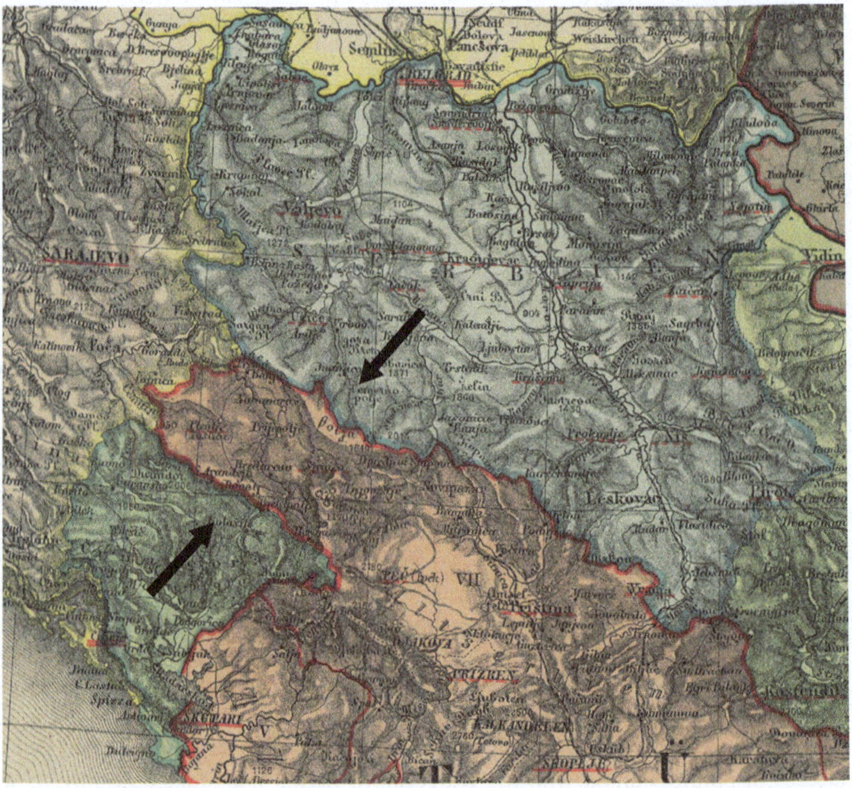

Fig. 5.10 The Sandžak landlock in 1903. *Source* Harvard Geospatial Library (Peucker et al. 1903). *The landlock existed in a native form since the beginning of the nineteenth century. The land strip narrows in stepwise manner with Montenegro and Serbia extending their sovereign areas northward and southward. Until its annexation by Austria–Hungary, the corridor was used as single land access to Ottoman Bosnia*

These elements demonstrate the permeation of territorial and national interests. Our primary approach implies studying southern Serbia, Kosovo, and (FY)ROM as a single issue. The last two *countries* have emerged on behalf of Tito at the end of WWII with two different statutes—federal republic versus autonomous region (Radan 1999). Before this date, none of them had ever been identified as a political-territorial unit. The Ottoman administration had adopted a basic "non-compliant" principle for its planning policy. The withdrawal of the Porte results in a complete disruption with the past. The Vilayets, for instance, used to embrace areas divided afterward between Yugoslavia, Albania, and Greece. The Vilayet of Üsküb–Skopje extended all the way to Plevlja and Novi Pazar. Unlike today, "the South" used to govern northern lands placed afterward under the influence of Belgrade. In Fig. 5.12, the current boundary between Serbia, Kosovo, and (FY)ROM finds no equivalent. Spatially defined by its border with Serbia, and

Fig. 5.11 Kiepert's map of Montenegro (1862). *Source* Zeitschrift für allgemeine Erdkunde (Kiepert 1862), Digizeit Göttingen University

divided between the Sandjaks of Prishtinë, Prizren, and Ipek, there is no unity in "Yugoslavian" Kosovo.

With the establishment of First Yugoslavia as an independent country in 1918, a new territorial order arises that breaks up with the Ottoman framework[4] (Trbovich 2008). The Vidovdan Constitution of 1921 establishes a unitary state and creates 33 oblasts transgressing the former administrative network. Similarly, the post-1929 *Banovine* division and toponymy systematically avoid referring to a previous spatial configuration, especially in the South (Fig. 5.13). Montenegro is thus integrated into a *Zeta Banovina*. The *Vardar Banovina* includes both Skopje and

[4]And with the post WWII Yugoslavian administrative reference frame.

Fig. 5.12 Administrative map of Turkey in Europe, Vilayets of Salonika, Monastir, and Üsküb, 1908. *Source* Museum for the Macedonian Struggle, Salonika (Carte des provinces de la Turquie d'Europe (vilayets d'Üsküb, de Monastir et de Salonique) soumises aux réformes 1908)

Prishtinë. Bosnia–Herzegovina is split into three banovine whose features have no prior reference. No trace of Kosovo is reported. All cards are reshuffled.

Half a century later, the project of Tito restores some spatial *coherence* to Bosnia, Herzegovina, Serbia, and Montenegro. Few historical outlines are recalled, although the southern *question* is left untouched. This leads to another explanation. A first landlock between Montenegro and Serbia had been mentioned. A second buffer zone comprising Kosovo and northern Macedonia deserves careful consideration, with Prishtinë and Üsküb as cities ensuring land control.

By the end of the nineteenth century, four national ambitions compete for control of the *Great Southern Serbia*: Albania's claims on Kosovo, Bulgaria's views over present (FY)ROM, Serbia's purpose of following its southward extension, and Montenegro's mainland appetites. Few people identify the southern limit of Kosovo with Bulgarian territorial ambitions. Observers commonly focus on the competition between Serbs and Albanians and forget that south Kosovo used to be seen as

Fig. 5.13 Administrative map of First Yugoslavia, 1929. *Source* V. Marinković (Marinković 1935)

a contact area between Bulgaria and Serbia. They prefer to ignore the disastrous military eagerness of the past. Although not part of the GIS database because of its unofficial status, a document emphasizes the past existence of this interface (Fig. 5.14).

The inter-Slavic confrontation opens the way to further delineations. The northeastern limits of current Kosovo match with the old Serbian–Ottoman boundary. Their historical legitimacy is quite grounded. But the southeastern frontier follows an unprecedented outline, mentioned only in two "fictitious" agreements:

- The Treaty of San Stefano had already adopted the same tracing for Bulgaria's boundary. It was never applied.
- Serbia and Bulgaria had signed the Treaty of Alliance while preparing the first Balkan War against the Ottoman forces. Facing both Albanian and Bulgarian territorial claims, Serbia had firstly chosen to give up its ambitions south of a line Veles-Ohrid, only contesting a transversal land strip with Üsküb at its center.

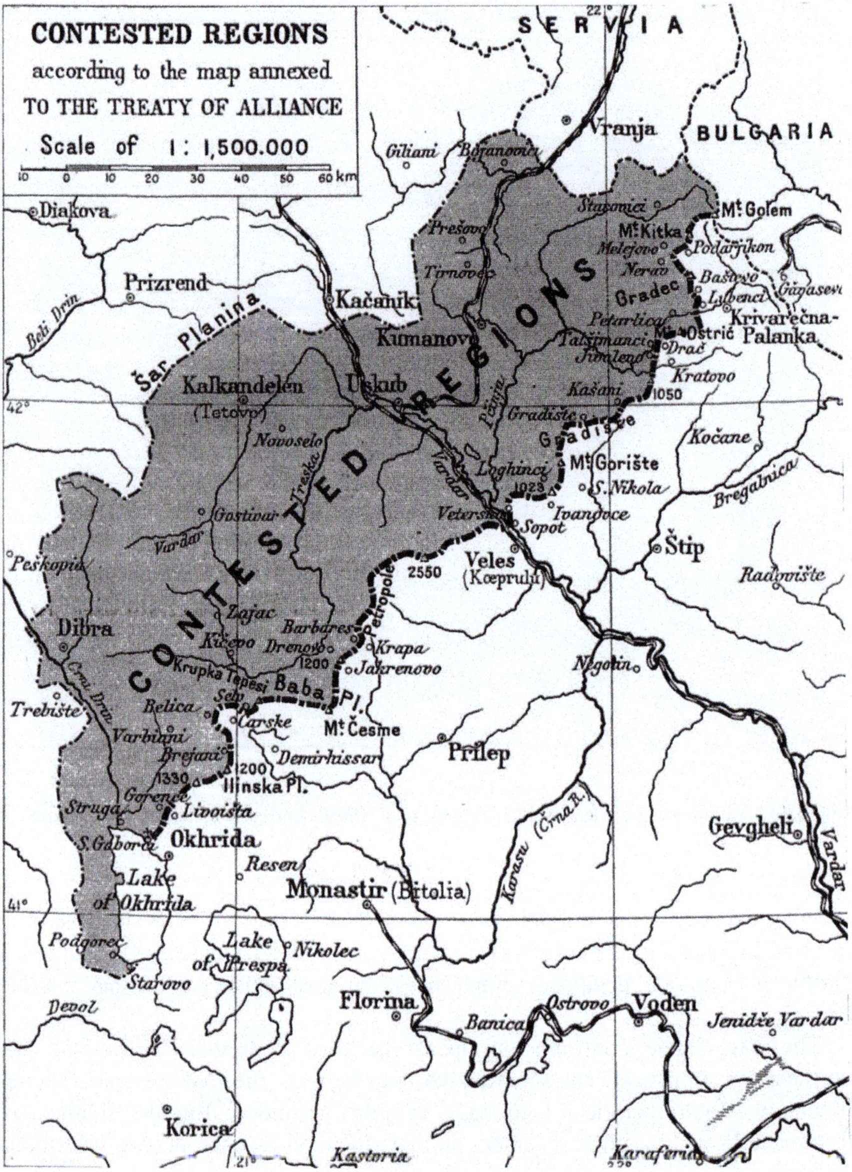

Fig. 5.14 Serbian-Bulgarian confines in 1912. Treaty of Alliance. *Source* Report of the International Commission to enquire into the causes and conduct of the Balkan Wars (Peace 1914)

The temporary border is reactivated during WWII when Bulgaria takes control of the districts east of a line Stara Planina–Šar Planina. There is no coincidence that Tito adopts the same boundary when fixing the limits of the *Republic of Macedonia*

—with a small nod to history. The “contested regions” were claimed by Bulgaria. They end up acknowledged by a Yugoslavian political body.

The Serbian southern ambitions are a known stumbling block. The *independence* of Kosovo in 2008 obviously reinforces the geographic containment of the country. It reduces its contact with (FY)ROM to a narrow strip of land between Vranje and Kumanovo. Figure 5.8 shows that this other *bottleneck* is nothing compared with the broad way connecting Albania and Kosovo among the physical “barriers” of the Albanian Alps and Šar Planina. Since the 1990s, Serbia never had such a distant sea horizon.

5.4 Dalmatian Far Ends

The first three “hot points” were part of the Ottoman Rumelia. The forthcoming cases involve the outside frontier of the Ottoman Empire, on its interface with Austria–Hungary and Russia—where the most stable boundaries concentrate. An old imperial resonance is added to any national development. The oncoming “second rate” spots are less reliant on international boundaries; they demonstrate the efficiency of a GIS database enriched with other internal *limits*. The *marchland* system is particularly relevant and needs further investigations (Fig. 5.15).

Studying blind spots implied paying attention to Dalmatia sooner or later. As already mentioned, this strip of land has long been secluded from the Balkan mainland until the creation of the State of Slovenes, Croats, and Serbs at the end of WWI. Figure 5.15 presents two worlds apart: The coastal area of Austria–Hungary faces the Bosnian hinterland whose fate is attached to eastern horizons. This land pattern remains unchanged until the Treaty of Berlin and the establishment of the BiH Condominium placed under Habsburg rule. Until the fall of the Empire, the Austrian confines live under a specific legal status and serve as a “cordon sanitaire” device against the troublemaking Turkish Empire: These are *Military Frontiers* (Militärgrenze).

The difference with Montenegro is obvious: This latter country achieves an early independence and gradually develops, with many intermediate cycles of spatial shrinkage–expansion. Bosnia–Herzegovina knows an almost reverse trajectory. Its western contour line[5] is remarkably steady. The same is true with Serbia: The Drina is an early demarcation stream. Most importantly, Bosnia–Herzegovina never experienced state independence until the Dayton Agreements of 1995. The “country” goes from an early position of Ottoman Province to another imperial dependency (Austria–Hungary), then joins the Kingdom of Yugoslavia, and later

[5]It follows the Sava, Una, Glina, and Korana rivers (northern border, including Cazinska Krajina) and then catches on the upper Una and Krka, which joins the Dinara range until reaching the seaside near Dubrovnik.

Fig. 5.15 Southern Dalmatia in 1805. *Source* Digital National Library of Serbia (Traux de and Berken 1805)

becomes part of the Yugoslavian SFR. Despite the Yugoslav Wars of 1991–1995, its outer limits remain almost untouched.

Dealing with Bosnia when studying Dalmatia could be misleading, although these areas share a common interface (Burdett 1995). Until 1918, Bosnia–Herzegovina

Fig. 5.16 Historical borders in Dalmatia, 1800–2015. *Source* Author's own GIS database, NASA DTM, SRTM Program, ESRI

was not entirely landlocked. It took advantage of two corridors providing access to the Adriatic at Neum and Sutorina—a situation that makes Bosnia a pictorial representation gifted with a tiny pair of legs. These are remainders of an early eighteenth-century decision of the Republic of Ragusa to cede two land strips to the Ottoman Empire at its north and south. The decision avoided being in contact with lands ruled by Venice. Until its annexation by France in 1808, the Republic of Ragusa was a closed area south of the Habsburg possessions. At that time, the Ottoman Empire controlled all mainland areas south of the Carpathian Mountain range and all coastal lands south of Herzeg-Novi (Fig. 5.16).

In 1815, after the Congress of Vienna, the Habsburg Empire takes benefit of the French Empire's defeat and reaches an extreme southern point at Bar, close to Shkodër's Lake. The Bay of Kotor thus joins a newly created *Kingdom of Dalmatia*, part of the crownlands. The Bosnian–Ottoman sea corridors of Neum and Sutorina continue to play a strategic role until the Treaty of Berlin. Between 1878 and 1918, both strips are maintained even though BiH has become an Austrian possession.

In 1948, after 30 years of Yugoslavian "melting pot" and "openness," Tito reestablishes the historic borders of Bosnia. However, the unification of *Boka Kotoska* (Bay of Kotor) with the federative republic of Montenegro deprives Bosnia of one of its "legs."

Despite a long isolation from its hinterland and a secured position inside Croatia's present boundaries, Dalmatia remains highly fragmented. The creation of Yugoslavia in 1918 puts an end to conflicting Italian–Austrian ambitions. Getting into each local dispute would not be appropriate, but one might argue that Fig. 5.15 reveals a rough division into five zones:

1. Dubrovnik's area is in the far south.
2. A medium zone spreads north of Neum, delimited by the Velebit Mounts and the Dinara range: It defines the former Kingdom of Dalmatia, with Split at its center.
3. The old Croatian space, part of the Austrian–Croatian Military Frontier, is bordered west by the Velebit and north by the Kupa River. From the Kupa, the boundary line reaches the sea south of the Risnjak Peak and leaves the district of Rijeka to other mainland areas. Under the Habsburg dominion, the Croatian Kingdom had a permanent access to the sea at Crikvenica.
4. Istrian and Slovenian possessions occupy the area stretching north of the Kupa's thalweg. Here, Italian and Austrian–Hungarian interests directly collide. The Habsburg Empire keeps an access to the Adriatic until 1918, mainly through the harbors of Rijeka and Trieste. The interwar period changes the balance of power. Italy officially gains Istria in 1918 and maintains its presence until 1947, leaving the inland areas to the First Yugoslavia.
5. The islands define a "space apart." The whole insular area belongs to the Austrian Empire until 1918 and returns to Yugoslavia after 1947. The interwar period allows Italy to gain Rijeka, Cres, Zadar, and Lastovo. WWI and WWII also assign to Rome many other coastal areas and insular possessions. The Italian withdrawal from the Dalmatian Islands is fully achieved in 1947—a historic event putting an end to the former "Italian" influence.

All travelers who have crossed Yugoslavia shall admit the importance of the Velebit-Kapela and Dinara ranges in physically defining frontiers. The Riviera, which begins at Rijeka and stretches until the end of the Velebit's maritime channel, east of Zadar, is a road traffic nightmare in summer. As usual throughout the Mediterranean mainland areas ignore seaside areas—and vice versa. The geographic "law" is quite explicit, with a possible exception in Central Dalmatia.

5.5 Wandering Frontiers Along the Lower Danube

Continuing along this path, the examination of the Ottoman outside boundaries leads us to the Black Sea coast and Lower Danube. This other maritime area has a profile completely different from that of Dalmatia. Whereas the Dalmatian hinterland seems to turn its back to the Adriatic, helped by an irregular relief, the Lower Danube is a natural compromise between land and water—whatever fresh or salted. Across open landscapes, connections are easier. Which landmark will bring support for political limits? (Fig. 5.17).

Fig. 5.17 Historical borders of the Lower Danube, 1800–2015. *Source* Author's own GIS database, NASA DTM, SRTM Program, ESRI

Chap. 3.6.1 has evidenced the indeterminacy of the Danube thalweg. The delta is a long-standing issue for mapmakers. In comparison with other past changes, the latest territorial adjustments decided by the International Court of Justice in 2009 sound ridiculous. The decision settles down an old dispute that had been interfering between the USSR and Romania since 1948 about the status of six micro-islands of the Chilia branch of the Danube. Obviously, the old imperial land uncertainty has given way to close negotiations between states who are eager to regulate their respective positions even down to the last details.

The essential is elsewhere. As opposed to other mountain districts, this lowland area evades the rules of "single use" boundaries. Most frontiers are spatially steady, although politically unstable. For instance, the well-known Dobrudjan Cadrilater has changed trusteeship no fewer than six times after 1800, and Budjak eight times, each of these districts integrating or leaving the same states: Russian Empire/Russia/Ukraine versus Romania (for Budjak), and Ottoman Empire/Bulgaria versus Romania (for Dobrudja) (Durandin 1979).

These examples deny any validity to the idea that border instability and mobility are exclusive characteristics of remote mountain reserves—supposedly engaged in low regulation. Three rivers are key-references: Danube, Prut, and Dnistr. Besides, other outlines rely on "human-based" demarcations. These latter are closely negotiated because they divide cheek by jowl rural communities and large agricultural areas extensively cultivated. The Moldavian–Ukrainian boundary created in 1947 is an example of such political border, as well as the Romanian–Russian outline of 1856, or the northern and southern successive limits of the Dobrudjan quadrangle. It is hard to imagine how such frequent border changes have decided the destiny of civilians caught in ceaseless military fights, without other option but to face the twelve official Russo-Turkish Wars—countryside areas are far more populated here than other Balkan mountain regions.

Romania's access to the sea is versatile. Locked in between Budjak and Dobrudja, the status of the Danube Delta seems more obvious: Its alleged "neutrality" displayed by few old maps finds a logical outcome in the recent creation of a natural protected area, with the support of Romanian and Ukrainian governments.

Significantly, the Lower Danube between Silistra and Galați is not a dividing line. Its unifying role seems exceptional for a waterway so often used in support of political divisions. It should be kept in mind, however, that until 1878 Wallachia and the United Principalities were formally placed under Ottoman suzerainty: The Danube has escaped a full boundary charge until lately.

5.6 The Marchlands of Banat

Northern Serbia is a last case study of former cross-imperial frontier. The area is shaped by long-lasting boundaries[6] backed up by less official outlines that have no equivalent downstream. Unlike Dobrudja, the Pannonian Plain displayed in Fig. 5.18 welcomes many internal frontiers of the Empires—short-lived Republics and local autonomous districts. Its fragmentation reaches peaks when adding semi-legal borders to the picture. All else being equal, flat plains do not pale into insignificance against other mountain areas.

Few people have heard about Banat as a geographic region. The name is today rarely applied to this part of the former Austrian Empire. Three countries occupy the historic lands of Banat: Romania (south of the Mureş and west of the Carpathian Mountains), Serbia (east of the Tisa), and Hungary (residual sector southeast of Szeged). Until the 1870s, a legal military status applies on half of Banat. Along this large frontier area, the governing place of Timişoara-Temeschwar lasts until WWI. On the opposite bank of the Danube, Belgrade assumes at that time an unusual position: Although straddling the border, the city becomes the official capital of Serbia in 1841, as a successor to Smederevo-Semendrin and Kragujevac. For almost 80 years, Belgrade leads to the Serbian destinies, its hillsides flooded by the Danube, and its suburbs occupied by Austria–Hungary across the riverbank (Fig. 5.19).

Both sides of the Danube and Sava have been military frontiers for centuries (Peacock 2009; Maier 2003): A series of fortified places punctuates the river streams for the greater enjoyment of modern-day visitors. The *Military Frontier* system on the Habsburg side is effective until 1849 in Transylvania, 1873 in Banat, and 1881 in Slavonia. The device plays an important role in structuring space and shaping landscapes. It is meant to remain in place as long as any Ottoman "pressure" would persist on the outside face of the Empire. The progressive independence of Serbia against the Ottomans leads to the expression of many other "self-governing" wishes after the revolution of 1848 (Fig. 5.20)

The different forms of Banatian political autonomy reveal the existence of disputed lands that resist the dismembering attempts of combined imperial–national forces as when creatures fancy they can survive their creators. Local autonomy is a relevant indicator of ambitious external appetites. In the nineteenth century, Banat is a complex area populated by a large number of ethnic and confessional communities: Romanians, Serbs, Hungarians, Germans, Croats, Roma, Czechs, Rusyns, Slovaks, and Krashovani. Such a diversity makes any border settlement problematic (Marin 2012).

Two episodes betray the local resistance to land fragmentation and unveil the geographic-political specific nature of Banat:

---

[6]Despite their graphic prominence, neither Sava nor Danube is separating river today on both sides of Belgrade.

Fig. 5.18 Historical borders around Banat, 1800–2015. *Source* Author's own GIS database, NASA DTM, SRTM Program, ESRI

- The self-proclaimed *Serbian Vojvodina*, created in 1848, leads to the establishment of the *Voivodeship of Serbia and Banat of Temeschwar*, preserved until 1860.
- In 1918, a *Banat Republic* is shortly declared in Timişoara. The fake state survives for only 15 days until the invasion of the Serbian army.

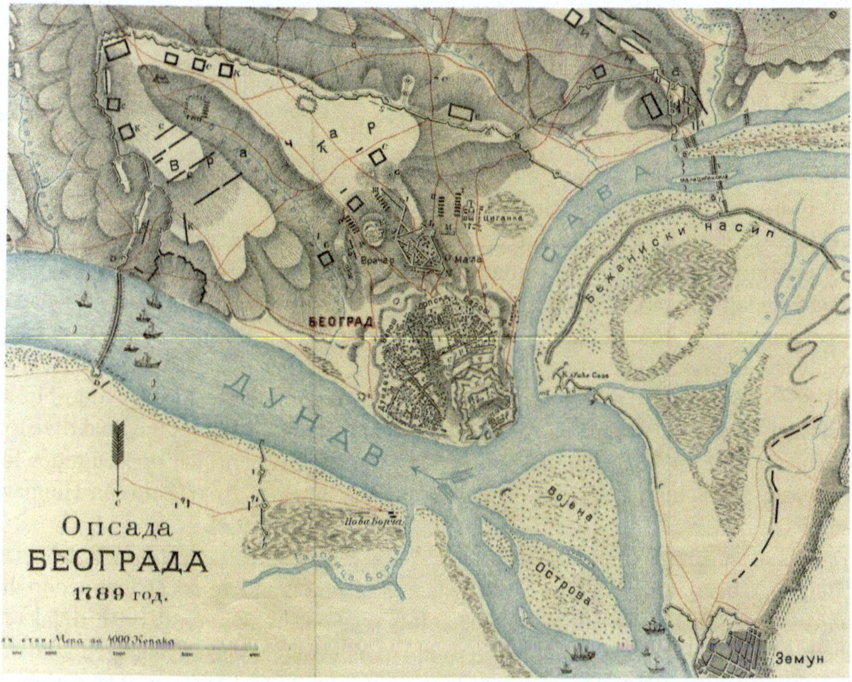

Fig. 5.19 Belgrade in 1789. *Source* Lithography, Digital National Library of Serbia (Опсада Београда 1789 год, 1887). *As a border fortress, Belgrade is a military construction, like many other edge cities. Although submitted to an almost exclusive military control, exchanges are made possible (bridges)*

Fig. 5.20 Banatian Military Frontier in 1856. *Source* David Rumsey Map Collection (Scheda 1856)

Banat is thus an old crossroads where competing forces meet. Other political "oddities" indicate that some *regional existence* persists:

- The *Baranja-Baja* short-lived Republic (1921).
- The autonomous *District of Brčko* (from 2000 onward).

In the nineteenth century, most internal and external borders were attached to river streams. With the Empires' withdrawal, nations stop using such landmarks. From 1918 to the early 1990s, Danube and Sava lose their prominence for other "unorthodox" demarcations. Settled in 1916 and revised in 1924, the Romanian–Yugoslavian border cuts Banat into pieces. Even the Mureş-Tisa river confluence is abandoned: The new outline carefully bypasses Szeged. Between Yugoslavia and Hungary, the boundary stretches west of Szeged and north of Osijek from 1915 to 1918. In 1918, it finally adopts part of the outline granted to the Voivodeship of Serbia, half a century ago. The new Yugoslavian frontier achieves the final division of Banat while splitting Baranja and Bačka. All this leaves further opportunities for European cross-border regions! The same occurs north of Arad, between Hungary and Romania.

This description naturally drives us back to the great question of Austria–Hungary's dismembering after WWI, when the Pannonian Plain is abandoned to the *unfortunate fate* of dismantling projects. It is noteworthy that, until 1991, thanks to the process of federal integration, the southern Yugoslavian lands managed to avoid the turmoil other trans-Danubian former crown lands had to face: The Ottoman areas have no monopoly on territorial indeterminacy.

Present Vojvodina is thus a late patchwork of Syrmian, Bačkan, Banatian, and Baranjan pieces of land. Its creation as an *autonomous region* within the Serbian Republic is dated 1945. During the interwar period, the Banovina of Belgrade covers a large trans-Danubian area, already breaking up with the old Ottoman–Austrian boundary (Fig. 5.13). After 1945, Belgrade heads a new administrative unit, the *capital district*, which steps over the Danube and contributes to lowering the dividing role of the river. Belgrade perfectly embodies the defeat inflicted to the old imperial demarcation.

This brief review would not be achieved without mentioning the Croatian–Serbian boundary (Klemenčić and Schofield 2001). The present outline is established by Tito as an inter-republic division immediately after WWII. Between the Hungarian border and Bačka Palanka, it follows the Danube and thus borrows segments of the Military Frontiers. South of Bačka Palanka, its position is fixed by the Yugoslav Wars, without prior historical and/or geographic references. For the first time in 1991, this demarcation is raised to an international level. Vukovar was in the middle of a powder keg. In hindsight, it comes as no surprise that this city has been plagued by the hardest fights of the Yugoslav Wars.

5.7 Cross-Looks on Border Districts

The transversal perspective on cross-border areas helps avoid different errors committed by classical segmented works. The analysis brings geographic consistency to lands heavily impacted by late fragmentation processes. Some may say it would be best to settle old scores and to ignore painful historical episodes. At a time of general questioning on the European internal and external frontiers, the criticism appears quite unfounded. The upper figures easily demonstrate how much border-related issues and Balkan geography are closely related.

These "hot points" exemplify constant interactions between neighboring states, administrative regions, and local communities. The Balkans are pervaded with countless "blind spots" and "in-between spaces" whose geographic features often withstand scrutiny. Against this background, our transnational approach was the only way to provide new insights into Balkan complexity without making the area an artificial patchwork of mutually ignoring lands—a self-serving falsehood.

At this point, certain biases could cast some doubts on the efficiency of the analysis. One might consider that building the GIS database from 1800 onward introduces an artificial break in the chronology of developments. The artifact cannot be avoided. Of course, the introduction of pre-1800 data could favorably complete and weight many assertions. The issue belongs to further research.

Another legitimate question should not be taken off the table: How could such a work on frontier issues avoid addressing ethnic-confessional concerns? There are many answers to this. It should first be noted that no panoptic Balkan map—*ipso facto* endorsing transnational perspectives—has never been published. "Global" mapping ambitions have not survived the collapse of the Empires.

An ex-post review of low-scale cross-national sheets demonstrates that many such mappings aimed to picture the ethnic-religious distribution of people. The willingness to investigate linguistic issues usually prepared further annexations. These were meant to sacrifice complexity and scale dependence. Rough and politically oriented maps have left painful memories. They deserved too serious objections.

Thus, as an answer to the initial question, any accurate transnational mapping of ethnic-confessional affiliations/identities is virtually impossible. The few available statistical data on this particular matter are usually considered truncated or biased. Furthermore, addressing ethnic and confessional issues across administrative units submitted to divergent collection systems is, by definition, unachievable project. Overcoming the problem is worthy on limited areas, but the price to pay is high: Bringing statistics to convergence is a hard task whenever and wherever data exist (Chap. 8).

Sequentially mapping boundaries was meant to provide new insights into Balkan issues while avoiding futile controversies. Some of these regions are still sensitive today, not to say highly volatile. Border debates never stopped poisoning people's life for centuries. From our perspective, "reactivating" a transnational approach at different scales was a way *not* to add fuel to the fire. The project precisely had to break up with the outdated mappings of the late nineteenth century whose aim was to help political and military authorities. As opposed to the enhanced indeterminacy of demography, maps based on *legal* definitions of political bodies are safer. For a similar reason, as an answer to former "engaged" mappings, any term with negative or positive value has to be banished. No border is "acceptable," "conformal," or "defective." This work focuses on actual frontiers, not on demarcations that "could be" or "could have been" (Peckham 2000).

References

Andree R (1895) Greece, with Crete or Candia. The Times, London

Burdett ALP (ed) (1995) Bosnia, Croatia, Serbia: historical boundaries, 1815–1945. Cambridge Archive Editions, Slough

Carte des provinces de la Turquie d'Europe (vilayets d'Üsküb, de Monastir et de Salonique) soumises aux réformes (1908). Istituto Geografico De Agostini, Rome

Demetriou O (2004) Prioritizing 'ethnicities': the uncertainty of Pomak-ness in the urban Greek Rhodoppe. Ethn Racial Stud 27(1):95–119

Dimitropoulos D (2008) Limites intérieures dans l'espace grec au temps de la domination ottomane. L'aspect géographique des administrations politiques et ecclésiastiques. Historical Review-La Revue Historique 5:239–253

Durandin C (1979) La Russie, la Roumanie et les nouvelles frontières dans les Balkans. Le cas de la Dobroudgea. Cahiers du monde russe et soviétique 20(1):61–77

Greek Army Headquarters (1919) Carte ethnographique de l'Epire du Nord en 1913. Ankyra, Thessaloniki

Guy NC (2005) Fixing the frontiers? Ethnography, power politics and the delimitation of Albania, 1912–1914. Stud Ethn Natl 5(1):27–49

Guy NC (2008) Linguistic boundaries and geopolitical interests: the Albanian boundary commissions, 1878–1926. J Hist Geogr 34(3):448–470

Kiepert H (1862) Das Fürstenthum Zrnagora oder Montenegro. Zeitschrift für allgemeine Erdkunde, 13-1862 edn. Reimer, Berlin

Klemenčić M, Schofield CH (2001) War and peace on the Danube: the evolution of the Croatia-Serbia boundary. Bound Territ Brief 3(3):60

Maier L (2003) Die Grenze zwischen dem Habsburgerreich und Bosnien um 1830. Von einem Versuch, eine friedlose Region zu befrieden. Jahrbücher für Geschichte Osteuropas 51:379–391

Manta EK (2009) The cams of Albania and the Greek state (1923–1945). J Muslim Minor Aff 29 (4):523–535

Marin I (2012) Contested frontiers in the Balkans: Habsburg and Ottoman rivalries in Eastern Europe. I. B. Tauris, London-New York

Marinković V (1935) Geografski atlas. Za gimnazije i druge škole Stefanovića, Belgrade

Milenkoski M, Talevski J (2001) The borders of the Republic of Macedonia. Bound Secur Bull 9 (1):81–85

Nasrullah M, Rüsdi M, Esref M (1907) Memalik-i mahruse-i şahaneye mahsus mükemmel ve mufassal atlas. Şirket-i Mürettibiye Matbaası, Istanbul

Peace CEfI (ed) (1914) Report of the International Commission to inquire into the causes and conduct of the Balkan Wars. Washington

Peacock ACS (2009) The frontiers of the Ottoman world. In Proceedings of the British Academy, 156. Oxford University Press, Oxford-New York

Peckham RS (2000) Map mania: nationalism and the politics of place in Greece, 1870–1922. Political Geogr 19:77–95

Peucker K, Maschek R, Bannwarth T (1903) Die Staaten von Südost-Europa: Rumänien. Türkei, Griechenland. Artaria, Vienna, Serbien, Montenegro, Bulgarien, Kreta, Europ

Radan P (1999) Yugoslavia's internal borders as international borders: a question of appropriateness. East Eur Q 33(2):137

Rumley D, Minghi JV (eds) (2015) The geography of border landscapes. Routledge, London, New York

Scheda J (1856) General Karte des Oesterreichischen Kaiserstaates. Expeditions-Bureau, Vienna

Traux de M, Berken J (1805) Festungen Dalmatiens und Albaniens nebst vorliegender Inseln. Belgrade

Trbovich AS (2008) A legal geography of Yugoslavia's disintegration. Oxford University Press, Oxford, New York

Turkish Prime Ministry İstanbul Boğazı'ndan Batı Trakya'ya kadar olan bölgenin haritası. General Directorate of State Archives

Vogel C (1908) Die Balkan-Halbinsel in 4 Blättern: Türkei, Rumänien, Griechenland, Serbien, Montenegro. Bulgarien, Justus Perthes, Gotha

Опсада Београда 1789 год, (1887)

Chapter 6
From Borderlines to Borderlands: A Perspective of the Frontier Issue in the Balkans

Abstract Balkan boundaries are not only linear artifacts. The areas impacted by their presence are covered by legal provisions that are thoroughly examined. Braving or enforcing regulatory frameworks, gunmen, soldiers, and policemen are powerful agents who determine the rules of a land planning system implemented without regard to, or in spite of, the local population. From one country to the other, military defense systems organize land and decide on territorial development.

Keywords Military defense · Banditry · Surveillance · Security · Extra-territoriality · Buffer zones · Colonization · Land control · Bunkers · Mines · Spatial simulation

People tend quite naturally to give credence to the idea that the destiny of political boundaries is connected to the profound development of fringe lands *unfairly* banished from the main decision centers. Arising from a classical core–periphery model, the marginal status of border areas has been established as a fundamental principle in applied geography.

In view of such conceptual archetype, the GIS database presented before may contribute to a sense of unease and confusion. It is now clear that any Balkan area is —or has been—a boundary district, at least for a while, over the past two centuries. It may even be said that narrow distances produce an everlasting *border effect* on any Balkan land. Whatever softened by the European integration of Bulgaria and Romania, the Peninsula has faced acute changes with the Yugoslav Wars. Between 1990 and 2008, the length of official international boundaries has increased from the 7 500 to 12 300 km (+4800 km, +63 %)—when including the new Bosnian entities and Kosovo.

How could "outer edges" still be considered a valid conceptual category? When boundaries spread across the world, is their structuring position further strengthened or weakened? Are the upper maps a cartographic mystification made of asynchronous and non-compliant data? Is land fragmentation a so-advanced process that it might even contradict the alleged power of central cities? All in all, will borders participate in the making of a *holistic* principle governing the whole Balkan geography?

© Springer International Publishing Switzerland 2017
R. Darques, *Mapping Versatile Boundaries*,
Springer Geography, DOI 10.1007/978-3-319-40925-2_6

Answering these questions is a worthy aim, although it would be ambitious to hope that this task can be achieved. As good inquirers, we will follow a first lead. Until now, boundaries have been treated as linear artifacts. The achievement of a *borderland* status requires that lines become surfaces (Burrough and Frank 1996). Such inflation implies environmental impacts on the surrounding landscapes, with different degrees of effectiveness depending on the distance to the borderline. In other words, the *deepness* of frontiers can be measured, directly or indirectly, which induces a change in thinking about the issue.

Identifying the "external projections" of boundaries and watching space dilations/retractions are perspectives underpinning the research initiative. The basic assumption is that the spatial imprint of borders is much more substantial than most studies centered on a "barrier custom" viewpoint make them out to be. The project implies enlarging the GIS database, which is a real challenge: To get deepness, casual "sideways glances" must acquire the dimension of an investigation method. This can be done in a number of ways.

6.1 Defining and Defending Borderlands

6.1.1 Uncontrolled Areas: Bandits, Gunmen, and Soldiers

If one considers that boundaries are open air theaters, who are the role players? Preparatory works are assigned to mapmakers and politicians before land monitoring is assigned to various state bodies. Two institutions directly or indirectly perform border surveillance. Military authorities come first (Kirsch and Flint 2011; Woodward 2004). They retain upper-hand control but require the assistance of civil administrators in traffic watch. Custom officers come second. They most often belong to tax administration and, occasionally, to police forces. These services work hand in hand. Former military officers make good custom officials. The review would be incomplete, though, without mentioning a third important group: smugglers.

Smugglers are a driving force in border areas. But we must not let ourselves be fooled. Contrary to common perception, smugglers and policemen do not belong to impervious groups of people. As first resistant fighters against the *occupying* Ottomans, outlaws figure prominently on the list of historic national leaders. Before young nations can build regular army corps, many such outlaws had a mission to maintain "order" and "control" along the newly created boundaries (Gavrilis 2003, 2008; Damianakos 2003; Hobsbawm 1969, 1959; Koliopoulos 1987). Balkan bandits are charismatic figures of popular culture, whatever name they are granted. *Andartes*, *Armatoles*, *Hajduks*, *Komitadji*, Chetniks, and *Kachaks* all inherit social and political prestige, to such a point that many of them have reached the highest governmental responsibilities—although or because their hands were sullied with blood.

Once attached to a red flag, or when confronted with other military occupations, they gain the status of *partisans*. Under this new skin, they often claim some clan leadership that roots their authority into local history (the Albanian *Bayraktar* is such a clan leader). *Outlaws* are among the most ambiguous and representative figures of the Balkan society (About 1861). Dealing with boundaries had to bring us back sooner or later to such central characters. In-between frontiers could not find representatives better than these shady fellows.

This is often forgotten, but land confines provide asylum to another fourth category of people: rural communities. In remote border areas, these are seldom composed of "peasants"—*raya*, i.e., poor farmers attached to the land, heavily taxed by powerful Ottoman landowners. Much more numerous are the seminomad communities engaged in traditional livestock herding, who used to follow seasonal housing.

Before they were forced to settlement by new national authorities, *Vlach* shepherds, for instance, used to travel freely across districts separated today by international boundaries, as between Romania, Bulgaria, Serbia, and Greece, with westward extensions to Bosnia and Albania (Trifon 2005; Weigand 1895; Schein 1975). The same applies to Karakachan and Yuruk pastoral nomads. Other ethnic groups or specific clans among them made every effort to keep the mountains untouched or closed to external influences, i.e., from the "plain." Some of them are famous: Mirditë and Dukagjin tribes along the southern Dinaric Alps, Pomaks in the Rhodope, Montenegrin tribes of the Maganik and Prekornica, Serbian *plemena* and *bratstva* of Herzegovina, Montenegro and eastern Dinara, Bosniak captaincies of the central Dinara, and Székely lands in the eastern Carpathian Mountains are a few examples.

Far from the semi-independence observed in mountain areas, peasant communities of the old Austrian–Ottoman marchlands face different constraints. Across these flat landscapes, people used to live hand in hand with armed troops in charge

Fig. 6.1 Keeping external borders (1): Bulgarian soldiers on top of the Belasitsa, facing Greece, WWI. *Source* www.lostbulgaria.com

Fig. 6.2 Keeping external borders (2): Turkish–Bulgarian border post, bridge across the Struma River, north of Gorna Dzhumaya (Blagoevgrad), 1900s. *Source* www.lostbulgaria.com

of maintaining public order, settled in permanent garrisons. Austro-Hungarian Military Frontiers back to back with Ottoman Pashaliks and captaincies: a military order presides over the destiny of civilians in edge places (Stein 2007), (Figs. 6.1 and 6.2). In the nineteenth century, ceaseless raids traditionally conducted across both sides of the boundary were still living memories. Land confines suffered the consequences of a special economic and fiscal status that worsened a situation of everlasting insecurity (Karpat and Zens 2004). The process of territorial integration induced by the era of national awakening pays little compensation. The end result can even be found particularly severe for old "rural autonomous communities".

What emerges from this? The lessons learned indicate that the anthropological approach is useless in helping to define borderlands on spatial criteria. Quite the contrary, the human-based perspective results in a blurred image where land confines and "borderline" communities mix. Soldiers, smugglers, and captains occupy the whole Peninsula with varying degrees of pervasiveness. Even the newly established capital cities have to manage in their suburbs the erratic behavior of quarrelsome gangs expressing "local" resistance against "upper" sovereignties (Figs. 6.3 and 6.4). All this confirms the results of Chap. 4 and validates the holistic dimension of border issues.

Spatially qualifying the impact of boundaries requires using other methods. The legal status of land needs to be given further study and debate. Because no reference to any Ottoman legal status of the marchlands could be found—with the partial exception of the semi-autonomous buffer zones based on tribal allegiances described, among others, by Ateş (2013)—and leaving the well-documented case of the Austrian *Militärgrenze* apart (Schwicker 1883; Rothenberg 1970; Horel 2009), we shall focus on the border systems promoted by nation-states.

Fig. 6.3 Smederevo and its fortifications in 1914–1918. *Source* Old postcard, own collection. *The edge city of Smederevo has developed as a fortress. The Ottomans officially leave the town in 1867. The place had been chosen to be the temporary capital of Serbia in 1806. A capital city exactly located on the national borderline: The paradox is glaring. Political intentions are obvious. Belgrade will inherit the same geographic characteristics*

Among the national military–political authorities, the need to define a special status is made possible by improved control means on land. Legal statutes are planning tools developed to enforce already settled boundaries. They are unsuitable for still-expanding borders that are nothing less than temporary demarcations.

Investigating such a sensitive topic is difficult. Applying research to legal and military defense frameworks throughout the Balkans is an enormous task, all the more so as access to classified information is almost impossible. By way of default, personal observations and field-based investigations will provide missing data and help remove inconsistencies. Altered by the outcomes of a series of cartographic revolutions, national laws have typically evolved toward an always improved conceptual and spatial accuracy in defining borderlands.

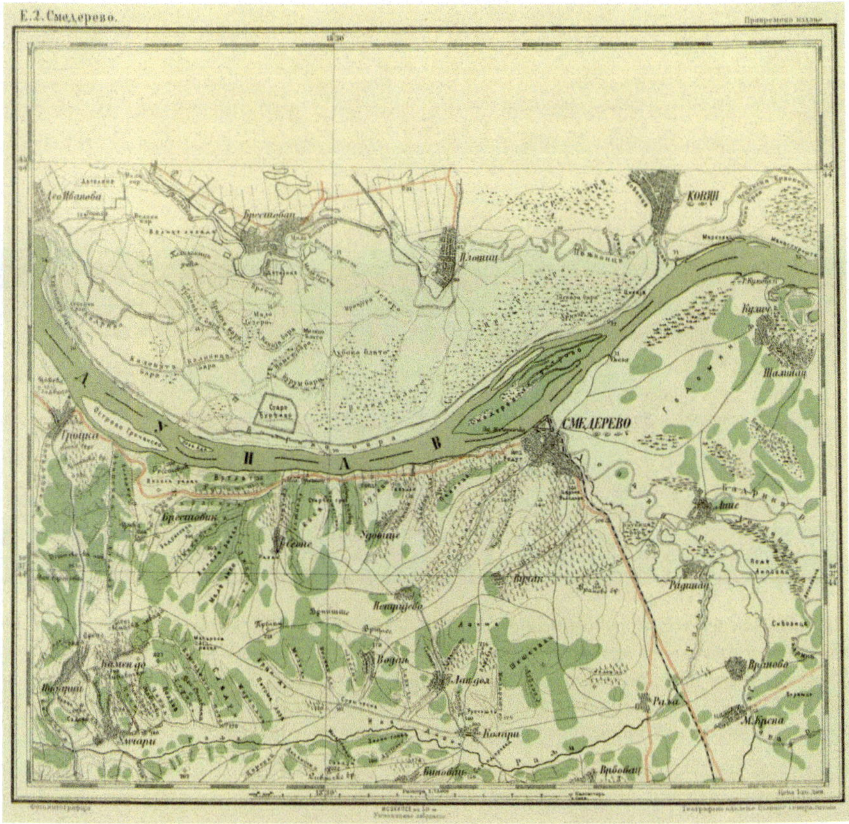

Fig. 6.4 Map of Smederevo and its surroundings in 1894. *Source* Military Topographic Map, 1894, 1/75.000, National Library, Belgrade. *This document is remarkable. Compared with other low-scale Ottoman—and even Austrian maps, its topographic accuracy is baffling. Its style is obviously German (Gartner and Ortag* 2010*). Swamplands spread everywhere on the outskirts of scattered military fortresses*

6.1.2 Greece's Security Planning

With the fall of the Iron Curtain, the sophisticated border systems developed in former socialist countries suddenly collapse. They could not survive the breakdown of the states. However studied by many anthropologists and renowned experts on international migrations, random works address the border issue from a military–geographic point of view. Exception to this are the writings of Labrianidis (2001, 1997). His much documented work remarkably clarifies the way the Hellenic frontiers and adjacent lands used to be managed during the 50 years of the Cold War. According to the author, who mainly focuses on the Pomak region, Greece first creates an official *surveillance zone* (επιτηρούμενη ζώνη) along its northern boundaries in 1936, during the dictatorship of General Metaxas. The strip extends across 15–45 km within the Greek territory, depending on the degree of tension of international relationships, topographical conditions, and military demands. This control area may be viewed as an "internal frontier" because it implies strict restrictions on land use and people's mobility. To put it otherwise, this military stranglehold organizes an *extra-territoriality*, which in turn results in the long-term underdevelopment of extensive areas.

With the provisions of law No. 376/1936, the Metaxas government officially defines for the first time "defense perimeters" divided into two sectors: the inner one called "forbidden zone" (απηγορευμένη ζώνη), exclusively reserved for military forces; the external one named "surveillance zone." Operating conditions are drastic:

1. Any entrance to the forbidden zone is prohibited to anybody without prior military authorization.
2. Military control also applies to the surveillance zone, with strict conditions:

   - The entrance and circulation of people and vehicles are limited and submitted to a compulsory procedure whose application is assigned to military and naval forces. Some categories of individuals can be denied access to the area. Local police stations thus deliver a *laissez-passer* equivalent to an internal passport. When duly presented at checkpoints (μπάρα), this document allows people to enter the area. Any circulation within the surveillance zone is also prohibited during the night.
   - Individuals living outside are not allowed to carry out a professional activity in the area without prior permission of the Commission of Military Security.
   - All foreigners (αλλοδαποί ή αλλογενείς) are forbidden to settle in. This interdiction may apply to those already installed.
   - Any change of residence inside the surveillance zone must be approved by competent authorities.
   - As for the status of land, any field, building, or piece of forest within the forbidden districts is effective propriety (κυριότητα) of the public domain, with a management delegation granted to the military authorities. A general expropriation is applied. Across the surveillance zone, forest areas inherit the same status. Private ownership persists, although it may be contradicted by unilateral military requisitions and expropriations. For self-reasons, the armed forces are authorized to cut down forests, destroy buildings, stop industrial production, and break off agricultural works. In such circumstances, private landowners are promised compensation.
   - The military authorities define the limits of the surveillance sector and set up ground signs: barbed wires, barriers, road signs, and markings.

Furthermore, a special identity card is delivered to local residents. This white ID paper has to be renewed every 6 months.

L. Labrianidis holds the idea that each local military commission could fix the spatial limits of the area under surveillance according to the presence—or absence—of ethnic minorities along the boundary. As shown in Fig. 6.5, the surveillance area reaches its maximum southern extent in 1958 and covers 15,100 km^2—11 %

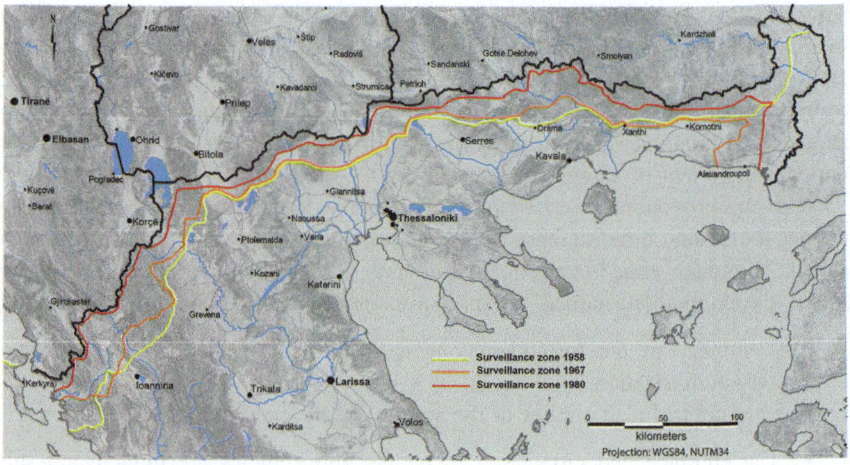

Fig. 6.5 Control areas in northern Greece. *Source* Greek Ministry of Defense, DTM NASA, after (Labrianidis 2001)

of Greece's land surface. Along the Albanian boundary, military forces closely manage former Çamëria and all districts plagued by the Civil War (1946–1949). Further east, other perimeters are placed under control. They include all former Slav-speaking lands neighboring Yugoslavia and the Pomak regions of Thrace, but eastern Evros retains a conventional civilian status. On an objective assessment, Jannina, Kastoria, Corfu, Drama, Xanthi, and Komotini are "preborder cities." Barely surviving this situation, they support the presence of large army camps, face demographic depletion, and suffer economic lethargy.

In 1967, the total surface placed under military control climbs to 16 250 km^2, with minor spatial adjustments, except for the Evros district finally included into the system. A few years later, in 1980, the area is reduced by half (6700 km^2). In some places, the surveillance zone almost merges with the forbidden area. This is the result of a change in the Greek foreign policy. After the fall of the military junta in 1974, C. Karamanlis and A. Papandreou organize an easing of political tensions.

This overall territorial *system* has maintained large districts under a biding status that could only narrow the scope of land development projects. Applied for more than half a century with varying degrees of commitment, this policy organizes the depopulation of entire regions. "Frozen" lands slowly replace agricultural fields and pasture areas, while forests develop. Such a security policy creates a long-lasting civilian "insecurity," with highly negative effects in the economic field, leaving apart the industrial and service activities involved by the presence of troops and police.

Because of this, countless villages had disappeared before a decent asphalted road came up to them. For the few places that have succeeded in surviving, there is no shortage of challenges. As a sign of bitter disappointment, generations of individuals and families leave the land they once knew without choosing a Greek destination: They prefer to move faraway, toward Europe, America, or Australia (Kolodny 1981, 1982). In front of such “devastated” landscapes, the present European programs promoting cross-border development are a *last chance*. Maybe too late.

After 1974 and under various governments, defense policy dogmas are gradually challenged and legal restrictions are softened. G. Arsenis, the Ministry of Defense, partially abrogates the system by the end of 1995 and withdraws the last internal checkpoints. Any practical impediment to the circulation of people is suppressed. The legal instrument, however, remains untouched. The Greek legislation still keeps effective today the legal provisions settled by law No. 376/1936. Thus, the system did not survive the fall of communism, but executive authorities still have the right to revive it by decree.

Nowadays, a series of “security” problems continues to impact the northern Greek confines, and it appears everywhere that border districts still remain under “special status.” From the 1990s onward, the development of mass migration movements has required the deployment of renewed control units to contain illegal entries. Therefore, on the basis of practical eventualities—“against their own will,” Greek borderlands still take on the function of buffer zone. South of the Rhodope range and along former Yugoslavia, the residual presence of non-official ethnic “minorities” implies maintaining a close eye on cross-border movements and traffics. Newspapers sometimes recount stealthy police interventions and mention the need for more accurate and extensive controls near borders. These spaces are still suspect, although apparently softened rules are applied.

Beside this military-inspired planning system, other relevant legal provisions apply to the Hellenic borderlands. The legislation currently in force keeps reference to “border zones” (παραμεθόριες περιοχές or επαρχίες). This mention can be found in a set of documents regulating territorial development, local administration, career management of public servants, land rights, and so on.

Historically, the terminology stems from the stage-by-stage expansion of the country. Through a process of territorial integration, upon annexation of new districts, soldiers and army officers were the first government agents sent to take possession of land. Civil administrators come second. The most effective and studied case concerns the “new provinces” of northern Greece gained in 1912. The legislative ordinance of 22–24/06/1927, for instance, defines the conditions permitting any immovable property acquisition. According to this text, all acquired or inherited land rights of foreign citizens on properties located in border zones have to remain intact. However, any new purchase by other foreigners is clearly prohibited. Such political decisions play a central role in determining the future of entire regions.

Background information is necessary to understand the outcomes of this ordinance. The decision comes after population and estate exchanges between former conflicting countries. At that time, Greece has no cadastre—the same is still true today. After the departure and arrival of countless people (Balkan Wars, WWI, and *Asia Minor Catastrophe* of 1923), thousands of square kilometers are left without clear ownership. With the settlement of hundreds of thousands of Greek refugees,

Greece becomes an extended homestead land. This law dated 1927 fixes the terms and conditions upon which the new *colonization* will take place—among other measures. "Border area" is a concept introduced within the Hellenic law as a support to the project of "internal conquest."

Surprisingly, the legal designation of "border areas" persists to this day. Law No. 1892/1990 reaffirms that no foreign physical or moral person is allowed to buy any real estate property in border places. Purchase is permitted to EU citizens after submitting a request to the Hellenic administration with the mention of the final use of the property. In 2011, such limitation has been abrogated to fulfill the EU legislation. Such apparently anecdotal evidence affects the daily lives of Greek people, though. It is known for long that all capable male individuals are sent alongside national boundaries when called up for service in the armed forces. Other applications of the *frontier* concept are less familiar. For instance, law No. 287/1976 establishes that all public employees seeking for a transfer or recruited to fulfill their missions within frontier districts—otherwise called "problematic regions, προβληματικές περιοχές"—shall be granted material and financial advantages as a compensation for their commitment.

"Border areas" are defined variously according to the period and application field (Figs. 6.6, 6.7 and 6.8). The law establishes the fundamental principle; a presidential decree decides on the terms and conditions of its application, on the proposal of different ministers (foreign office, defense, agriculture, economy). The decree can be modified at any time.

In 1927, the spatial extent of the *paramethories periohes* is at its maximum. About half of the country is listed in the *frontier* category. With such a degree of internal penetration, applicant areas hardly deserve being called "border districts." Obviously, strategic purposes and other planning projects interrelate and reinforce each other in view of managing the integration of new provinces. Priority is granted to land confines. Already placed at a clear disadvantage because of their spatial fragmentation, the islands are included in the consolidation scope. They are seen as *fragile components*.

Old Greece versus New Greece, center versus periphery, and mainland versus islands: Geographers recognize the keystone role of Athens in laying down the necessary conditions for its own urban "global" expansion. A long-term decay of remote lands is planned in full transparency. In fact, the Hellenic authorities institute a policy based on a powerful spatial hierarchy strictly deriving from a consummated divorce with the rest of the Balkans. Military and civil officials perfectly coordinate their efforts to keep only half of the country under "normal" status. Anticipating what the course of discussions will be, Enver Hoxha's Albania appears to be far less exotic country.

The old legal framework is maintained until the end of the Cold War. In 1990, law No. 1892 burns the bridges with the past. The text promotes land and economic development through a broad array of measures. Frontier areas undergo a critical shrinkage, without disappearing. Salonika and Halkidiki are excluded from the "protected" zones, as well as many islands, including Crete. At last, the second Greek city gets out of its straitjacket. It is interesting to note that Santorini and the

Fig. 6.6 Greek border areas in 1927. *Source* Law of June 22–24, 1927, Greek National Printing Office. *Border zones are marked in yellow. The black outlines are recent administrative divisions. The Dodecanese is annexed in 1947*

islands facing Anatolia are granted with special provisions: Officially, no foreigner is allowed to buy estates in half of the most popular Greek tourist resorts (without special permission)!

In fact, the list of districts presented in Fig. 6.7 is not exhaustive. For instance, the official text does not mention the northern prefecture of Drama, even though this sector still belongs to the core military defense system in 1990, doubtlessly. It seems clear that the initial law of 1927 stands for the reference model of any further internal frontier definition. Its legal provisions have been applied with more or less accuracy, following relevant developments in international relations, local context, economic purposes, and financial needs.

Figure 6.8 displays the last official provisions. Surprisingly, despite an appeased Balkan context and the European integration of Bulgaria, border areas broaden out. Obviously, this is no longer a question of national defense. Law No. 3731/2008

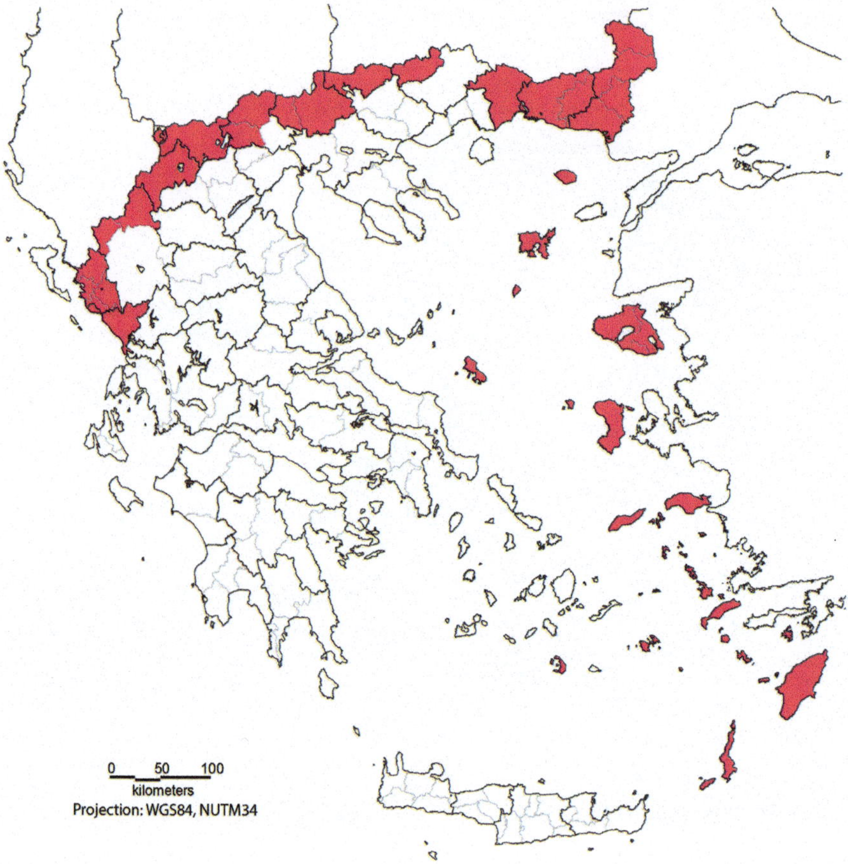

Fig. 6.7 Greek frontier areas in 1990. *Source* Law No. 1892/1990, Greek National Printing Office

discusses the management of police forces. Relevantly, it supplies a priority support to the Bulgarian and Turkish borderlands that face major migrant flows. There is also a lot hidden between the lines.

The border system turns into an economic instrument. Territorial cohesion has to be reinforced to help borderlands face an unfair competition. Undeniably, the long-lasting military freezing did not prepare the local markets of northern districts and eastern islands to compete with other Bulgarian or Turkish low-cost enterprises. In order to reverse the trend, the military system has to change into a planning advantage for restoring the balance with Bulgaria and Turkey—both countries enjoying convenient exchange facilities of people and goods with the European space and other good transport infrastructures to help their development.

The EU integration of Bulgaria could have changed the Greek marchlands for the better. It has sometimes produced worsening. In 2011, the National Confederation of Greek Trade sounds the alarm about a new plague: It reveals to the press that about

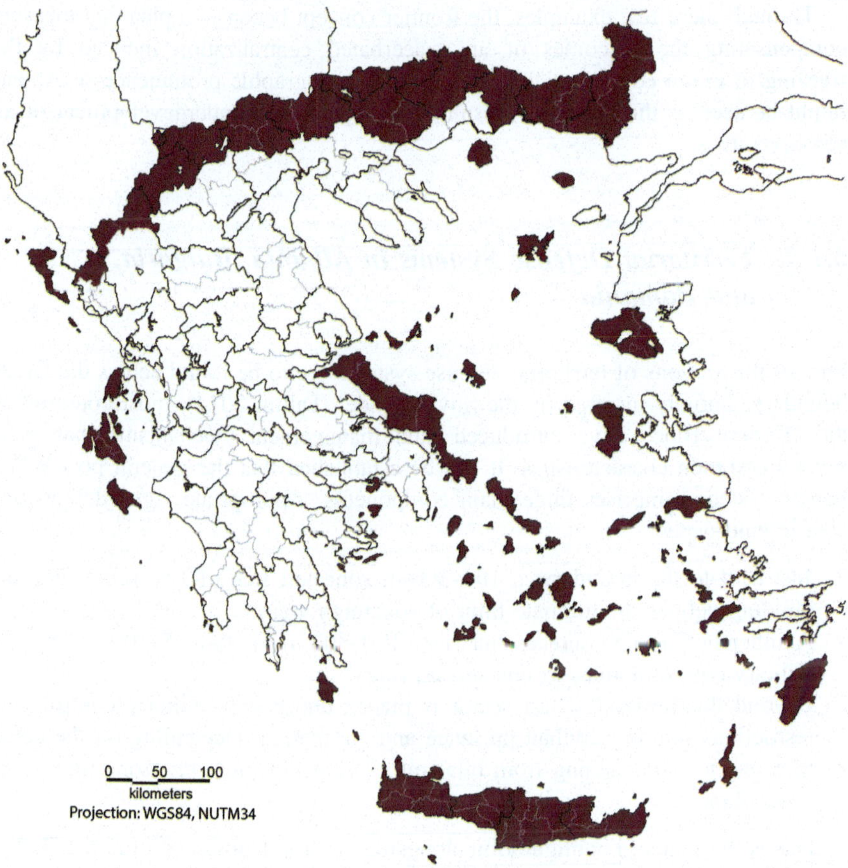

Fig. 6.8 Greek frontier areas in 2008. *Source* Law No. 3731/2008, Greek National Printing Office

1500 Greek enterprises have moved out to settle in neighboring countries, because of the crisis. Regularly crossing the boundaries for shopping, villagers and city-dwellers of the frontier zone have made the local trade go from bad to worse. The estimated hemorrhaging risk is high: The confederation surmises that 600 000 people go abroad for shopping each month (1/20th of the Greek national population)! Many goods are concerned: clothing, footwear, drinks, cigarettes, fuel, medical services, and car repairs. Households engage in such consumption practices to cover basic weekly needs. One in four enterprises would have closed since the beginning of the crisis.

Land confines obviously retain unsafe position. Taking action against this *unfair game*, the Greek government intended to define "special economic zones" to preserve business communities. They would thus find a partial solution to their remoteness and increase their competitiveness against "advantaged" partners across the boundary.

Through these few examples, the frontier concept becomes a planning measure compensating the outcomes of an exacerbated centralization induced by the freezing of entire edge districts. Therefore, the geographic prominence of Athens might be seen as the "positive" result of a programmed underdevelopment of all sensitive areas.

6.1.3 *Territorial Defense Systems in Albania, Bulgaria, and Romania*

Further illustrations of territorial defense systems are to be found across the Greek boundary. Initially inspired by the Soviet model (Dullin 2014), most countries of the "Eastern Bloc" have introduced land management mechanisms that share several common characteristics. It can be established that the general post-WWII border system comprises three main components. Appropriate signs define each area in landscapes:

1. Just next to the boundary, a 100–300-m zone left free of any vegetation and building defines the ultimate limit of sovereign areas.
2. Another buffer zone spreads out from 100–300 m to about 5–10 km. A strict military control applies to this second ring.
3. Beyond this horizon, a last sector is placed under semi-military control with restrictions limiting civilian presence and circulation. Depending on the countries and tensions arising from bilateral relationships, this last zone may reach 25–50 km.

This is the standard configuration alongside the Iron Curtain. Within the Balkan Peninsula, the Bulgarian–Romanian and Moldavian–Romanian borders are the only ones to escape such regulatory frameworks.

It is well known that, as soon as the early 1950s, the political crisis between Eastern and Western blocs has led to the powerful enforcement of all boundaries. The authorities establish and apply extensive security policies. Hundreds of thousands of people are pressed to leave their homelands and join temporary camps before enforcing the urban growth through national industrialization programs. The geographic "cleaning" creates large areas of natural wilderness. The vast majority of Balkan states adopt such measures although local situations might be widely divergent.

(1) Albania is an extreme case. Under Enver Hoxha's initiative, the whole country finds itself gradually covered with a stupendous amount of bunkers (Figs. 6.9 and 6.10). The entire earth changes into a besieged place left in the hands of military authorities. The number of bunkers is estimated to be more than 750 000 units. No region has escaped the proliferation of these "mushrooms" whose planned construction begins immediately after WWII and reaches a

Fig. 6.9 Bunkers at Gjirokastër's citadel, Drino valley. *Source* R. Darques, 2004

high in the 1970s and 1980s. Few Albanians would confess to having escaped the general enrollment necessary for bringing such a project to its end.

Officially, the system is created in the early 1950s in reaction to repeated external acts of agression and raids (especially from Greece). Later on, the suspension of diplomatic relations between Albania and Yugoslavia would lead to the same results in the north. In such a small country, the concept of "border area" loses its proper meaning and becomes a universal land management principle. Civil spaces and military reserved areas permeate each other (Woodward 2004). Of course, the mixing degree varies with the distance to the boundary. The system is not just about remote mountains. It applies to any village and city suburb, even in low-lying areas. As a human response to this general landscape "bunkerization," people are placed under strict surveillance. To control any movement, public officials edit resident cards, family passports, and other authorization papers.

In response to this situation, it comes as no surprise that, at the end of the 1980s, the final defeat of the regime produces opposite outcomes: The military overcontrol gives way to a political–geographic undercontrol. Forced settlements and close supervision of any travel are replaced by massive and spontaneous migration movements. The political and territorial anarchy of the 1990s arises from the deregulation caused by a general land planning "failure."

Fig. 6.10 Defense systems, Derviçan, north of Kakavijë, Drino Valley. *Source* R. Darques, 2004

Little information about the legal status of the Albanian borderlands has been found. Such elements have long been kept secret. Furthermore, the border regime has significantly changed across time—the Internal Affairs and Defense Ministries have assumed jurisdiction alternately or simultaneously. According to Georgia Kretsi's works about Konispol (Kretsi 2007), the decree No. 1535/1952 officially defines two adjacent zones: The first one (*brez kufitar*, border strip, from 500 m to 2.5 km wide) is placed under an exclusive military control; a second area (*zonë kufitare*, border zone) stretches over 40 km in the hinterland.

As in Greece, albeit in a more systematic manner, people were asked special authorization to travel during decades. The checkpoints took position along the main transport routes. The Ministry of Defense used to publish a list of permanent border battalions and assigned great many troops to the surveillance. All people living within the area had to be duly registered and their movements were strictly controlled.

In 1964, Enver Hoxha orders the installation of an electrified fence along the *brez kufitare*—the so-called *kloni*. As compensation for such constraints, public authorities grant material and financial advantages to workers engaged in cooperatives. In the end, only "accredited citizens" could get on living in the area; the others were called to leave home. G. Kretsi, as an anthropologist, notices that physical and mental barriers encouraged the general imprisonment feeling of the Albanians, especially for those living close to the border (Sintès and De Rapper 2006). All visitors report the general atmosphere of fear and paranoia, even in Albanian heartlands (Champseix and Champseix 1990).

(2) In Bulgaria, the surveillance is organized after 1946. The government makes the decision to assign permanent armed troops to the external protection of the country in "official" response to illegal crossings observed along its boundaries. The security system improves its coherence after the signing of multilateral agreements with Romania (1950), Greece (1955), Yugoslavia (1955), and Turkey (1967) aiming at preventing any incident. The *tightening* of the Iron Curtain comes next to the agreements: The accuracy of demarcations is improved, and permanent security facilities are installed. Finally, the country finds itself entirely surrounded by barbed wires, starting from the Black Sea and ending with the Yugoslavian border districts (Figs. 6.11 and 6.12).

The decree No. 36/17.11.1988 is a late text regulating the legal border regime. Picking up post-WWII provisions, it defines a first forbidden zone (*забранена зона*) 10–300 m large followed by a frontier zone (*гранична зона*) 7 km wide on the Greek boundary and 2 km wide along Yugoslavia. In certain places, the distance may increase up to 15 km. As in other countries, the "deepness" of the protection depends on the importance of the estimated external "threats." The existence of minorities is another criterion to consider in defining appropriate planning policies. The southern contact region with Greece and Turkey is equipped with thousands of land mines. The same access restrictions apply as before: limitations to the entrance of external visitors, edition of special transit papers, and residence authorization. Local dwellers are recruited to help the surveillance. As related through interviews with Bulgarian local entrepreneurs and representatives, any acquisition of fields or buildings by foreigners was forbidden until 2007 and the introduction of the European legal framework.

As in Greece, the present Bulgarian legislation keeps a specific reference to "end border territories or regions" (*крайгранични територи или райони*) in a series of official documents—leaving aside the European provisions and transboundary development projects that are core management policy today. For instance, the ordinance No. 3 of April 3, 2005, about "transport network improvement" defines a limited number of border districts for its implementation (Fig. 6.13). The 56 selected *obshtini*[1] that are to be recipients of program funds cover 24 600 km^2–

[1]For further information on administrative networks, please refer to Chap. 8.

Fig. 6.11 Barbed wire fence at Nova Lovcha, Greek-Bulgarian border, Blagoevgrad Province. *Source* Darques 2005

22 % of the Bulgarian national area-, leaving aside the border districts already equipped with public transport facilities.

In Bulgaria, some national contingencies hamper the march to a "successful" borderland management: As opposed to Athens, Sofia itself might be regarded as a

Fig. 6.12 Entrance to the border zone, south of Gotse Delchev. *Source* Darques 2006

Fig. 6.13 Border zone *obshtini* according to the Ordinance n°3, April 3, 2005 (public transport). *Source* www.lex.bg, author's workout

border city because its distance to Serbia is no more than 50 km. Consequently, its suburbs are potentially impacted by restriction measures. This close vicinity is not unconnected with the distinct legal status afforded to Yugoslavian borderlands. Incidentally, the far-distant districts of Dobrudja, Thrace, Rhodope, and Vidin share an enforced spatial marginality.

(3) In Romania, the first communist period organizes a complete reversal of the official planning policies. A "vacuum" security doctrine replaces the colonization spirit associated with the integration of new regions after WWI. Border districts have to be drained from any potentially conflicting population. A "sanitary" versus security area is established. Decree No. 200/1956 and law No. 678/1969 stipulate that both military units and villagers supply safety. The frontier zone (*zona de frontiera*) concerns all local communities adjacent to the state boundaries, with the exception of the Black Sea coast. Border guards are assigned a jurisdictional area extending 30–40 km beyond the border. Within this area, they are authorized to install any facility and apply any security measure they might find necessary. People entering the border area must report to the local office of the militia within 24 h. The same procedure applies to workers. Of course, special identity and residence certificates are edited. Any night trip is prohibited within a 2-km land strip next to the boundary. Farming activities are closely regulated.

Paradoxically, the post-communist law No. 45/1990 extends almost all legal provisions. It is not until 1992 and law No. 56 that a significant change occurs. This text reduces the border zone to a 20-km strip. Many access restrictions are abrogated. Local militias are dissolved. The external Romanian districts thus recover a "normalized" status that comes with the new freedom of movement. In 2001, the border zone gains ten more kilometers (it reaches 30 km) and includes the Black Sea coast, although the Danube zone is limited to 10 km only.

Ultimately, the Romanian borderlands include extensive areas Fig. 6.14. Until 1992, the 40-km land strip defines a global surface of 88 400 km^2—the Black Sea coast is excluded from the calculation. This represents 37 % of the country and one-third of its population. The numbers speak for themselves: They put the frontier issue in its proper place, even concerning a large Balkan state.

As well as Sofia, Bucharest may also be assimilated to a border capital, with many other leading Romanian cities. Compared to the Hellenic case, the geographic framework is reversed. The Carpathian Mountains are central. The country has limited sea boundary, when Greece has restricted continental contacts. Most frontiers stretch in the middle of lowlands, whereas most northern Greek boundaries belong to sparsely populated mountain areas.

In 1992 and until 2001, the Romanian border strip is reduced to 20 km. The area decreases to 22 % of the national territory before increasing again after 2001. With a current surface of 64 200 km^2, it involves about 700 *comune, oraşe,* and *municipii*, and 25 % of all Romanians. Playing with words, it could be said that military confines are not a *side issue*, with or without the Schengen rules.

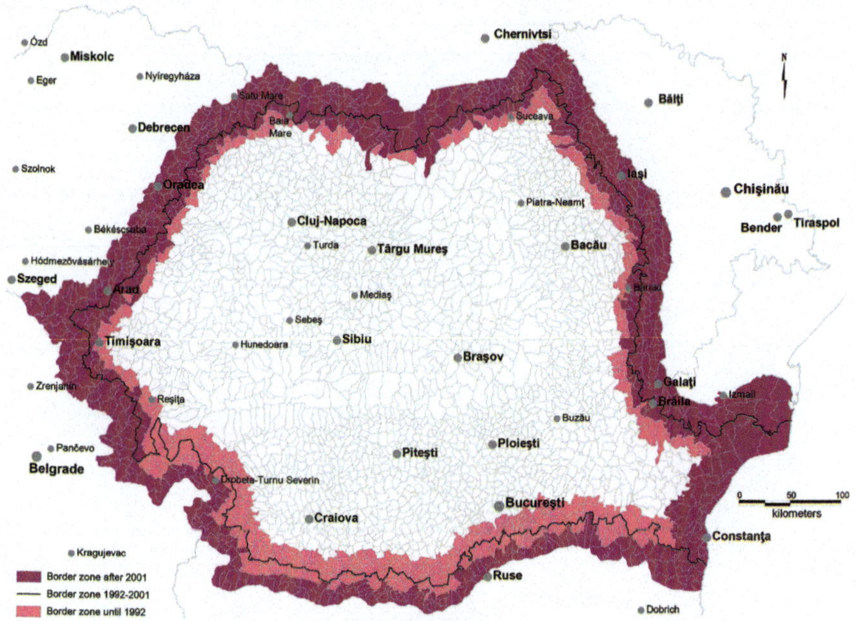

Fig. 6.14 Romanian border areas. *Source* Monitorul Oficial

(4) Let us return to the Albanian case and adopt the same mapping approach as in Fig. 6.14 (Fig. 6.15).

Once applied to Albania, the 40-km zone initially introduced by the Soviet delegates de facto creates the "besieged citadel" depicted by so many authors, including I. Kadare. The brown regions around Berat, which also include part of Mirditë, would be the only places escaping a border status and its downsides. This picture promotes a few mountain districts almost entirely depopulated as leading communities of "Free Albania". Close to Tiranë, the heights of the Dajti would be a safe haven preserved from military pressures. All this is fictional, of course.

In the Balkans, the expected easing effect of diplomatic relations following "the fall of the Berlin wall" is not clear. The communist collapse has not borne the predicted "revolution." Obvious practical changes are not accompanied by a deep revision of legal provisions. Most laws regulating the current "soft borders" still use the same old terms as before. Furthermore, research has shown that "borderland" is a rather elastic notion. Without mentioning the administrative, legal, and practical proceedings recently imposed by the European Union to promote harmonization, fringe areas are generally run as *flexible* territories.

Finally, beyond the screen of an "East and West" confrontation, national territorial defense systems echo each other. Military measures define an 80-km-wide band covering both sides of the same boundaries. Small countries cannot put up with such pressures without experiencing major geographic outcomes.

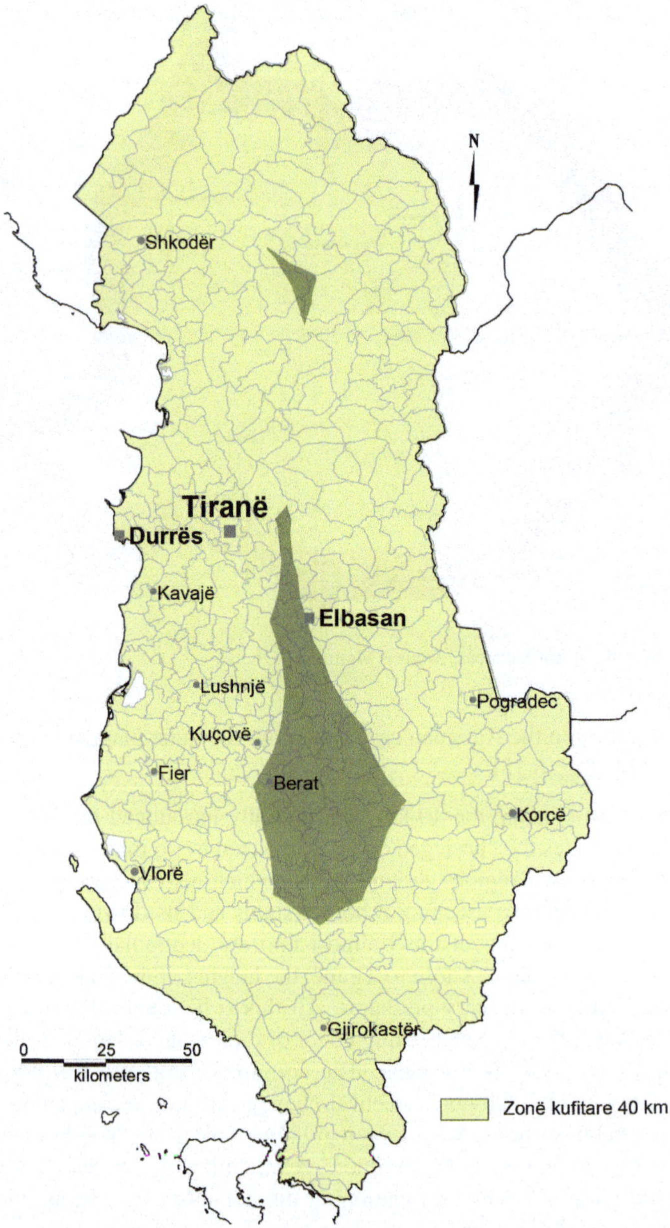

Fig. 6.15 Albanian border zone, socialist period (40 km). *Source* ESRI, Gfk Regiograph, author's own workout

To go even farther in this approach, it must be noted that most "border worlds" have experienced considerable downgrading after 1990. The dismantling agricultural cooperatives and small-scale industries has led to an abrupt rural withdrawal. The barren fields of the communist era have been proudly replaced by wildlands. Between 1997 and the early 2000s, the signature of the *Mine Ban Treaty* allows the implementation of demining programs close to the Bulgarian, Albanian, Greek, former Yugoslavian, and Turkish borders. Over the course of the project, the European integration of Bulgaria almost immediately follows its "material" liberation.

The overall number of removed mines after the treaty may have reached 1.5 million units in Greece, 1.6 million in Albania, one million in Bulgaria, and one million in Croatia (especially in Slavonia) (Human Rights Watch 2001)! Boggling amounts: In 1997, after the political crisis caused by the falling down of the financial pyramids, Albania has to face anarchy and lawlessness. State control over land is virtually nonexistent, to such a point that armed gangs replace national security forces. The army finds itself in such a degraded situation that its higher authorities are obliged to negotiate with the "rebels" to recover stockpiles of weapons. That is what happens to 80 tons of antipersonnel mines stored in sealed tunnels near the village of Lazarat (district of Gjirokastër).[2] The total national ammunition and armament surplus was estimated at 200 000 tons at that time. The Balkan "demilitarization" process obviously opens a window of opportunity for the illicit trade of arms and light weapons.

Of course, all Balkan borders do not fit to such a clear profile. The Yugoslav Wars especially lead to new militarization processes, some of which are still going on. Bosnia and Herzegovina was considered in the early 2000s as the most heavily mined country in Europe. An extensive land *glacis* used to cover the former "separation zone"—a territory 1100 km long and 4 km wide. About 10 % of the whole Bosnian space was suspected to be mine-contaminated. Rural geography and natural protection here change meaning. Easy transport and travel freedom are not self-evident in the Balkans, even today (Fig. 6.16).

"Open-border" policies are far from systematic, even in EU member states. In 2011, before the *refugee crisis*, Bulgaria—through its Council of Ministers—decides to build a fence along its 210 km boundary with Turkey. The official goal is to prevent any straying animal to cross the line between both countries, for sanitary reasons. Finally, the Bulgarian government would have come back over its

[2]The settlement shall become famous in 2012 when an international audience "discovers" the scandal of extensive cannabis fields spreading throughout the countryside. Lazarat becomes the *European capital of cannabis.* Strengthened by prosperous narco-trafficking activities, local authorities and "farmers" prioritize resistance against repression and manage to maintain a local-based extra-territoriality against the security forces mandated by Tirana until 2014. In June of the same year, heavily armed police forces take control of the "lawless" area. They defeat cannabis growers supplied with machine guns, rocket-propelled grenades, and heavy mortars. Illegal plantations are destroyed under the scrutiny of television cameras. The income generated by this traffic was estimated to have reached about half of the national Albanian GDP. It seems, however, that the issue is not fully resolved. Plantations would have been transferred to other parts of the country. In June 2015, violent clashes between police and residents were still reported in Lazarat.

Fig. 6.16 Mine situation in Bosnia as of 2004. *Source* BH MAC, www.bhmac.org

decision. Greece intended to do the same along the Evros riverbank. After different backtracking and dithering episodes, Athens decides not to bring the project to its end, at least temporarily. With the refugee crisis, the entire dynamic has changed toward a general rebordering process. Bulldozers digging trenches and soldiers setting up barbed wire fences have become common practices recently. Official statements reject the idea of a *new Iron Curtain*, even though the deployment of such facilities in the landscapes might bring up unpleasant memories.

Sometimes, contradictory information is disclosed. Debordering operations may be completed almost simultaneously with rebordering projects. In 2010, at Romania's conflicting northeastern boundary, 360 km of barbed wires dating from the early communist period has been removed alongside the Prut. This decision came before the vote of law No. 10/2010, which defined the terms of a legal cross-border small-scale traffic. According to the Moldavian–Romanian agreement of 2009, the Moldovans living close to the Prut have the right to enter Romania up to 30 km without visa. The administrative units partly located in this 30-km band, whose territory extends to 50 km from the boundary, shall be considered *border communities*. The project mainly aims at improving transboundary relationships hindered by decades of blockade (Marcu 2009) (Fig. 6.17).

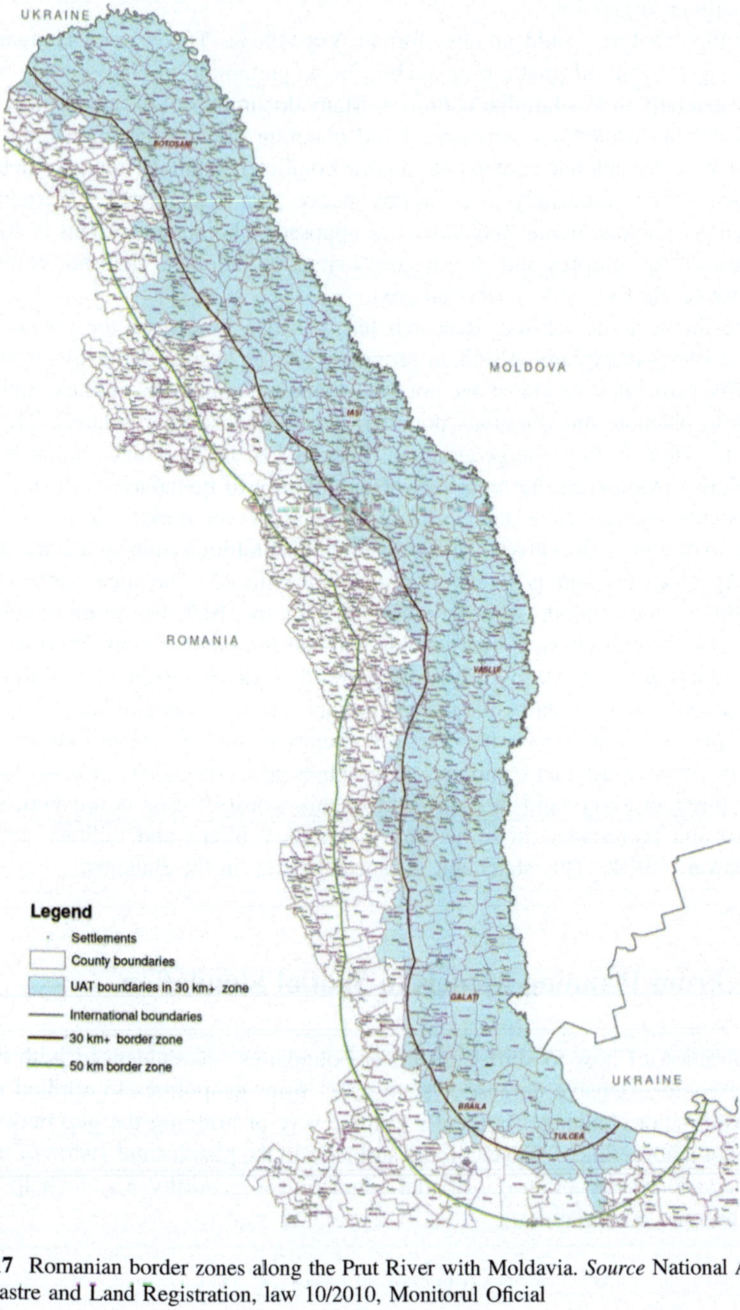

Fig. 6.17 Romanian border zones along the Prut River with Moldavia. *Source* National Agency for Cadastre and Land Registration, law 10/2010, Monitorul Oficial

These examples show that microscale geography sometimes prevails over macroscale logic. From enclosure to opening and back to defensive fences, the "border game" seems unpredictable. This points to the importance of a deep historical analysis in assessing how local circumstances may eventually decide on international injunctions.

Further analysis could address former Yugoslavia, Turkey, and Hungary, but gathering reliable information on such a theme pertaining to land security is difficult, especially in less familiar countries. Many documents of the communist period could not be located and accessed. Land planning policies are often more determined by international agreements and/or conflict resolutions than by deliberate choices. This is especially true for tiny states. Are Kosovo, (FY) Macedonia or Montenegro able to define *heartlands*, as opposed to borderlands? This is doubtful. Because of its complex and still unsteady situation, attempts shall not be made to investigate the former Yugoslavian area.

The theme is not closed. Research has, at least, evidenced the prevalence of frontier issues in explaining Balkan geography. As an answer to the *uncertainty* and *mobility* principles validated by our first mapping approach, military authorities prefer to promote unambiguous boundaries and enclosed borderlands. There is a basic paradox. In fact, the geographic "freezing" of land confines might be interpreted as a reaction to the mobility versus ubiquity of boundaries shown before.

Mapping military areas and camps scattered across the Balkans would have been useful to complete the survey. Unfortunately, this ambition runs up against national security obstacles and is likely to remain unachieved. The idea comes spontaneously to one's mind: Almost everywhere in the Balkans, from remote rural districts to central cities, military fields pervade the environment. Who has never bumped into military "dead-ends" during a walk or driving excursion, within cities, in the middle of the countryside or next to a marvelously wooded area? Any border geography is also military—not war—geography to such an extent that we can say with R. Woodward that "military geographies are everywhere; every corner of every place in every land in every part of this world of ours is touched, shaped, viewed and represented in some way by military forces and military activities" (Woodward 2004). This statement is especially true in the Balkans.

6.2 From Planning Models to Spatial Simulation

The question of how to shift from linear boundaries—as evidenced with the GIS database—to extensive defense systems, i.e., from geopolitics to applied geography, is a major challenge. Is there a graphic way of bridging the gap between the fiction of a borderline deprived of any surface and the phantasmal vision of frontiers that pervade whole countries? Spatial simulation is a worthy way to help borderlines become borderlands.

Although any map can be interpreted as an approach to reality, i.e., a "simulation," the next mappings might be seen as attempts to circumvent ineffective conventional picturing. Figures 6.18, 6.19, 6.20, 6.21, 6.22, and 6.23 are fictional documents based on a process of data harmonization among states.

The GIS database presented in Chap. 4 is used to complete the project. A first series of maps deals with the most recent situation. A second sequence brings historical resonance to the initiative. The mapping exercise excludes maritime boundaries, shorelines, and island units, even though these are usually included in official borderland definitions. The focus on mainland areas leads to a fall by half of the border imprint. However, this choice does not compromise research results.

The simulation assigns three possible radii to border *corridors*: 10, 20, and 40 km. The shorter the distance agrees with "low-water" periods, the wider it matches with "high-water" levels. The varying dimensions of corridors might also be interpreted as an expression of the balance of power between military authorities and civil administrators.

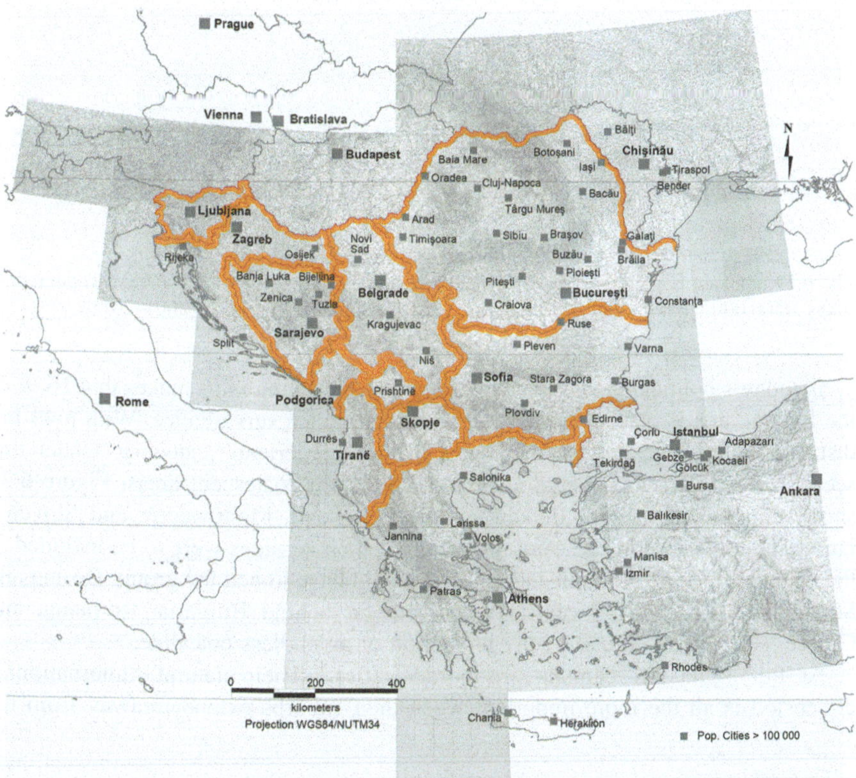

Fig. 6.18 Balkan borderlands in 2015—10 km area. *Source* Russian Military Topographic Maps 1/500,000, author's own GIS database, ESRI, population censuses (2001–2011)

Fig. 6.19 Balkan borderlands in 2015—20 km area. *Source* Russian Military Topographic Maps 1/500.000, author's own GIS database, ESRI, population censuses (2001–2011)

In today's political–geographic environment, a 10 km radius makes that 16 % of the Balkan area is placed under strict army and police surveillance. With a 40 km distance, half of the Peninsula belongs to the "borderland" category. Under this scenario, land corridors merge and eventually form coalescent areas. Figure 6.20 shows entrenched western Balkan countries. Kosovo, Montenegro, and Slovenia ignore the true sense of the word *heartland*. If their seasides were to be included in the mapping process, Croatia and Albania would have joined the group. Bosnia and Serbia hold some "free" central space, but far behind Bulgaria, Romania, and Greece—the latter "enjoying" the protection of an endless coastline.

At this scale, all capitals are border cities.[3] Their natural "international" dimension is all the more important when their suburbs extend faraway from the

[3]Athens would remain outside, despite its position close to Piraeus and its water front.

Fig. 6.20 Balkan borderlands in 2015—40 km area. *Source* Russian Military Topographic Maps 1/500.000, author's own GIS database, ESRI, population censuses (2001–2011)

countryside. The reverse perspective is equally relevant: Over the course of national expansions, capital cities are built on a model of citadels. Such projects are made possible through the development of effective defense systems (Table 6.1).

The same simulation applied to historical delineations produces astounding although predictable results. A 10-km border imprint covers 50 % of the Balkan area. When the 40 km limit is reached, the entire Balkan Peninsula is submerged. For redundancy reasons (overlapping zones), the surface gain is not proportional to the increase of border length, but inversely exponential.

Figure 6.23 clearly demonstrates that borders are ubiquitous. Central Wallachia and Western Moldavia are the only regions to ward off bad luck.

Because of their advanced fragmentation, Balkan lands are shrinkable. When boundaries move outward, the released districts are "rewarded" with a strict control. Open spaces are exceptions. The legacy of such "inward" spatial construction is twice and unambiguous:

Fig. 6.21 Balkan borderlands, 1800–2015—10 km area. *Source* Russian Military Topographic Maps 1/500.000, author's own GIS database, ESRI, population censuses (2001–2011)

- The power of *local* political forces is obvious and can be viewed as a natural outcome of former semi-independences. This explains many present political tricks.
- On the other hand, *national* authorities demonstrate their own capacity for action through powerful centralized administrations, developed military systems, and dominating capital cities.

Caught between hostile "enemies," *regional* identities and political intermediate powers[4] are almost nonexistent. Spatial simulations perfectly illustrate the importance of border issues in the Balkans. Land control, surveillance, and security topics

[4]The term "regional" refers here to middle-scale territorial communities as opposed to national and local authorities.

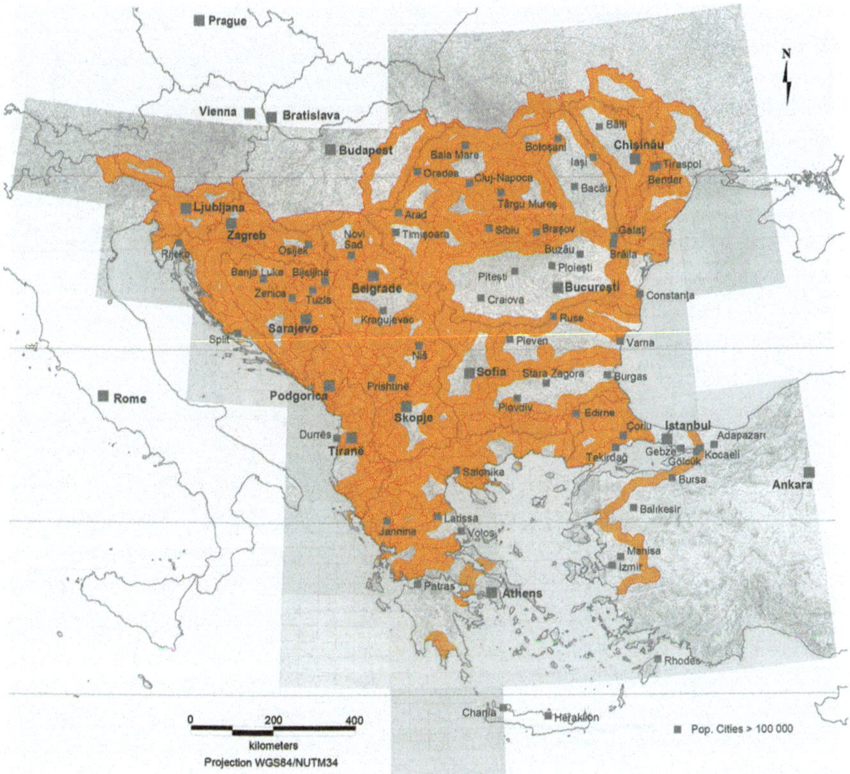

Fig. 6.22 Balkan borderlands, 1800–2015—20 km area. *Source* Russian Military Topographic Maps 1/500.000, author's own GIS database, ESRI, population censuses (2001–2011)

infuse all facets of planning policies—directly or indirectly. This makes perfect sense for small states fully assuming political–military sovereignty as opposed to other microstates placed under the "protection" of major neighboring countries (Bartmann 2002; Catudal 1975). The development of cross-border cooperation over the past two decades does not contradict this principle. As the experience of recent rebordering decisions has shown in a backlash against uncontrolled migratory flows, land control and cooperation policies are more closely linked than is generally thought. A new change in scale should enable to go even further into the reflection beyond clichés.

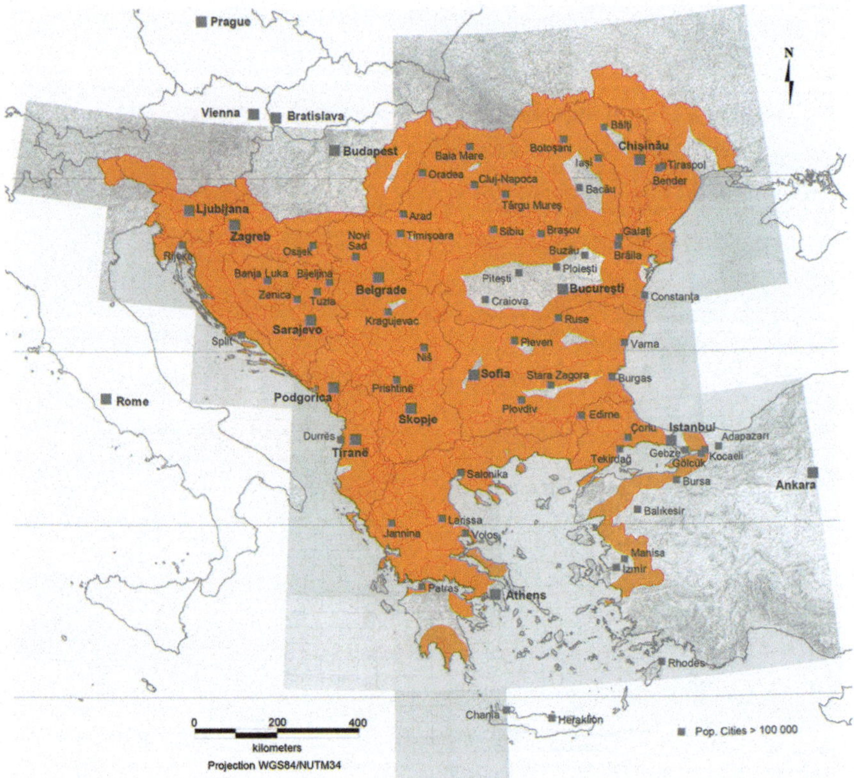

Fig. 6.23 Balkan borderlands, 1800–2015—40 km area. *Source* Russian Military Topographic Maps 1/500.000, author's own GIS database, ESRI, population censuses (2001–2011)

Table 6.1 Measured area of continental borderlands

		Present map		Historical map[a]	
Number of segments		31		1 138	
Boundaries' length (km)		9 373		41 875	
Boundaries' strip (one side)	**10 km width**	123 700 km^2	*16.2 %*	490 200 km^2	*50.9 %*
	20 km width	221 200 km^2	*28.9 %*	688 100 km^2	*71.5 %*
	40 km width	375 000 km^2	*49.0 %*	856 000 km^2	*88.9 %*
Balkan total area		**765 000 km^2**	**100 %**	**962 600 km^2**	**100 %**

[a]Including today's boundaries
Source Author's own GIS database

References

About E (1861) Le roi des montagnes. Hachette, Paris

Ateş S (2013) The Ottoman-Iranian borderlands. Making a boundary. Cambridge University Press, Cambridge, pp 1843–1914

Bartmann B (2002) Meeting the needs of microstate security. Round Table 91(365):361–374

Burrough PA, Frank AA (1996) Geographic objects with indeterminate boundaries. GISdata, vol 2. Taylor & Francis, London-Bristol

Catudal HM (1975) The plight of the Lilliputians: an analysis of five European microstates. Geoforum 6(3–4):187–204

Champseix E, Champseix J-P (1990) 57 boulevard Staline. Chroniques albanaises, La Découverte, Paris

Damianakos S (2003) La Grèce dissidente moderne. Cultures rebelles. L'Harmattan, Paris

Dullin S (2014) La frontière épaisse. Aux origines des politiques soviétiques (1920–1940). EHESS, Paris

Gartner GF, Ortag F (eds) (2010) Cartography in Central and Eastern Europe. Springer, Berlin

Gavrilis G (2003) Territoriality, boundaries, and conflict management: escalation and restraint along the Greek-Ottoman boundary. In 2003/08/27/2003 Annual Meeting, Philadelphia 2003. pp 1–27. doi:apsa_proceeding_935.PDF

Gavrilis G (2008) The Greek-Ottoman boundary as institution, locality, and process, 1832–1882. Am Behav Sci 51(10):1516–1537

Hobsbawm E (1959) Primitive rebels. Studies in archaic forms of social movement in the 19th and 20th centuries. Manchester University Press, Manchester

Hobsbawm E (1969) Bandits. Weidenfeld and Nicolson, London

Horel C (2009) Soldaten zwischen nationalen Fronten: die Auflösung der Entwicklung und die Militärgrenze der Königlich-ungarischen Landwehr (Honved) in Kroatien-Slawonien 1868–1914. Akademie der Wissenschaften, Vienna

Human Rights Watch (ed) (2001) Landmine monitor report. New-York

Karpat KH, Zens RW (2004) Ottoman borderlands: issues, personalities and political changes. University of Wisconsin, Madison

Kirsch S, Flint C (eds) (2011) Reconstructing conflict: integrating war and post-war geographies. Ashgate, Burlington

Koliopoulos GS (1987) Brigands with a cause: brigandage and irredentism in modern Greece, 1821–1912. Clarendon, Oxford

Kolodny E (1981) L'émigration macédonienne vers l'Allemagne Fédérale: le cas de Neokaisaria (Piérie). Méditerranée 1:21–34

Kolodny E (1982) Samothrace sur Neckar. Des migrants grecs dans l'agglomération de Stuttgart, Institut de recherches méditerranéennes, Aix-en-Provence

Kretsi G (2007) Verfolgung und Gedächtnis in Albanien. Eine Analyse postsozialistischer Erinnerungsstrategien. Balkanologische Veröffentlichungen, Band 44. Harrassowitz Verlag, Wiesbaden

Labrianidis L (1997) The impact of the Greek military surveillance zone on the Greek side of the Bulgarian—Greek borderlands. Bound Secur Bull 7(2):82–93

Labrianidis L (2001) 'Internal frontiers' as a hindrance to development. Eur Plan Stud 9(1):85–103

Marcu S (2009) The geopolitics of the eastern border of the European Union: the case of Romania-Moldova-Ukraine. Geopolitics 14(3):409–432

Rothenberg GE (1970) Die österreichische Militärgrenze in Kroatien 1522-1881. Verlag Herold, Vienna

Schein MD (1975) When is an ethnic group? ecology and class structure in Northern Greece. Ethnology 14(1):83–97

Schwicker JH (1883) Geschichte der österraichischen Militärgrenze. Prochaska, Vienna

Sintès P, De Rapper G (2006) Composer avec le risque: la frontière sud de l'Albanie entre politique des Etats et solidarités locales. Revue d'études comparatives Est-Ouest 37(4):243–271

Stein ML (2007) Guarding the frontier: Ottoman border forts and garrisons in Europe. Tauris Academic Studies, London-New-York

Trifon N (2005) Les Aroumains, un peuple qui s'en va. Acratie, La Bussière

Weigand G (1895) Die Aromunen. Ethnographisch-philologisch-historische Untersuchungen. Über das Volk der sogenannten Makedo-Romanen oder Zinzaren, Barth, Leipzig

Woodward R (2004) Military geographies. Wiley-Blackwell, Malden

Chapter 7
Spaces Under the Microscope: Looking Closely at Border Fluctuations Through Remote Sensing

Abstract The availability of Landsat images allows the close investigation of a wide variety of districts at high scale. In this chapter, borderlands are seen as places to live. People must deal every day with practical problems caused by territorial demarcations. Field experiences are used in combination with satellite imagery for explaining sensitive spatial systems.

Keywords Satellite imagery · Environment · Tourism · Natural parks · Forest · Water management · (De)Collectivization · Agriculture · Informal economy · International trafficking · Land reform

Although recent borders exist through dedicated and readily identifiable facilities "forming landscape," older frontiers used to follow planning models based on the principle of environmental "overnaturalization" versus "undernaturalization." Human deserts—be they green, rocky, or liquid—had to assume the role of barriers. Studying borderlands thus implies to take a real interest in a wide variety of geographic objects. Are these remote spaces capable of maintaining local specificities against normalizing forces? How singular are frontiers when seen up close?

In answering these questions, geography has special expertise. At first, research has to overcome a major obstacle: Field investigations cannot claim to be exhaustive. This is even enhanced by the access restrictions imposed by the authorities; but addressing past landscapes is also problematic and supposes to develop a passion for archaeology. Unfortunately, people's lives are time-limited.

Satellite imagery allows multiple holes to be filled. Unlike aerial photographs, remote sensing imagery equally covers the earth's surface. It also offers a distinct advantage: Legal interdictions do not apply. The width of swathes enables comparative works on a large scale. Repetitive data acquisitions open the way for change studies. Furthermore, satellite sensors produce objective information, which ensures consistency in further data treatments.

Of course, the remote sensing approach will not clarify issues pertaining to human geography or, if so, incidentally. The methodology employed is closer to the domain of "physical" sciences. A powerful instrumental filter restricts opportunities

© Springer International Publishing Switzerland 2017
R. Darques, *Mapping Versatile Boundaries*,
Springer Geography, DOI 10.1007/978-3-319-40925-2_7

for observing human-based phenomena. Environmental impacts shall be placed at the center of focus. Choosing the right scale, i.e., the most adapted spatial resolution, is the point. Without disposable financial resources, free online imagery sources have been used. The USGS Landsat data are products one never can find a substitute. Such tools enable medium-scale analyses. Some VHR imagery would have been useful on an ad hoc basis to identify military structures or observe places of high interest. Unfortunately, Google Earth is of unequal help: Its QuickBird imagery is randomly available and of variable quality.

In order to prevent future frustrations, it should be remembered a basic principle of land observation: The eye of any observer located at the surface of the earth has few common characteristics with a multisensor scanner placed in orbit at an altitude of 700 km. This applies not only to acquisition modalities and spectral discrimination capacities, but also to landscape perception. Administrative and political outlines have no "environmental equivalent" that might be identified from space, with few exceptions resulting from the artificialization of certain open areas.

Ordinarily, boundaries are discreet. Military facilities have been designed to remain hidden. Abandoned or underutilized lands are quickly gained by spontaneous vegetation or remodeled through erosion and sediment deposits. Several months are enough to achieve the victory of nature against any artificial human construction. The point is not to play a game of landscape archaeology. We shall also avoid the pitfall of presenting "earth from space" documents for purely visual/esthetic purposes, with slight geographic interest, although much rewarding these pictures might be.

Thinking along these lines will continue, particularly through the observation of selected border systems. Focusing on present-day landscapes is a good way to question the complex reality of frontiers. Sustained efforts allow performing targeted investigations on interfaces that are "obvious" geographic objects as opposed to older boundaries, *lost in translations…*

7.1 The Greek–Turkish Interface of Lower Evros-Maritsa-Meriç

Symbolically, the lower Evros is a "final frontier"—the ultimate step of the Ottoman–Turkish withdrawal from the Balkan Peninsula. In the southeastern part of Fig. 7.1, the Bulgarian–Ottoman boundary stretches along the line Enos-Midia, as it used to be after the Treaty of London (05/30/1913). This position remains effective until 08/10/1913. In the first times of its establishment as a boundary, the Evros River divides two state entities that are no longer visible on local maps: Bulgaria and Ottoman Empire. Bulgaria legally keeps its positions on the right bank of the river until the Treaty of Neuilly in 1919. Greece then replaces Bulgaria, whose access to the Aegean Sea is abolished. The boundary briefly disappears between 1920 and 1923, when the Hellenic Kingdom launches its troops toward Turkish Thrace and Asia Minor, with the *Great Catastrophe* bringing an end to its ambitions. The Treaty

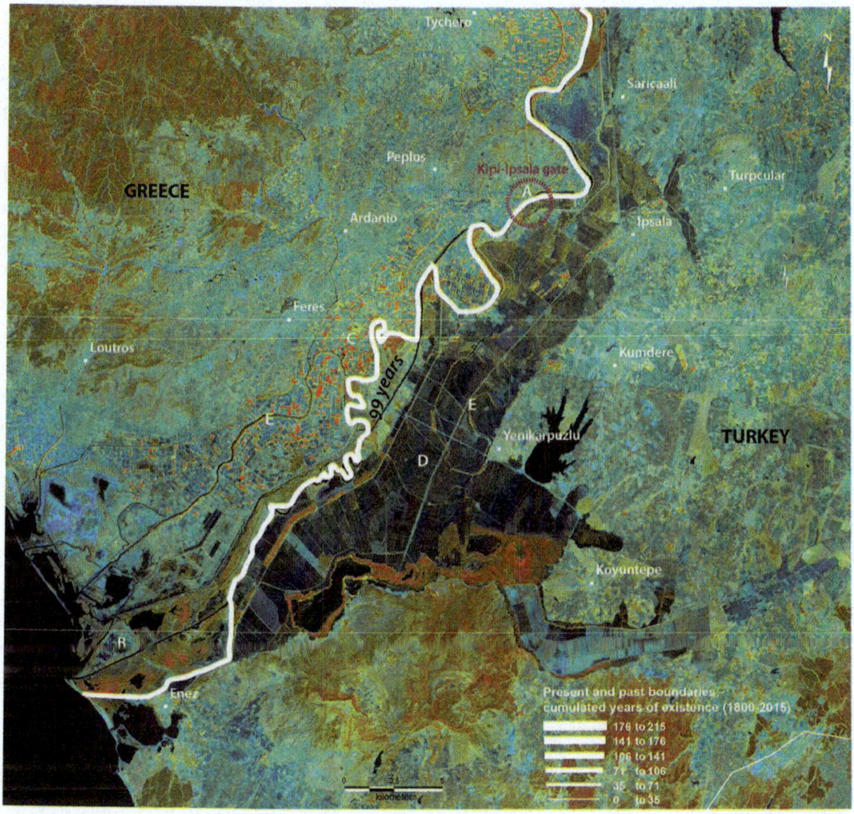

Fig. 7.1 Hydrographic complex of lower Evros-Maritsa-Meriç. *Sources* USGS, mosaic of two Landsat ETM images, 182/31, 06/14/2000, 182/32, 06/01/2001, WGS84/NUTM34, NDVI over Pan, author's own GIS database

of Lausanne re-establishes the Evros-Meriç thalweg as a demarcation. All in all, the main boundary segment cumulates 99 years of existence.

Today, the area is an external European border regarded as sensitive. Greek border guards keep the whole perimeter under strict surveillance. Their main mission is to contain migration flows concentrating on this gateway as best they can. The Kipi-Ipsala Bridge (A) is a main eastern entrance to the European space.

The ETM Landsat images of Fig. 7.1 date back to 2000 and 2001. A 15-m ground resolution is required in order to offer the best possible viewing experience at the desired scale. With such data available, the Evros system evidences its complex nature. Human settlements and activities leave few parts of the riverbed "untouched" (B).[1] North of Enez, the floodplain is almost totally devoted to land farming, with a great imbalance between Greek and Turkish sides. The Hellenic

[1] The *Evros Delta National Park* is declared in 2006 only.

part is made of regular plots situated on drained soils provided with irrigation (C) mainly assumed by pumping systems (Béthemont and Sivignon 2002).

Obviously, lower farmlands contrast with the upper slopes dominated by a series of villages. Rice-growing lands concentrate on the Turkish side of the floodplain (D). Hundreds of patchy plots align along dozens of channels in charge of supplying water to communities of end users through gravity-assisted irrigation. South of Sarıcaali, the overall surface liable for controlled irrigation and flooding can be estimated to 17 000 ha—not counting the Koyuntepe branch. All other Turkish arable lands of the hinterland are cultivated in drought conditions. The contrast between flooded fields and dry inland areas is impressive—an observation made possible by the NDVI algorithm.

Figure 7.2 provides an overview of the same territory a few years before the Balkan Wars' *big bang*. The general scheme is quite different. The Gala Lake, northeast of Enez, is much larger than today. In fact, the area under water roughly corresponds to the National Park of Gala Lake, created in 2006, which covers 3 000 ha. With its widely disseminated puddles and swamps, the valley is still

Fig. 7.2 The lower Meriç in 1902. *Sources* Harvard Map Collection (Erkânıharbiye-yi Umumiye 1902)

repellent in 1902. All settlements are located on safe hillocks, a few kilometers away from the river.

The 1923 boundary divides the delta into two political entities. Although destinies have parted, both sides share two common characteristics:

- With the exchange of populations, both Greek and Turkish borderlands change into asylum for refugees. During the 1910s, anticipating future upheavals, an early flood of Greek migrants from eastern Thrace settles down on the western side of the Evros. In contrast, a large number of Muslims—including Pomaks—originating from different parts of the Ottoman Rumelia find refuge on the eastern bank. With such a change in geographic determinants, local conditions are deeply impacted. The settlement of thousands of people creates new needs for land occupation.
- Emerging from an old isolation, the delta area gains a new status of "land to master." Soil reclamation and land needs result in major public works beginning in the 1920s. Dams, digs, and other irrigation ditches increase the availability of resources previously left to the whims of nature. However, none of these facilities have succeeded in eliminating flood risks. In 2005, two consecutive spring floods cause severe damage in the countryside, leading to further interventions. In 2006, the Turkish authorities decide to build a new dam at Koyuntepe.

The Evros delta and its uplands reveal glaring contrasts. On the one hand, human pressure on natural resources implies maintaining a great number of sensitive facilities. On the other hand, because nature protection is now a major public concern, restoring permanent or temporary flood areas has become an essential objective of "good territorial governance."

Such measures are late variations of a basic principle laid down by the boundary settlement: closely mastering space. When international officials arrive in those far-off Ottoman lands in 1926 (Backer 1926) with a view to define the new demarcation, they directly oblige the neighboring states to exercise tight controls in the long term on *uncertain* lands (E). When economic and strategic ambitions collide, competing authorities have no choice but to find arrangements despite political tensions. The process is less burdensome across borderlands that are subject to lower human pressure.

7.2 Unfriendly Mountains of the Greek–Albanian Confines

On the opposite side of the Peninsula, the Greek–Albanian interface takes us back to much more "savage" landscapes. Here, any human presence is evasive. The Ionian side of the Balkans, east of Corfu, is built as a natural amphitheater, starting from the seashore and ending at around 2 000–2 500 m across the Pindos—the backbone of Greece. From Sagiada to the Black Stone at Peak Gramoz Grammos, the distance is only 110 km as the crow flies. Between both points, a series of longitudinal rocky escarpments define cross-border valleys interrupted by intermediate mountain ranges and more random plateaus. Traditionally, north–south communications are much easier than transversal east–west journeys. As has already been said, during the Ottoman era "southern Albania" and "Greek Epirus" belonged to the same Vilayet of Jannina.

Figure 7.3 presents the most southern part of the contact region between Albania and Greece. Before its partition, the area used to be called *Çamëria-Tsamuria* and welcomed mixed Albanian–Greek populations, both Christian and Muslim. Territorial ambitions never ceased to poison bilateral relations since the Balkan Wars: Tirana aimed at recovering present Thesprotia (besides Igoumenitsa), and Athens demanded the return of "Northern Epirus." Against this background, the boundary plays a demarcating role. Unlike what happens in lower Evros, which keeps men and activities in close physical proximity, mountain areas have fostered the development

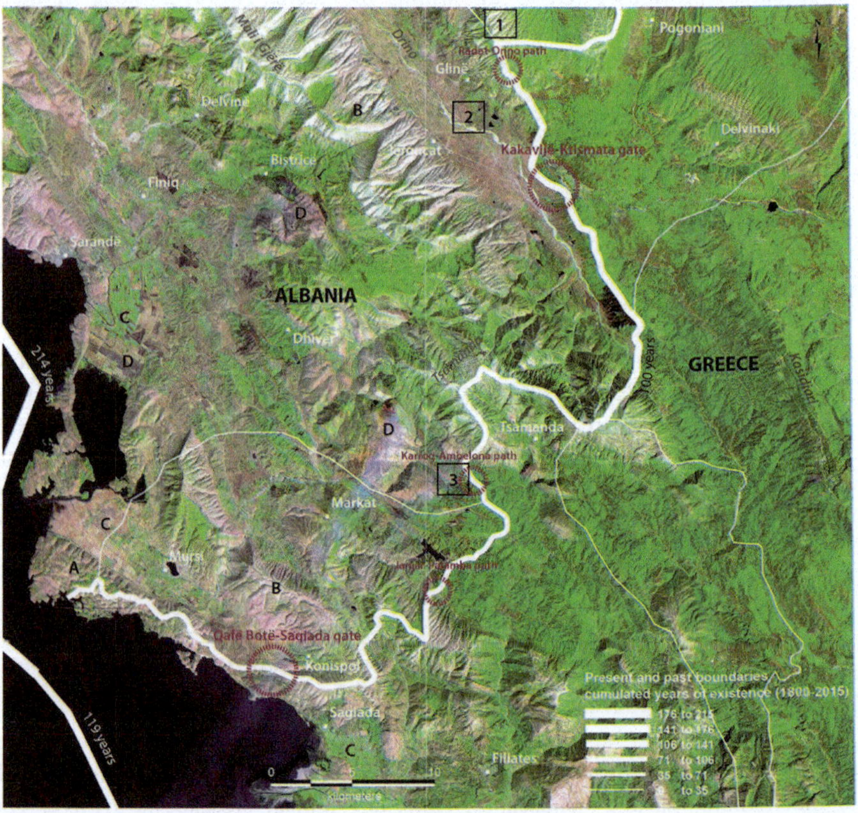

Fig. 7.3 Edge lands between Albania and Greece. *Sources* USGS, mosaic of two Landsat ETM images, 185/32, 08/22/2000, 185/33, 08/22/2000, WGS84/NUTM34, HSI visualization of bands 541, author's own GIS database

of a vast buffer zone. As one of the "hardest" sections of the Iron Curtain, the boundary had to apply a splitting force on lands and people as long as possible.

Settling the boundary was a considerable challenge to overcome. At the beginning of the twentieth century, Corfu and the Heptanese already stay apart from mainland Greece. With 214 years of existence, the border stretching north of Corfu Straits is one of the oldest of the region. The number drops significantly to 119 years in its southern Greek part. Until lately, the division remains clear between the insular domain open to external influences and foreign kingship, and the Ottoman mainland.[2]

In its present shape, the Albanian–Greek boundary cumulates 100 years of existence. The first period of Albanian independence is difficult to describe because no accurate mapping has come into being prior to the 1920s. By the end of the first Balkan War, Albania officially appears on maps as an independent entity. The Protocol of Florence fixes its position at the end of 1913, but this latter is immediately contested. Just before the beginning of WWI and during 2 months, Northern Epirus is declared an independent Republic, afterward incorporated in Albania as an autonomous region. The borderline is officially left untouched when the Albanian Principality becomes an Italian Protectorate. At the Paris Peace Conference (January 1920), part of Northern Epirus returns to Greece. The protests reach their target when Albania recovers all of Northern Epirus at the end of 1921. The 1910s are thus operating under the auspices of a legal territorial *uncertainty*. Kingship is defined by shifting occupation zones. From 1921 onward, this part of the Albanian–Greek frontier remains unchanged.

The situation on the ground is unambiguous:

- The past Ottoman land management system unites lands and determines "interstate" corridors. For instance, an ordinary administrative limit divides the area ruled by Jannina and the Sandjak of Delvinë. Without known precise position, it roughly follows the Drino Valley or Mali i Gjërë and reaches Mount Tsamanda before continuing until the heights of Souli.
- After the unsteady decade of the 1910s, the area is split by a SW–NE line. The demarcation is complex. The boundary crosses mountain ranges and cuts hydrographic basins. Mount Tsamanda is clearly a reference point between both emerging "worlds." The position is easy to receive. South of it, the situation is more confused. A surprising "end of the earth" stretches west of Sagiada: The narrow strip, completely depopulated, stops at Cape Stillos (A). Obviously, this was done on purpose. Such a geographical singularity allows Corfu and mainland Greece to face each other, which results in a limited Albanian control on the Straits.

From space, today's boundary is clearly visible, despite obvious complex geographic systems. The 541 ETM bands overlay reveals significant landscape contrasts between the Greek and Albanian parts, not least in respect of vegetation

[2]This long existence does not avoid diplomatic disputes about its exact location across the sea.

Fig. 7.4 Greek–Albanian borderlands, north of Ambelona, Tsamanda mountain range. *Source* R. Darques, 2002

covers. Political choices, military interventions, and planning programs have determined the emergence of diverse landscapes on both sides of the boundary.

Greece's up-country seems devoted to wild nature and forest. This is not surprising, given that the Ionian coast is well supplied with rainfall. However, rural underdevelopment, withdrawing pastoral practices, human emigration, and field decay provide optimal conditions for the development of overabundant vegetation. The facts are clear. In the surroundings of Kakavia, travelers are often stunned by the luxuriant woods of the Hellenic slopes as opposed to the bare rocky soils of the Albanian side. Such recursive observations reveal divergent systems of land planning in the long term (Fig. 7.4).

Many factors might explain these disparities:

1. Military defense systems play an obvious role. Greece's protection is assumed through a "green belt." Albania would have used a reverse "bare soil" policy. With limited vegetation covers, the confinement is made easier: Any dissident willing to leave Albania would have faced increased difficulties in escaping the border guards' scrutiny.
2. Contrasting population densities may be seen as a countereffect of the military expulsion of the Çams from present Thesprotia by the Greek forces in 1944. In this operation, dozens of villages have been reduced to rubble; the private

properties of these "Albanians" remain frozen up to now. Many of those Çams took up residence across the border, in the district of Sarandë. The desertification of the Greek northern lands and the enhanced rural development of the Albanian *Far South* were at an early stage.

3. The area has a karstic profile. Acid and poor soils are common outside the alluvial plains, polje, and other narrow basins that stretch between mountain ranges (A and B). Erosion plays a decisive role in altering landscapes when vegetation vanishes for whatever reason. Usually, low reliefs are badly drained and submitted to regular flooding. They alternate with bare hills, where limestone outcrops give birth to scattered vegetation. Overpasturing used to be a main cause of the forest draw back. During the Ottoman rule, the lower arable lands belonged to local *beys* heading *çiftliks* with the help of servant farmers, whereas free—and poor—villages spread across mountains had livestock breeding as a main occupation.
4. During the Cold War, diverging planning policies probably provide the necessary conditions for such landscape emergence. While most people in northern Greece join the main Hellenic cities or choose abroad destinations, Hoxha's views on rural collectivization aim at developing farms even in outlying districts. Overpasturing practices and cultivation over newly gained arable plots (in terraces, for instance) reduce "natural" areas to a minimum. Intense competitive land pressures early degrade settlements and farming conditions (Blanc 1963). The transition phase of the 1990s leads to a pressure drop. The setback is impressive: Without any care and submitted to savage destruction, rural communities cannot avoid a severe worsening of their situation. Erosion finds the necessary conditions for increasing its effects. Irrigation systems are destroyed, leaving few plots possibly watered (C). Such land use patterns contrast with the irrigated fruit fields and orange groves of the lower Kalamas, next to Sagiada. Badlands and wastelands spread almost everywhere. In accordance with local constraints, 50–80 % of the population leave their homelands between 1991 and 2001, although land property rights have been transferred to private hands (Darques 2004).
5. As evidenced in Fig. 7.3 (end of August 2000), many farmers obviously use post-crop burning and sartage techniques on a regular basis, although such agricultural practices are risky at this time of the year. These badly supervised operations prevent the reconstitution of soils and facilitate erosion as autumn begins.

To sum up, the rural desertification of the Albanian borderlands meets an already advanced Greek relinquishment. This is usually the fate reserved for land confines. The Albanian–Greek frontiers thus participate in a "standardization" process.

As "relics" of the past, the facilities devoted to military surveillance evidence the fact that regular armed forces did not consider "normality" as a useful concept until recently. Against the background of a dead border placed under close surveillance during decades, Fig. 7.3 and the QuickBird imagery provided by Google Earth allow one to discover how cross-border exchanges work today. Until the communist fall, Kakavijë-Ktismata used to be the second largest Greek–Albanian checkpoint behind Kapshticë-Krystallopigi, east of Korçë. Lately in 2007, a third southern gate opens between Sagiada and Konispol. The EU provides funding for the construction of a wide access road 10 km off the crossing point. This initiative brings significant benefits: A car journey Sarandë-Igoumenitsa (45 km) took approximately 4 h until then.

The biggest problem today is not any more the political enclosure of Albania; it stands in the poor condition of the road network. There is almost everything to build, or rebuild. The Google Earth imagery points toward small crossings at Karroq-Ambelona, Janjar-Palamba, and Radat-Orino: These are potentially trafficable tracks. Of course, it is needless to mention the great number of footpaths joining both countries that can be detected by satellite views. Illegal migrants are regular users of such *monopatia*. These networks criss-cross fields and forests. From the Hellenic custom officers' view, the control mission is impossible to achieve, considering the extent of the area to supervise. Behind the pyramids' first line, their attention focuses on certain strategic positions: bus stations, bridges, and major roads.

The abundant vegetation in Greece prevents us from observing the remaining military facilities. Most of them have been dismantled by the end of the 1990s. In contrast, the importance of the Albanian defense system combined with a lower canopy enables the observation of military installations and fortifications inherited from the Cold War (Figs. 7.5, 7.6, and 7.7, corresponding to the three black squares of Fig. 7.3).

North of Kakavijë and southeast of Libohovë, the Greek–Albanian boundary reaches the altitude of 1 700 m. The Bureto mountain range overhangs the Dropull Valley, which stretches 1 500 m lower down. On the eastern side, the height

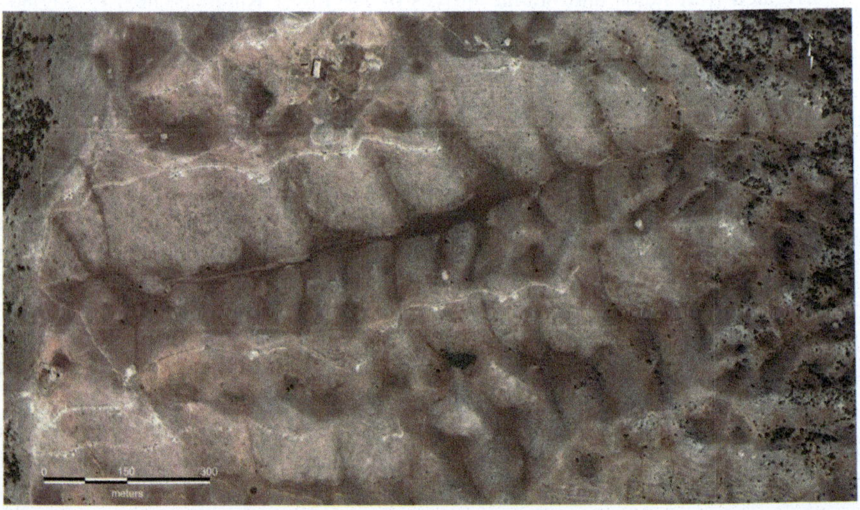

Fig. 7.5 Southern Buretos chain: military watchpost and defense system (square 1 on Fig. 7.3). *Source* Google Earth imagery, 09/02/2011

Fig. 7.6 Dropull Valley: bunker defense lines (square 2 on Fig. 7.3). *Source* Google Earth imagery, 07/20/2010

Fig. 7.7 Ksantos-Pavlla Valley at Karroq: barrier blocks (square 3 on Fig. 7.3). *Source* Google Earth imagery, 08/08/2010

deviation between the peaks and the Kikomosë stream is "only" 1 000 m. The panoramic view on both Greek and Albanian borderlands is breathtaking. Bare rocks swept across by the wind make a perfect watching position. The surrounding heights and slopes used to be reserved for extensive forms of pastoralism. During the Cold War, the army settles down never to leave again this highly strategic place.

Considered to be the main corridor for a potential foreign invasion, the Drino Valley and its southern part (called Dropull) are a large military playground. Spreading over kilometers—at least until Gjirokastër, various defense devices pollute the environment. The extent of the areas assigned to farming are very limited. Between the villages of Glinë and Jergucat, arable plots left uncultivated still welcome ranges of bunkers waiting for potential aggressors. Figure 7.6 covers seventy hectares and receives about hundred visible bunkers of different types—not counting the facilities covered by vegetation and those still kept underground.

Figure 7.6 epitomizes the ambivalence of such remote districts: Military facilities closely mingle with other standard agricultural practices. Once constructed, the bunkers stand in the middle of living cooperative fields. Besides, the workers "invited" to help build these military facilities were the same as those employed in state farms. Army engineers and farm technicians have shaped landscapes. The great depression of southern Albania has derailed all past plans. Military and rural withdrawals end up together as they had begun, side by side.

East of Karroq and on the side opposite of Ambelona, the river Ksantos-Pavlla crosses the boundary under the guardianship of Mount Tsamanda. This watercourse takes its source in Greece and runs through Albania before flowing into the sea, south of the lake of Butrint. Although the place is only popular for a few shepherds and hunters today, the Albanian army considered the path as a strategic one. Just beyond the boundary line, the valley broadens for a while between steep slopes. The area reaches more than a hundred hectares entirely covered with antitank and infantry blocks (Fig. 7.7). Land is made sterile and shall remain so until future dismantling, whose cost might be prohibitive. Meanwhile, only sheep and goats enjoy the *peacefulness* of this area.

Many more examples could be quoted. It might be objected, though, that these border systems belong to a gone by era and that this work pertains to military archeology. In fact, military forces are one among many ways of expressing geographic remoteness. War provides the necessary conditions for generating spatial marginality. Peaceful times do not expunge the past, but preserve what has been achieved.

7.3 Nature Protection and Tourist Flows Between Montenegro and Albania: The Lower Bojana-Bunë

The northern confines of Albania conform to another organization pattern. The Albanian–Montenegrin boundary is 287 km long. Its lowest part (32 km) comprised between the lake of Shkodër and the Adriatic Sea lies in the middle of an atypical region. As opposed to the stepped relief of Epirus, this area is a sedimentary zone only interrupted by hillocks. The contrast with the Albanian hinterland (Malësia) is striking (Figs. 7.8 and 7.9). Between the snowy peaks of the Albanian Alps and the awaited beaches of the coast, geographers would expect a

Fig. 7.8 Drin and Bunë confluence south of Rozafa Castle (Shkodër). *Source* R. Darques, 2002

Fig. 7.9 Shkodër's plain and Malësia. *Source* R. Darques, 2002

Riviera associating both advantages of easily watered soils and sunny climate. Here, communities and cultivated areas could harmoniously coexist.

Although not entirely unfounded, this scenic view does not stand up to adverse environmental conditions. At the foot of sheer inland mountains, rambling watercourses and temporary lakes spread across sedimentary plains. Unpredictable rivers define large dejection cones before bursting into the lower valleys they regularly dismantle. Then, swamps alternate with damp fields gained by abundant vegetation. Environmental conditions thus create landscapes that are unique within short distances. The violence of the elements is stunning—as in many other Mediterranean regions.

Forty kilometers long and 15 km wide, the lake of Shkodër-Skadar is separated from the seashore by a land strip (Fig. 7.10). The Rumija range forms the upper part of this band; it reaches an altitude of 1 500 m. Following a perpendicular path, the boundary imposes a cross-sectional division of space. The demarcation begins at

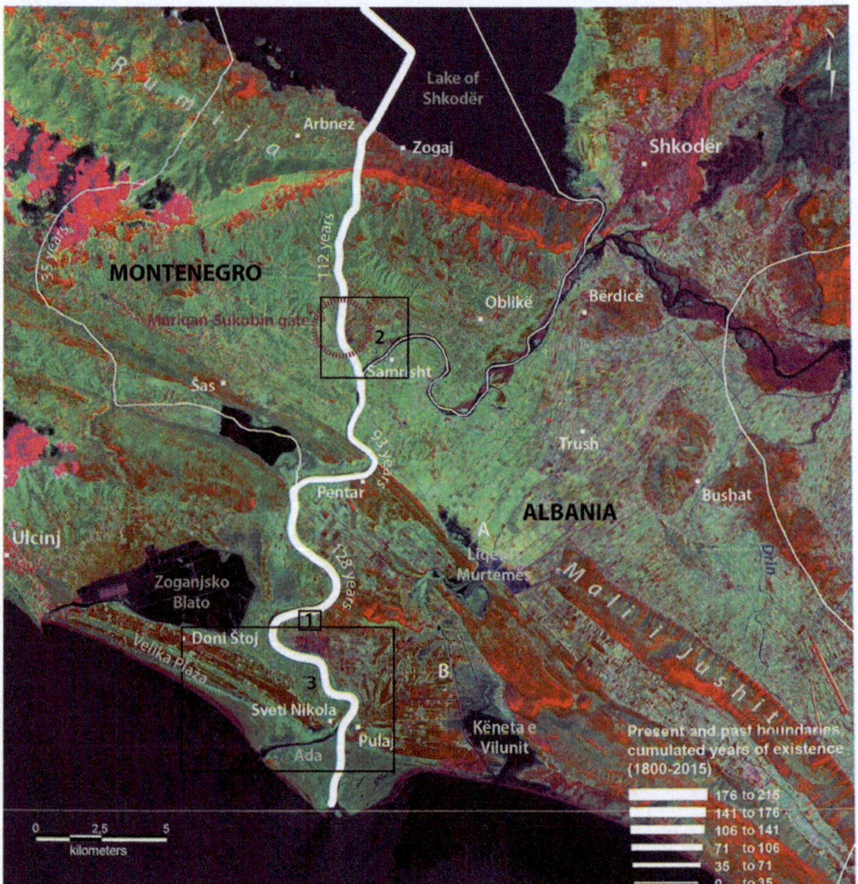

Fig. 7.10 Bojana-Bunë hydrographic complex. *Sources* USGS, mosaic of two Landsat ETM images, 186/31, 08/13/2000, 187/31, 07/19/2000, WGS84/NUTM34, Tasseled Cap transform over Pan, author's own GIS database

the Bojana-Bunë delta and follows the river until the village of Samrisht. Immediately after leaving the thalweg, it climbs across the Rumija and ends up joining the lake.

Up and down the mirroring waters of Shkodër Lake, a complex watercourse network causes visible changes to landscape forms and shapes. South of Shkodër city, three rivers join:

- Regulated by the dam of Vau i Dejës,[3] the *Great Drin* comes from the east;
- The *Kir* runs from the Dukagjin Mountains. Its watercourse meets the Drin just east of Shkodër city;
- The *Bunë* River has its source in Shkodër Lake.

The hydrographic system is complex and unsteady. In 1858, a great flood of the Drin changes the watercourse position. The old way headed to Lexhë. After the disaster, a second thalweg discharges into the Bojana. Such hazards illustrate the difficulty attached to geocoding old maps in the region. Human settlements and facilities are unreliable, and neither are natural reference points even when appropriately positioned. Figure 7.10 shows the new (post-1858) location of the watercourse and Fig. 7.11 the older one.

Studying the settlement of the boundary in this area takes us back two centuries ago, much earlier than before (Evros and Epirus). The basin of Shkodër is an old contact region between rival political forces. The far south of the Austrian Empire meets the Ottoman *Far West*. Uneasy ground conditions and far distances from the main urban centers weaken the land control capacity of both mastodons. Between these giants, a tiny state imposes its presence, trapped behind its mountains. From its early narrow positions, Montenegro will not stop expanding in all directions.

Until 1809, all lands shown in Figs. 7.10 and 7.11 belonged to the Ottoman Empire through the powerful *Pashalık of Işkodra*. At this date (Treaty of Schönbrunn), the former Venetian possessions of Kotor are legally transferred to the Illyrian Provinces of the French Empire. A few years later, in 1815, the Congress of Vienna allows the Austrian Empire to replace French authorities. After the defeat of the Ottomans against the Russian army in 1829–1830, Montenegro extends its hold to the south until meeting the Austrian possessions of the seaside. The layout will not change until 1878.

In 1862 (Fig. 7.11), Kiepert's map presents a typical land pattern of the late Ottoman period. First of all, it should be noticed that, until 1878, Montenegro is still semi-independent from the Ottoman Empire. The autonomy of the principality, supported by Austria, builds on chieftains that maintain armed forces able to resist the Ottoman troops. Riots, ambuscades, and other struggles are "normal" everyday events.

There is no significant difference in character between the highlands of Montenegro and those of Malësia, even though the latter are fully integrated into the Ottoman Empire and reported by Kiepert as "Albanian." Land distribution is

[3]Formerly *Mao Zedong*, (Sivignon 1977).

Fig. 7.11 Extract from Kiepert's map, 1862, region of Shkodër. *Source* Zeitschrift für allgemeine Erdkunde (Kiepert 1862), Digizeit Göttingen University

based on clan organizations and subdivisions, rather than topography. Significantly, districts are defined by clan names (Doll 2003). The Austrian authorities ignore this hinterland, focusing their attention—like further north—over a narrow strip of land consisting of seaside cities and harbors.

The rivers Morača and Bojana are not landmarks: Most boundaries avoid their thalweg and prefer to adopt "non-compliant" routes. Nestled in a meander of the Zeta River, Spuž astonishingly remains Ottoman, a few kilometers north of Podgorica. It is only after 1830 that Montenegro succeeds in accessing to Shkodër *Sea*, failing for a time to reach the Adriatic. The extremely complex ethnic–confessional background of the area will not be discussed. In this respect, it is sufficient to point out that the Sublime Porte uses social mixing as a way, among others, to preserve its power.

High Albania This name immediately recalls adventure novels, exotic trips of unconscious or brave travelers (Durham 1909; Lane 1923; Schon and Galaty 2006). Doubtlessly, Shkodër is the head city of the whole region. Neighboring the boundary is an old thing around here. Cetinje, Bar, Budva, and Podgorica are border places for ages. And yet, the concept of "border" loses most of its meaning when one considers the mobility of the human groups that had pastoral activities as their main source of income. M. Reinkowski argues that, across these Balkan confines, the notion of sovereignty falls into two parts: The Ottoman Empire officially defends its districts, but its administrative bodies accept the practical inability to uphold the law (Reinkowski 2003). Who cares about frontiers? Who could control the bandits and rebels that are known to chop off the head of their enemies? Seen from Istanbul, Montenegro is a real "Far West," a frontier of wilderness erected against "civilization." At the beginning of the twentieth century, bullet shooting was reported as one of the first causes of death in the region (Kaser 2012).

Until the fall of the Austrian Empire, the access of Montenegro to the sea is blocked north of Bar. After the Treaty of Berlin and until 1918, Austrian garrisons surround the country: Bosnia and Herzegovina has become part of the Empire, and the Sandžak of Novi Pazar is occupied. Montenegro has no other option but to look south and encroach upon the Ottoman area. The Bunë becomes a borderline after 1878. Halfway, the demarcation leaves its thalweg to reach the lake of Shkodër. Making a detour to the west, it comes to include the communities of Šas and Arbnež. More than half of the Great Lake belongs to the independent country from then on.

A second period begins in 1913. The Balkan Wars authorize a temporary Montenegrin extension until the suburbs of Shkodër city, although the demarcation has no official existence. The independence of Albania will not completely change the border location. The Protocol of Florence fixes the new delineation west of Zogaj. According to the London Pact of 1915, Montenegro gets southern Dalmatia and northern Albania. Cetinje extends its territory to include the lower Bojana and Drin, until Lexhë. The Paris Peace Conference re-establishes Albania in its former legal limits depicted by the Protocol of Florence, whereas Montenegro becomes part of First Yugoslavia.

The reverse occurs during WWII. Although submitted to the Italian domination, Montenegro suffers a temporary territorial withdrawal on behalf of Albania, which "recovers" Ulcinj. Finally, despite the parenthesis of the World Wars, the boundary has remained relatively stable during the last 100 years. However, the current ethnic structure of both Montenegro and Albania keeps traces of these past wanderings.

Back to Fig. 7.10. From Ulcinj to Lexhë, through Shkodër, a vast triangular area of unlikely lowlands stretches over 60 000 ha. For the new Albanian authorities established after 1912, and afterward for the communist *nomenklaturë*, the whole perimeter was a land to master. Albanians, Serbs, and Montenegrins, as proud mountain folks, traditionally overlooked those remote places dominated by freshwater vegetation with little interest. Taking all possible steps to control the implacable clans that had succeeded in resisting for centuries to the Sultan's armies from their upper villages, the first Albanian administration and—later on the party—had to conquer a new frontier (Gruber and Pichler 2002; Rugg 1994). The great socialist project shall find a relevant application field across the Bunë and Drin low-lying lands. The workforce necessary to carry out such enterprise will precisely come from the highlands: The process of collectivization finds an ideal playground to demonstrate the effectiveness of official planning policies. This is made easier for them by the fact that the planned works concern sparsely populated areas (low densities, wide unproductive *çiftliks*). Breaking down social and political resistance for the benefit of the socialist project and strengthening border control are a twofold objective.

Starting in the 1950s, a series of large state farms and cooperatives take possession of the Albanian side. The collectivization radically alters landscapes and development models: Major public works free all swamps and drain water out of the permanently flooded areas (A and B points in Fig. 7.10). The marshlands are cleared out of any riparian vegetation. Because of the abundant waters coming from upstream, the riverbed has to be straightened and its deviations limited. Dams built across the Drin allow the establishment of a general water conveyance system. No land reclamation would have been possible without domesticating the upper watercourses and limiting the devastating effects of floods. These efforts boost agricultural productivity. The results achieved are impressive as regards the level of economic development of the first national period.

Getting a clearer picture of the "revolution" underway in the *Ultësira Shkodrane* (lowlands of Shkodër) implies comparing seldom available data. These lowlands are made of several adjacent plains:

- *Fusha i Mbishkodrës*, north of the city, along the lake.
- *Fusha e Nënshkodrës*, south of it.
- *Fusha e Bunës*, along the river of the same name.
- *Fusha e Zadrimës*, east of the old Drin course.
- *Fusha e Kakarriqit*, a long strip of land caught between both flysch hills of Mount Kakarriqit and Renc, east of the delta.

These geomorphological and hydrologic units are, of course, disconnected from the administrative network. From the few available statistical data, special attention shall be paid to the district (*Rreth*) of Shkodër as defined during the socialist period. This latter covers all lower plains mentioned earlier, plus the highlands of Malësia e Madhe and Dukagjin. The districts of Pukë, Lexhë, and Mirditë stay apart from it (Table 7.1).

The district of Shkodër faces a real revolution. In 1927, according to Selenica (1928), the area holds 68 500 inhabitants, of whom 23 800 belong to Shkodër city. Arable lands cover 5 500 ha in all, plus 500 ha of vineyards. The annual production reaches 66 000 quintals of corn, 28 000 of wheat, and 21 000 of tobacco. Other vegetables and fruits for human consumption complete the set. Fishing is another significant activity: 6 000 tons of fish are annually caught. The lowlands provide 90 % of the total cultivated area. The livestock is more equally distributed: 38 000 sheep, 7 000 goats, and 5 000 lambs wander the outskirts of Shkodër, whereas 24 000 sheep, 12 000 goats, and 4 000 lambs pace the mountain slopes. Fig and plum trees are by far the most developed shrubby plantations. The evidence will thus suggest that populated and poorly productive uplands differ significantly from

Table 7.1 District (Rreth) of Shkodër: some indicators on agriculture production, 1927–2001

		1927	1956	1971	1990	1998–2001
Rreth	*Inhabitants*	**68 500**	**116 100**	**171 500**	**241 500**	**222 087**
Shkodër city		**23 800**	**38 600**	**56 500**	**81 800**	**83 600**
Cultivated area	*Area (ha)*	5 500	28 800	34 800	45 600	19 600
Vineyard		500	500	1 100	1 500	221
Corn	*Production (tons)*	6 600	9 500		32 400	
Wheat		2 800	5 700		30 700	
Tobacco		2 100	2 800			
Rice			700			
Potatoes		7	1 500		6 200	
Sheep	*Heads*	38 000	110 000		110 000	72 224
Goat		12 000	72 000		78 400	48 027
Cattle		7 000	39 000		42 600	54 500
Total fruit trees	*Trees*	**240 300**	**333 900**		**757 000**	**56 709**
Fig		22 000	138 000			17 962
Plum		67 000	49 000			15 077
Mulberry		64 000	29 100			
Olive		1 600	25 000		346 000	1 925
Pear		11 500	19 000			4 032
Apricot and peach		4 000	13 000			

Source (Drejtoria e statistikës 1959; Selenica 1928; Drejtoria e statistikës 1973; INSTAT 2000; Drejtoria e statistikës 1991; INSTAT 2004)
Discontinued and/or unpublished statistical series and altered collection methods produce blank fields. The district (Rreth) definition changes across time. The limits adopted during the socialist era have been chosen. They include the *Nënprefekturat*/subdivisions of Shkodër, Malësia e Madhe, and Dukagjin. Pukë, Lexhë, and Mirditë sometimes belong to the district. These areas have been removed from the accounts when necessary. When such subtractions were impossible, blanks remain.

the wealthier areas around Shkodër city, which benefit from arable lands and terraced farming along foothills.

In 1956 (Drejtoria e statistikës 1959), the situation has drastically changed. The agrarian reform engaged in the early years after the war starts to show positive results, although the cooperative sector is at an early stage of development. Most lands are still placed in the hands of private owners that carry out the greatest share of production. Visible progress has been accomplished through the process of land reclamation. Arable areas increase from 5 500 to 29 000 ha! The statistical analysis reveals a country that reinvents its productive potential. In parallel with the demographic growth of the coastal plains, the entire farming system is revised. The "new Albanian peasant" is proud to introduce original products: Potatoes are now harvested, and 700 ha are reserved for rice cultivation. Thousands of fruit trees are planted with preference granted to fig, peach, apricot, and pear. Olive groves also develop.

By the 1960s, the collectivization of land takes off. Cooperatives replace private farms everywhere. Provided with better-performing equipment and facilities, state farms reinforce the public domain. Land reform is achieved. A few Western travelers and scientists describe this *new world* during the 1970s. Their stays in Albania are strictly supervised by Foreign Affairs intermediates. The image spots they bring back are significant (Sivignon 1977). M. Sivignon briefly describes the state farm "Perlat Rexhepi," south of Shkodër. In the mid-1970s, this farm controls 3 000 ha, of which 1 800 are actually cultivated. With such dimensions, the production unit may be assimilated to a new working "village." Its buildings and facilities spread across open landscapes. The lower Bunë changes into a colonization area. Here, though, human presence was evasive before the war. No community exceeded thousand souls.

At the end of the 1980s, the socialist system is at its acme. The latest figures doubled the amounts shown in the mid-1950s in virtually every area, including registered population. The clearing of badlands and forests has reached a maximum and will not go further. Farming productivity is at its peak. The number of fruit trees hits record levels (around 800 000 units). Thanks to extensive irrigation, those plantations produce massive outcomes for national consumption or export. The former "desert" has become a "garden"—at least if official data do not lie…

The socialist collapse brings a "backward" revolution (Civici 2010). The whole system falls down in a few years. A whole political and ideologic world disappears. Figure 7.10 provides a truncated image confirming the latest statistics. Between 1990 and 1998, the arable area is divided by more than two. The cultivated area declines steeply and comes back to pre-WWII rates. Half of the vineyards are destroyed. Sheep and goat flocks decrease by a quarter. The most significant damage supported by rural communities is still the eradication of fruit plantations: From the 760 000 trees registered in 1990, only 56 000 remain in 1998, according to the agricultural censuses (7 %)! A population drain follows the desertification of the countryside. As a consequence of the reforms, plots and dwellings return to private hands. Many of them end up being abandoned and revert into a wild state. Migration is the only answer left to people, especially to young workers.

Figure 7.10 dates back to this downfall. The *tasseled cap transform* applied to ETM datasets is usually helpful for identifying agricultural practices and crops. On the Albanian side, the "forward to the past" process is ongoing. Purple, blue, and dark blue colors reveal water surfaces and damp soils. Obviously, hundreds of hectares have returned to their "natural" condition, i.e., before the installation of a drainage system. Although taken at the end of August—theoretically the driest period of the year, Liqeni i Murtemës and Këneta i Vilunit are expanding areas. Around Trush and Bushat, but also along the Kakarriqit Valley, water irrigation facilities are still operational. The collapse is obvious although the damage is limited as regards the conditions governing other southern Albanian districts. In these northern confines, the population experiences a decline of 20 % over the decade—only. Some collective facilities have "survived" the crisis. Land privatization opens onto an acute territorial fragmentation. Narrow strips of 20 by 200 m replace older plots. Deserted and occupied parcels alternate—at least wherever technical facilities do not need collective maintenance for keeping the earth farmable. Abandoned fields permeate the whole plain.

Figure 7.12 perfectly illustrates the change. Lagina e Re ("the new") is a place next to the Bunë riverbanks. This is not a "village" in the usual sense of the word because no central place exists. The main buildings are those left by the former collective farm. A dwelling hamlet occupies the southern part of a flysch hillock. Land has been sliced into strips. Little cottages appear on new private estates bordering roads. Few residual orchards spread around hamlets half asleep. Plots that are too distant from housing settlements are left on their own. A few kilometers southward, old paddy fields gained by salt bring back memories of a "glorious" past. The area is entirely thought of as a frontier, not only against nature, but also against the Yugoslav neighbor. Almost all fields receive bunkers or other defense blocks.

The present boundary is readily identifiable from the sky. Figures 7.13 and 7.14 avoid useless commentaries. Two worlds coexist, only separated by a thin river. The Albanian openfield contrasts sharply with a Montenegrin land division that has obviously escaped reform and reparcelling. All fields gained over scrubland of the right bank belong to the old villages of the upper limestone piedmont. Forest clearing is partial and performed on the basis of private uses, leading to a patchwork of small plots surrounded by hedges. Higher productivity is clearly not a credo in this countryside of Ulcinj. Small family farming is the rule, with a relatively wide variety of crops: Cereals, market gardening, tree plantations, and personal vineyards

Fig. 7.12 Lagina defense system and rural development (square 1 on Fig. 7.10). *Source* Google Earth imagery, 08/22/2009

Fig. 7.13 Samrisht: hedged farmlands and openfield (square 2 on Fig. 7.10). *Source* Google Earth imagery, 08/22/2009

Fig. 7.14 The Bojana delta (square 3 on Fig. 7.10). *Source* Google Earth imagery, 08/22/2009

are characteristics of mixed activities quite reluctant to external markets. Outlying lands are allocated to fodder production and pasture.

Emerging from dissimilar pasts, Albanian and Montenegrin agricultural sectors suffer today from the same low productivity. Although farming activities in the district of Ulcinj have never reached the industrial dimension observed across the boundary, "farm" as a production unit is to be reinvented on both sides of the Bojana.

Figure 7.13 mentions the checkpoint of Sukobin-Muriqan, which is the only customs post south of Shkodër. The gate was opened in 2002. As usual in remote areas, the advent of such facilities has a strong impact on local life. Until 2002, the only bridge to cross the Bunë was located just south of Shkodër; it was covered with wood and single way. With the new border gate, international traffics and circulation constantly develop. Local newspapers report that 3 000 people would travel to Montenegro or Albania every day. Shopping in Shkodër is advantageous for Montenegrins living in Budva, Bar, and Ulcinj.

Suddenly, Shkodër regains hope and dreams of recovering its past position of head city. Its fate seems close to that of Edirne. Both towns are major Ottoman centers lately "sentenced" to keeping state confines, with attendant negative consequences. For them, any indication of greater openness toward neighboring countries shall be taken as a sign of possible revival. For now, the Albanians are free to travel to Dubrovnik or Ulcinj and access their tourist facilities, although local prices remain high for them. The organic relations with the Albanian communities of southern Montenegro slowly recover from decades of blockade.

The essential is elsewhere. More remarkable are landscape contrasts alongside the delta of the Bojana (Fig. 7.14). To the general collapse of the Albanian side, southern Montenegro opposes an entirely different model of development where tourism plays an essential role (Grandits and Taylor 2010). Territorial disparities are not likely to change in the near future. Located 5 km south of Ulcinj, Velika Plaža is a remarkable place. Its sandy beach stretches over 12 km—a morphological anomaly compared to the abrupt rocky shorelines that Dalmatia and Montenegro typically offer. Tourist complexes constructed on lands owned by military authorities are not unusual in the Balkans. The distances are short, and the lack of space is obvious. Above all, any development opportunity has to be valued. Velika Plaža is situated right next to an old branch of the Bojana. The alluvial deposits form a coastal dune 1.5 km wide bordering the 1 500 ha of Lake Zoganjsko, a great part of which is used for saltworks.

The northern part of Velika Plaža is economically more developed. With its concentration of tourist facilities, the area may be considered as a suburb of Ulcinj. In order to allow further developments, hygrophilous woods and vegetation have been cleared along a strip of land until Sveti Nikola. The expansion goes all the way back to the 1970s: Hotels and dwelling units have no other function but to meet tourist needs. New apartment buildings and villas have been introduced recently. After the independence of Montenegro, the latest tourist development program has declared the prevailing position of Velika Plaža. Focusing its attention on the reported 1.1 annual million tourists of the country, government and local authorities are seeking private investors to improve the reception capacity of Ulcinj and its surroundings. Plans are impressive. Should they eventually apply, a wall of concrete would arise from the ground. Of course, environmental considerations are taken into account. This is the least warranty potential investors might grant to their future cash-rich clients at least in formal terms.[4]

For now, the fate of Ada Bojana Island is already altered. The area is theoretically included in a nature protection reserve. Despite this, dozens of illegal restaurants have been created on the banks of the Bojana. They build an informal resort center. Agreeable terraces over piles welcome a large number of clients keenly interested in enjoying famous fish dishes while mirroring onto the turquoise blue waters of this "end of the world" planned to become a paradise for nudists. The least one can say is that the boundary has lost its threatening power. Streams of careless tourists now replace heavily armed soldiers.

7.4 Syrmia: A Middle Land Between Danube and Sava

This new case study takes us on the opposite side of the Dinaric Alps. Syrmia is a historical region wedged by the Sava and Danube rivers, expanding from Vukovar to Zemun. Slavonia is to the west and Banat to the east. Environmental conditions differ significantly from that of previous examples, for many reasons:

- This is a "polygonal" border zone, interconnected with Bosnia, Serbia, and Croatia. After 1918, with the settlement of First Yugoslavia, the area gets rid of its old boundaries, although these had weighed so much in local history. Only recently did Syrmia come back to the hard materiality of sovereign perimeters and political divisions, after the Yugoslav Wars.
- This "new" international border zone is an old contact area between Austria–Hungary and the Ottoman Empire. Such "heavy" demarcation has left deep traces up to this day. Depicting the diversity of landscapes resulting from such "German" versus "Ottoman" influences is useless; they jump at anybody's eyes. Architecture, land patterns, farming habits, and even culture are directly concerned: This is self-evident for backpackers, despite the fact that remote sensing tools partially fail to reflect such legacies of the past.
- Terrain and topography are depressed. Geographers and travelers used to the cliff areas and slope declivities of the southern Balkans are usually impressed by such land availability. When compared to Rhodope, Pindus, or the Dinaric Alps, the large and open Danubian Basin is "another world." Among huge sedimentary deposits, rivers fight their way forward hoping to reach a far-eastern sea, wandering, changing location, and leaving behind them large swamps requiring further domestication before human settlements become possible. Even Mediterranean influences are declining, giving way to the development of vast mainland forests—wherever tree clearing has not been achieved.

[4]See http://albaniamakelaar.nl.

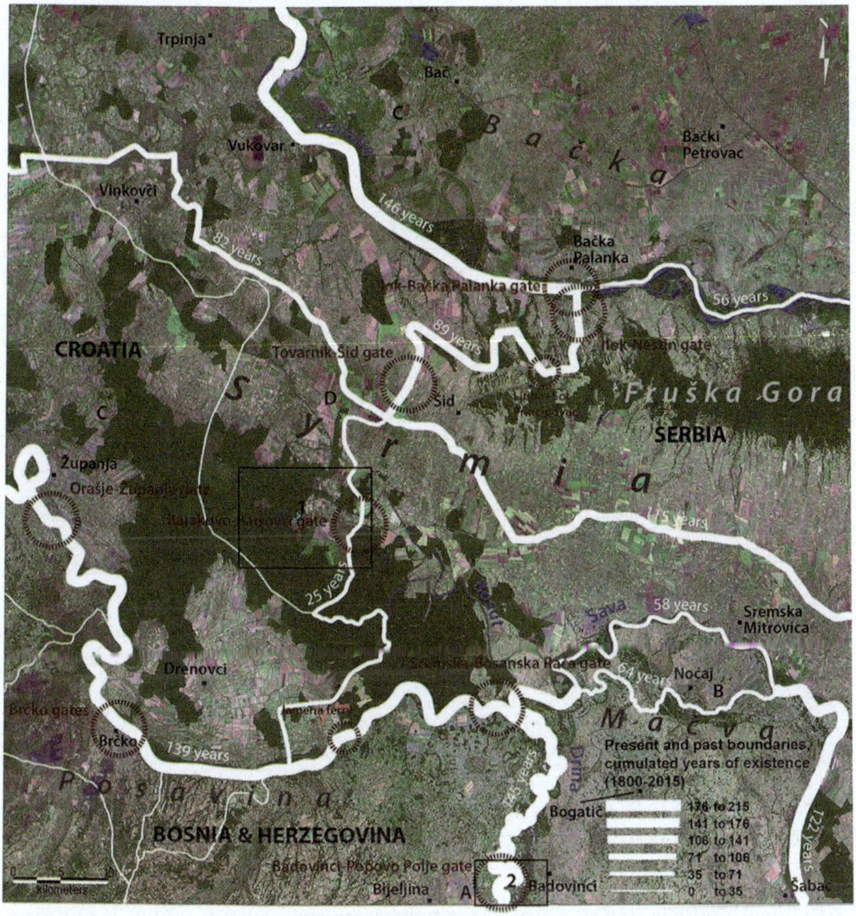

Fig. 7.15 Syrmia's border and buffer zone. *Sources* USGS, mosaic of two Landsat ETM images, 187/29, 08/20/2000, 187/28, 08/20/2000, WGS84/NUTM34, "Pleasing image" algorithm based on a modified Brovey Transform, author's own GIS database

Such even and open landscapes seem a priori badly compatible with any land fragmentation, as opposed to mountain areas. And yet, this region is the most broken up district investigated so far. Figure 7.15 shows a longitudinal cross section of the contact zone between the last Dinaric slopes and the Danubian Plain.

This land is highly contrasted. Danube, Sava, and Drina appear as stable boundaries that transcend the ages. They have been playing a pivotal structuring role for the three-fourths of the past two centuries. In reference to the classification of Chap. 4, two kinds of limits can be distinguished: Sava and Drina embody the past Ottoman and early Serbian boundaries; the remaining northern demarcations are often internal divisions of the Austrian Empire (Fig. 7.16).

Fig. 7.16 Syrmia in 1856 according to J. Scheda: a Military Frontier. *Source* David Rumsey Map Collection (Scheda 1856)

Political outlines define a west–east "stratigraphy" of land, which is consistent with local topography. However, two vertical alignments between Croatia and Serbia thwart the overall landscape pattern. Both countries are recent products of the Yugoslav Wars. The thinnest line represents the Western limit of *eastern Slavonia, Baranja and western Syrmia*, defined as a *Serbian Autonomous Oblast*, afterward part of the *Republic of Serbian Krajina* between 1991 and 1998. The second cross-cutting line marks the current border between Croatia and Serbia—a legacy of Tito's internal division of the Yugoslavian Federation. The northern part borrows an old demarcation line between the Kingdom of Slavonia and Hungary, plus the Voivodeship of Serbia. The shared section between Šid and Jamena has become an international de facto boundary in 1991, with the Croatian declaration of independence.

Parallel to the Sava and Danube, at mid-distance between both, a line fixes the upper limit of the *Slavonian Military Frontier*. Surprisingly, the remote sensing view of Fig. 7.15 clearly evidences past delineations as a living heritage, as if Yugoslavia had been a seventy-year parenthesis. In older documents, Syrmia seems loosely peopled. Mitrovica, Vukovar, and Brčko form a triangle whose center is occupied entirely by swamplands and forest. In the middle of this green and blue set, the Bosut River stretches its languorous meanders. This first greenbelt covering the Ottoman *march* land is doubled by the Fruška Gora range, north of Mitrovica, and its monastic bodies. The Military Frontier thus appears as a wide buffer zone. North of Palanka, another wooded and swampy area extends over kilometers along the Danube riverbed. Thus, Vukovar controls a corridor spanned by a major circum-Austrian road. Slavonia as a kingdom subdued to the Empire is here no more than 10 km wide—a rather curious situation evidencing the strategic importance of this land strip.

Across the Sava, on the Ottoman side, other rules regulate land occupation. The utmost projections of Majevica Planina allow better water drainage, hence a higher population density. Along the Sava, some boggy areas persist. The *Kitor Blato* occupies wide fields east of Lešnica and south of Mitrovica—a swampland of more than 50 000 ha surrounded by forests. The confluence zone of the Drina and Sava is almost empty of any human presence. To sum up, natural elements have no difficulty in helping the late imperial marchlands play their repellent role.

When Croatia declares its independence in 1991, its eastern border with Serbia has almost no historical background. Unlike previous examples, this current boundary is impossible to detect through remote sensing images. The Yugoslavian landscapes did not care administrative divisions between constituent Republics. Anyone who ever travelled across socialist Yugoslavia remembers flawless transitions between Republics.

Thus, it comes down to this: The most "obvious" land divisions are the oldest ones. They are sharpened by natural thresholds. Between the small cadastral allotments of Posavina and the openfield of Bačka, two farming worlds stand in front of each other: "old" and "new" Yugoslavian versus Serbian areas. The colonization of the northern areas is not only based on land reclamation. It also relies on property transfers induced by annexations. The old German and Hungarian estates that had succeeded to the Ottoman land pattern have been transferred to public and private domains in turn on a step-by-step basis after WWI. Any transfer of sovereignty provides a unique opportunity to adjust land tenure and structure (Lampe and Jackson 1982).

In 1935, Peyre describes the exact area under study (Peyre 1935). His report comes after the great land reform of 1919, which declares the general expropriation of all prestigious Habsbourg estates, with other Church properties, and organizes their transfer to private hands. The author accounts that 60 % of the ground belongs in 1935 to little landowners in Bačka and 70 % in Srem (Syrmia).

All settlements made of miserable houses built along straight roads are outcomes of the colonization project. Of course, land conquest and reform actively lead to broader instability. Abandoned fields of forced migrants, violation of rights, savage illegal settlements, and wide population movements participate in a large-scale restructuring. Derived from the old powerful agricultural *domains* or settled by local authorities after some state incentive, new cooperative farms play an increasing role. This emerging rural order lays down the conditions for a demographic explosion.

In its turn, the socialist reform might be seen as a “second phase” of agricultural modernization, after the massive troubles and devastation of WWII. The overpopulation and underproduction of the Yugoslavian rural areas were a considerable challenge to overcome—however “softened” by the German and Hungarian migration and/or expulsion. Obviously, few lands could be gained in western mountains or in the Mediterranean limestone areas. The aim had to be reached—once again—by increasing the agricultural productivity of the Pannonian Plain. During the first 20 years of Tito’s era, the growth often occurs in “experimental” conditions, i.e., across colonization lands where large cooperative units develop. They settle in a local context already open to such experiences. The socialist agricultural *kombinati* adopt the form of a vertical economic integration, rather than horizontal.

The Drava and Sava plains, with Vojvodina, receive dozens of thousands of migrants originating from the central uplands.[5] The conquest is performed by drainage and forest clearing as was already the case in the nineteenth century (points A to D of Fig. 7.15). The large oak wood, east of Županja, is thus reduced into scraps. Entire bogs disappear. They end up replaced by a fertile openfield over loess soils dedicated to cereals and industrial plantations, in the middle of which old meanders are the only traces left by past river wanderings.

Another assumption is that the Yugoslavian postwar “order” has not resulted in a global revolution as in other socialist countries. As Michel Roux argues (Roux 1979), the socialist sector occupied a limited place at the end of the 1970s: It owned only 15 % of all cultivated areas in Yugoslavia, because of an unachieved collectivization policy. Little peasant property persisted in most places. The old rural world had not vanished. Countless private estates opposed a fierce resistance to the collectivization. Vojvodina appears as a special case: Its social sector gathers 40–50 % of all agricultural areas at that time.

The situation has not much changed since then. Checking the validity of the landscape pattern evidenced in Fig. 7.15 implies gathering data from the Serbian Statistical Office on six municipalities: Bač and Bačka Palanka occupy the northern Danubian lands; Šid and Sremska Mitrovica shall represent the middle area of old Syrmia; Bogatič and Šabac are municipalities of the northern Mačva district. Because of few available statistics in Republika Srpska, the exercise has been limited to the eastern part of the spatial window. Data collection leads to meaningful results (Table 7.2).

After many years of planned privatization (Uvalic 2010), Serbia still appears as a double-sided country. The process of sequential transfer from the social to the private sector, through family holdings or capital sharing, has significantly changed land distribution and use. Family farms have increased their already important weight.

[5]Such migration toward the plain was also characteristic of the nineteenth century, when the Serbs originating from the Ottoman Empire found refuge in the northern Danubian districts.

Table 7.2 Utilized agricultural area (ha) in 2010, Serbian cross section of *Syrmia*

		North/Bačka		Middle/Syrmia		South/Mačva	
		Bač	Bačka Palanka	Šid	Sremska Mitrovica	Bogatič	Šabac
Population		**14 676**	**56 438**	**34 957**	**80 881**	**30 262**	**118 040**
Agricultural population		**2 987**	**6 461**	**6 683**	**11 939**	**16 473**	**25 324**
Agricultural land		**26 899**	**48 281**	**41 588**	**56 523**	**30 698**	**60 325**
of which %	Family holdings	*79,2*	*76,0*	*85,8*	*85,0*	*99,7*	*98,0*
	Companies and cooperatives	*20,8*	*24,0*	*14,2*	*15,0*	*0,3*	*2,0*
Arable land and kitchen gardens		**23 971**	**45 145**	**37 980**	**51 423**	**28 038**	**54 123**
of which %	Family holdings	*87,2*	*80,4*	*89,2*	*85,3*	*100,0*	*98,3*
	Companies and cooperatives	*12,8*	*19,6*	*10,8*	*13,7*	*0,0*	*1,7*
of which %	Cereals	*32,5*	*54,4*	*58,5*	*67,3*	*69,5*	*64,7*
	Industrial crops	*65,0*	*35,2*	*34,3*	*19,5*	*6,0*	*3,0*
	Vegetable crops	*1,4*	*6,5*	*4,4*	*7,3*	*8,8*	*10,0*
	Fodder crops	*1,0*	*3,5*	*2,7*	*5,3*	*15,4*	*18,4*
	Orchards	*0,3*	*1,1*	*2,1*	*3,2*	*5,5*	*5,9*
	Others (vineyards, pastures, and meadows)	*10,7*	*4,6*	*5,9*	*6,3*	*3,8*	*4,9*

Source SORS (Statistical Office of the Republic of Serbia 2011)

These microholdings own 80–90 % of all agricultural lands in 2010 in Bačka and 100 % in Mačva. Bringing a general enforcement of land division, the privatization law of March 2001 has planned the liquidation of the social sector. The share of large landowners would have been divided by two in the countryside of Vojvodina, even though the investigation on such a subject might deserve further research.

M. Roux's comments come back when analyzing the agricultural production. Along its gentle slopes, the "old" Serbia is still "obsessed" by cereals as was the case in the nineteenth century. This monopolistic position is only softened by a substantial animal production that entails maintaining plentiful fodder crops. Industrial and specialized crops belong to the Pannonian Plain. Here, history finds long-term continuity and legacy.

This Balkan area has remained split until 1918. With the emergence of Yugoslavia and for 70 years, it has enjoyed the advantages of territorial integration and open space. The early 1990s and the Wars initiate a new confinement era. In 2012, the trigon Bačka Palanka-Badovinci-Županja counts no fewer than ten border crossings (Fig. 7.15). When compared with the Albanian–Montenegrin confines and their single checkpoint south of Shkodër, the contrast is striking. Adopting a radically anachronistic position, one could say that, besides the war parenthesis, the newly independent states had no choice: The spatial integration and economic fabric inherited from the socialist period would not be terminated without terrific effects on local geography.

The creation of new boundaries poses a number of problems. Located on the edges of the Danube or Sava, cities that used to develop across both riverbanks now face critic conditions, although their new border position might bring some economic compensation: Bačka Palanka and Brčko are suddenly cut off from their trans-Danubian and trans-Savian suburbs. In the long term, this is no surprise. Brčko used to be an Ottoman bridgehead… one century ago. Despite this, which fate would be reserved for a border city of 40 000 inhabitants surrounded by a predominantly Muslim neighborhood and claiming to be the biggest "harbor" of Bosnia in front of Neum?

Brčko is a double bridge: With its two road connections with the Croatian side of the Sava, it welcomes two checkpoints in its center. The other "bridge" is strictly continental: The municipality ensures a seamless travel between the two "wings" of Republika Srpska. Brčko is a geopolitical crossroads of the utmost strategic importance, whatever obscure the place might have been during the socialist period. The Serbian military forces early identify the Pasovina linchpin as a central place for carrying out the planned annexation of all territories populated by Serbs. According to the 1991 census, Brčko city welcomes 55 % of Bosniaks, 20 % of Serbs, and 7 % of Croats, and another 13 % group declaring its Yugoslavian identity. The outskirts are more varied, Serbs, Muslims, and Croats sharing different villages. With the war, the Serbian troops enter the district and target urban spaces to provoke mass emigration. The place has to face devastation. By 1994, the military frontline divides the former *opština* into two parts: Federated Muslim–Croat soldiers secure the southern one; the northern side, including Brčko city, is in the hands of Republika Srpska (Jeffrey 2006; Farrand 2011). The *Inter-Entity Boundary Line* (IEBL) becomes the main structuring axis of the district.

The Dayton Agreements initially awarded the area to Republika Srpska, but the place was so disputed that the resolution of the subject has been left open for further negotiation. Sarajevo appeared as the symbol of Bosnia under war, and Brčko is the symbol of peace achievement. Without legal definition, the district sinks into a "no-place" position in the middle of the Balkans. The US army deploys its troops and manages the zone of separation along the IEBL (Crampton 1996). A "Texas road" is established on the right side of the Sava. It replaces the Yugoslavian highway, now inaccessible because of its location in Croatian and Serbian territory. A second road axis connecting Tuzla and Orašje sees a considerable growth in traffic. The units of the *Office of the High Representative* (OHR) in charge of the surveillance give its name to the "Arizona road."

In this backcountry of Brčko heavily affected by the war, soldiers and refugees live side by side awaiting a political solution to serious everyday problems, among which unemployment is surely the most acute. In 1996, in such a deregulated context, a spontaneous market develops beside a NATO military checkpoint, 15 km west of Brčko city. Placed under the *Implementation Forces* (IFOR) protection, the famous "Arizona market" enjoys an immediate success. Countless wooden barracks emerge from grounds that the US troops need to sweep from their mines. The landowners who used to grow wheat and pastures rent their plots to shopkeepers. These are built without any facility support.

No state, no taxes. The market becomes a place of "everything to sell" where criminal activities prosper irrespective of the ethnic or national origin of the actors. Somewhere between Ottoman bazaars and *Wild West* models, parking lots and restaurants, opened brothels, and bars develop alongside little shops. Such illegal trading places are frequent all over the Balkans. The initiative quickly reaches large dimensions because of favorable circumstances. To the international institutions in charge of building peace, Brčko becomes an example of cohabitation to preserve, made of military supervision, economic restoration, and inter-ethnic appeasement. Brčko is a "free city" of black market.

In 1999, the *Arbitral Tribunal for Dispute over Inter-Entity Boundary in Brčko Area* decides to create a special autonomous district of the same name. The Republika Srpska considers the case as a violation of the Dayton Agreements. The former *opština* is now a kind of *condominium* ruled by both delegates of the Republika Srpska and the Federation of BiH, with the assistance of a US supervisor. Officially, the Bosnian sovereignty applies to the district; in practice, it works like an autonomous land. The legal framework that allows political and administrative functioning is extremely complex (Karnavas 2003).

A return of the expelled populations is organized with some good results. Despite an impressive daily turnover, illegal traffics still tarnish Brčko's reputation. Prostitution, refugee traffics, drugs, and exchanges of stolen or illegal goods blatantly flourish. In 2000, in reaction to an international alert, both entities decide to clean up the Arizona market. To wipe out harmful memories, night clubs and other ugly bars close, hoping that the number of stalls and shops evaluated to 2 500 would not eventually fall. Monumental and clean warehouses replace rough roads, canvas roofs, and savage on ground arrays. A corporate tax law enters into legal force in January 2001. All shops are submitted to licensing, fees, and taxation. The market shall eventually move to provide better material facilities and security for owners and clients.

The new market reopens in 2004, next to the old one, thanks to private investments. The free trade place becomes a regular out-of-town mall welcoming flows of tourists mixed with other neighboring buyers originating from Slavonia, mainly. New rules come into force: The transactions take place only daytime. During summer, the mean attendance is evaluated at 20 000–50 000 daily customers. A parallel return of refugees is organized between 2000 and 2004: Their number now reaches around 19 000 in the whole district. Standardization seems under way, while Brčko enters the political science literature as a case study of successful international intervention occurring in a highly conflicting inter-ethnic context.

The story goes on. In December 2011, the government of Republika Srpska finally decides to give up the IEBL as a spatial delineation. This decision is highly significant because both entities now officially recognize the borders of Brčko district. Territorial integration is making significant progress, although many difficulties remain. Such resolution makes any international supervision obsolete.

A withdrawal of the international community would leave, however, immature institutions in charge of the district. The forces at work are still unsteadily balanced.

The association of Brčko "experience" with the emergence of new boundaries is no coincidence. Some argue that local conditions are unique—therefore, Brčko would be a territorial *anomaly*. This statement is of limited scope. Black markets, extra-territoriality, borderlands, military presence, fictitious state apparatuses, and ethnic and confessional intermingling make up for a typical cocktail the Balkans have the knack of enjoying.

Brčko is the latest expression of "savage" districts knocking holes in classical geographic models. It exemplifies the way an informal "game" can breach and eventually serve as an alternative to standard economy. The rules of urban planning are severely questioned by spontaneous local initiatives. Furthermore, military presence is not only the outside face of powerful state interests; it may also create bubbles of deregulation. Such reasoning puts back to back local and global powers always wrestling against each other.

Another way to better understand the success of this "mushroom implantation" is to put it into historical context. Most Savian settlements used to be imperial border colonies, on both sides of the river. Brčko is the most recent outcome of an epic history. North of it, the enclave of Orašje inherits almost the same characteristics. Both localities develop mainly in the 1800s, 1830s, and 1860s, when uprisings force the Muslim population to leave the new Serbian lands of Šabac, Užice, and Belgrade. These migrations strengthen a few late Ottoman military positions and fortresses.

North of the Sava, the colonization is mainly dated 1873. At a time when the legal status of the Military Frontier falls, the economic activities are released from trade restrictions. These edge lands had been used to welcome soldiers and refugees from both sides after each Imperial War. As long as garrisons have been maintained on these marchlands, the open settlements of the countryside have remained limited in number. Through a paradoxical union, the process of national expansion breaches the old military "order" and gives way to a civilian control. The agrarian colonies spread across lands suffering from limited access facilities. Economic development follows. A green light is given to forest exploitation and other industrial investments. Županja and Lipovac build their growth on forest clearing. The settlements follow simple plans: A series of individual houses align along a main street. Private and public buildings adopt wooden architecture. Family homes are built with private gardens. When needs arise, the model is duplicated, as in Lipovac (Fig. 7.17).

This last community assumes a new border situation after 1991. For all villages and cities included in Fig. 7.15, the 1990s are nightmarish years. The war wreaks havoc on the region. The immense destruction and human suffering inflicted on the people leave whole districts to rebuild. The Bajakovo-Batrovci border crossing, between Croatia and Serbia, first opens in 1994. The event might appear insignificant to external eyes. It is a revolution, indeed. The highway between Zagreb and Belgrade was a flagship project of Federal Yugoslavia. Developed in the 1950s, it had allowed millions of tourists to reach Greece or Turkey by car. The last

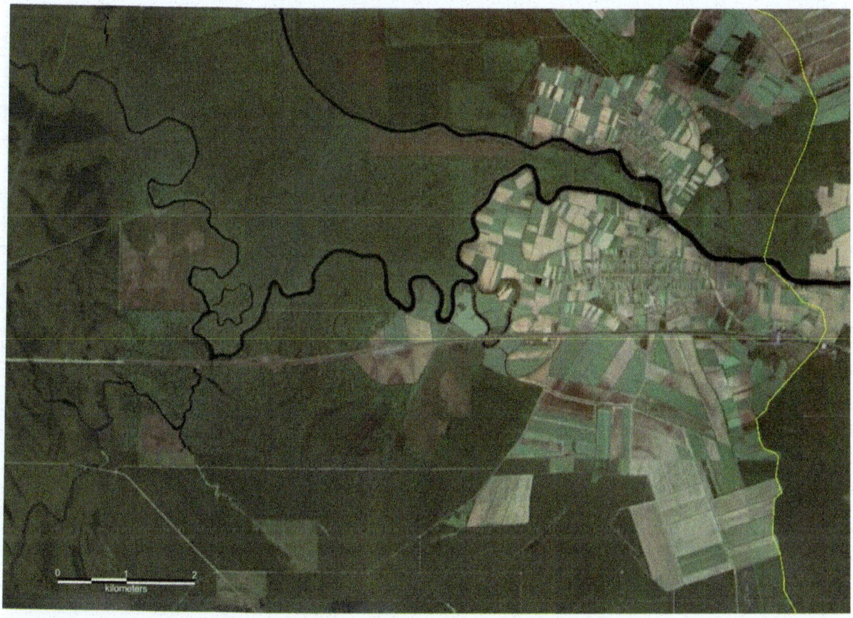

Fig. 7.17 Pannonian frontier and forest at Lipovac. *Source* Google Earth imagery, 04/15/2009

modernization works were achieved between 1984 and 1987. The Yugoslav Wars interrupt land traffic, thereby allowing the development of ferry lines departing from Italy. In 1994, the reopening of the highway is thus a symbol as well as an answer to actual needs. Growing traffic flows follow the progressive normalization of international relations, which leads in turn to the renovation of custom facilities in 2006.

Former Yugoslavia has a wealth of unsettled and contested borders. In the vicinity of Lipovac, the boundary line is entirely new and needs proper spatial definition. But that is without counting the vagaries of the Drina, a river that is reluctant to apply political decisions. Against such unwillingness, people have no other option but to master nature.

When J. Scheda publishes his map of the Austrian Military Frontier (Fig. 7.16), he pictures the boundary close to the city of Belina (Bijeljina), whereas the old village of Popovi immediately abuts the Drina. The author draws interlacing river branches that form a series of islets dispersed among a broad riverbed. Free meanders make up a basin almost 5 km wide. Figure 7.18 shows Popovi and its circular form to the left, opposite to the linear village of Badovinci, on the right side. The western meanders have been conquered by farming, and the international boundary has moved 2 km further east. The main river stream is still submitted to spatial wanderings. Thus, the current boundary sometimes coincides with the river course; some other times, it deviates across sediments and forest areas.

Between Bijeljina and Badovinci, the Pavlović Bridge is readily recognizable. Completed in 1992, the new engineering work spans across the Drina, between

Fig. 7.18 Lower Drina, east of Bijeljina: boundary and uncertainty. *Source* Google Earth imagery, 04/10/2011

Serbia and Republika Srpska, supplementing the 50-km blockade between Loznica and the E70 highway, north of the Sava. The financial support granted to the project comes from a Serbian businessman originating from Popovi but living in Chicago. The ambitions of the billionaire go far beyond this small project. They lie somewhere between philanthropic work, business investment, and the assertion of a prevailing local power against defaulting state authorities. The Pavlović couple intends to weigh on local geography by creating a new city, just next to the bridge and border crossing. "Slobomir, the city of freedom and peace[6]" rises from the ground in 1996. According to the first project, the built area would finally reach 45 ha on the left side of the Drina. Prospects have already accredited an extension of 300 ha, and plans call for the construction of suburbs on both sides of the Drina. The complex would be a mixture of resort center and trade platform. Under the guidance of the "Pavlović Kula" (tower), a "Pavlović International Bank" is operating beside the "Slobomir Pavlović University," next to a church, a hotel, and other sports facilities. All this makes an awkward ensemble positioned in the middle of the countryside. "Radio-Televizija Slobomir" is in charge of the advertising.

Notwithstanding the originality of the project, the place is of strategic importance. Popovi and Bijeljina stand on the road connecting Belgrade with Republika Srpska and Brčko. This way prevents a Croatian transit. Whatever future might be

[6]www.slobomir.com.

reserved for this urban work, the Pavlović ambitions are not so far from what has happened in Brčko. Isolated places benefit from "unwarranted earnings" associated with a "boundary effect" producing self-reinforcement.

In 2003, albeit more conventional, another project emerges. Two kilometers west of the Pavlović Bridge and city, the resort center "Etno Selo Stanišici" (Ethnic Village Stanišici, from the name of its founder) opens its doors to the public. The theme park is dedicated to the Serbian historical heritage. Building on its commercial success, the program constantly grows until it reaches considerable proportions. The monastery of Saint Nicola is established within the estate and consecrated in 2006. Many buildings of historical character are transferred from various places of former Yugoslavia.

The symbol has incredible strength. An area recently submitted to violent struggles has succeeded in becoming a tourist center devoted to Serbian folklore, just next to the boundary. Obviously, the borderlands of Republika Srpska are in search of perspectives. New futures often arise from open opportunities. The next few years will teach us if local geography has been permanently altered. Unlike other central places, international confines easily adopt original development trajectories. The fight between David and Goliath continues: Local chieftains still aspire to build kingdoms. Sometimes they succeed. The importance of "local personalities" is striking in the Balkans. Here, maybe more than elsewhere, economic interests, family networking, and political clientelism collide, and the resulting colorful mixture may be more successful than expected.

7.5 Between Ferizaj and Skopje: Post-Yugoslavian Crossroads

This new example brings us back to former Yugoslavia. Leaving one of the oldest boundaries of the region, the following considers a border system that has emerged in the 1990s. A series of basins, plateaus, high valleys, and mountain chains structures the area caught between (FY)ROM, Kosovo, and Serbia. The window covered by Fig. 7.19 has been Yugoslavian since the inception of the first kingdom. Before that date, this land entirely belonged to the Ottoman Empire. As a consequence, this strategic crossroads went through a process of abrupt fragmentation lately, from the early 1990s onward. After more than two centuries of uninterrupted territorial "cohesion," one would expect to observe the limited outcomes of recent independence in terms of spatial impact. Alas, the autonomy of local communities is deeply rooted in cultural tradition.

Figure 7.19 presents a 3D view that brings information about the geomorphological setting. This provides the opportunity to observe how "high Macedonia" interacts with the neighboring spaces. Skopje is the closest Balkan capital to its national boundary. The city is no more than 15 km from the first border crossing with Kosovo. An impressive barrier "protects" the old Üsküb: the Crna Gora/Malet e Karadakut and Šar Planina/Malet e Sharrit, respectively, reach 1 700 and

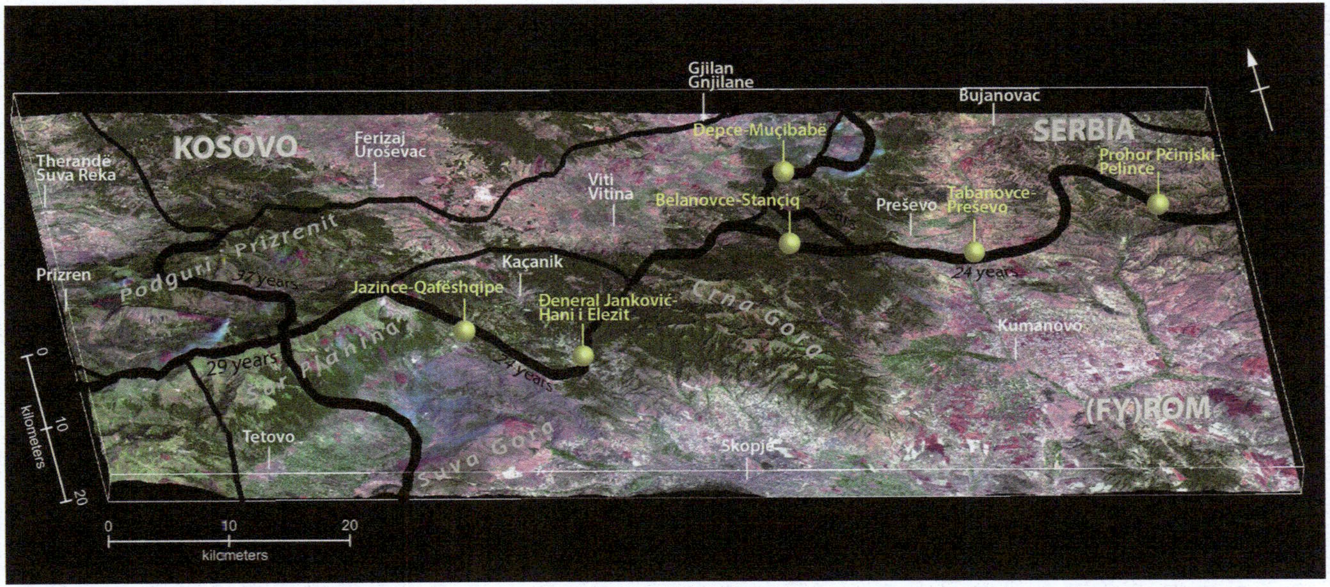

Fig. 7.19 3D view of Skopje and its crossroads. *Sources* USGS, mosaic of two Landsat ETM images, 185/30, 08/22/2000, 185/31, 08/22/2000, WGS84/NUTM34, NASA SRTM data, modified 741 visualization, author's own GIS database

2 500 m. These form a wooded and rocky impediment to any north–south communication. Located on the Vardar River benches, Skopje is at an altitude of 240 m, while Prizren is at 400 and Uroševac 340 m. Between Kosovo Polje and Metohija/Rrafshi i Dukagjinit, the difference in elevation is small. Everywhere around, slopes and inclines are impressive: Low-lying valleys and intensively farmed lands run against wide interfluves made of forests, bare soils, and rocky slopes affected by violent erosion. The Landsat ETM imagery clearly shows the existence of poles apart.

The position as crossroads of transit routes is clearly evident. Reaching Serbia and the *Old Serbia* (another denomination of Kosovo) requires climbing major passes. Two ways are open:

- A first road borrows the Kaçanik gorge, a narrow path overlooking the Lepenac River. This way crosses the Šar Planina and connects with the Vardar valley downstream. The second route adopts a steeper profile while reaching Tetovo directly.
- The main corridor leading to Belgrade follows the Južna Morava valley—a watercourse that takes its source a few kilometers north of Kumanovo. Here is where the famous E75 highway connecting Salonika and Belgrade is to be found.

The area holds another major road. The southernmost way that is to link Kosovo Polje and Prizren—and eventually Albania—stretches along the northern slopes of the Šar Planina.

A brief overview of the existing hydrographic network provides a better understanding of local geographic contingencies. The Metohija Basin is westward-oriented; the Beli Drim watershed and its many tributary streams flow out to the Adriatic Sea. Its waters cross the Albanian Alps before reaching Shkodër through the famous path of Morinë. On the contrary, running across Kosovo Polje, the Ibar, and Sitnica rivers look east. Through the Južna Morava, their waters flow into the Danube before emptying into the Black Sea. The Vardar Basin is a last southward-directed ensemble. The Vardar/Axios River meets the Aegean Sea west of Salonika. Thus, watershed junctions perfectly illustrate the crossroads position of the area under study, although watercourses seldom serve as boundary lines.

The 741 picture allows for an understanding of geographic frameworks. Kosovo, Metohija, and the upper Vardar Valley are isolated districts dominated by mountain chains. The area is densely populated. Low productivity, industrial underdevelopment, advanced mining activities, large workforce sectors engaged in farming activities, demographic surplus caused by high birth rates, and endemic unemployment are long-term characteristics of this Yugoslav *Mezzogiorno* (Roux 1995).

In terms of population, this *Central Balkan* is predominantly inhabited by ethnic Albanians. The eastern limit of the Albanian settlements trespasses the “boundaries” of Kosovo and Crna Gora to include the district of Preševo. All Macedonian municipalities close to the border are in the same situation, especially around Tetovo. However, this general assessment does not resist a scale change. “Majority” is here a concept of low significance. Rural communities and urban districts present reverse characteristics. Declared ethnic and confessional attachments find unstable statistical transcriptions. They vary in time and space. Regarding this subject, the post-1981 population censuses are unreliable. The post-socialist or late Yugoslavian surveys are tarnished by statistical biases arising from boycotts, time discordances, and diverging classifications. Embracing a still changing situation is always difficult, especially when three political entities are concerned. To put it otherwise, describing local contexts is the most efficient strategy for picturing the ethnic–confessional mix.

Prizren and Skopje are good examples. The diversity of populations is high in both cities, although with varying degrees. Mixed communities form multifaceted places. Besides the Muslim Albanians and Orthodox Macedonians, other more or less visible populations must be added: Catholic Albanians, Orthodox Serbs and Montenegrins, Gorani and Torbeshi, Romani, Ashkalis, Vlachs, Kosovan and Macedonian Turks, and Bosniaks. The list is non-exhaustive. Such diversity is clearly inherited from the Ottoman period; although questioned for long (Roux 1995), it has survived the Yugoslavian "interlude."

Getting deeper into the matter is useless. A mixed population is a characteristic of many Balkan lands. Coming back to Fig. 7.19, the specific situation of the Macedonian crossroads is more readily understandable. Unlike what happens so often in other Balkan areas, people have learned to deal with the boundary only recently. This is a substantial difference with Bosnia, Thrace, and Epirus.

The first GIS classification did not consider the former Yugoslavian cross-republican limits as relevant boundaries expressing local autonomy. As a consequence, all linear features drawn in Fig. 7.19 are fewer than 40 years of age. It must be noticed that the northern boundary of (FY)ROM suffers from a rough delineation. Unfortunately, we found no way of retrieving accurate data from the international commission in charge of fixing its exact position.

What was the northern Macedonian situation prior to 1991? As already explained, this *Central Balkan* area used to belong to the late Ottoman possessions of Rumelia. Kosovo, Macedonia and what shall become present Albania were parts of a vast interconnected whole. A few authors note the recursive links between this area and Istanbul, in terms of political affiliations, military participation, and migration (Clayer 2006). The permanent Albanian Muslim presence has played a central role in building and maintaining effective relations with other Ottoman central places, despite the pressure of emerging nations.

With the Treaty of Berlin (1878), Serbia and Montenegro officially escape the *Turkish* dominion, while Bosnia and Herzegovina and the Sandžak of Novi Pazar join the Austrian crownlands. To the east, the Principality of Bulgaria extends its boundaries until the upper Morava. From this date on, Kosovo and upper Macedonia leave their open horizons and find themselves gifted with geographic "myopia." The most Southern Serbian lands stop just near Vranje. The new Serbian–Ottoman demarcation line defines the northeastern limit of what will become Kosovo.

Within the Empire, administrative divisions clearly stress the predominance of Üsküb on all districts in contact with Serbia, Montenegro, and the Sandžak (Fig. 5.12). The city rules a *Vilayet of Kosovo* whose borders have no equivalent today. To the west, the *Kazas* of Gusinje, Djakova, and Lajme encroach on present Albanian lands as they include the adjacent valleys and mountains of Beli Drim. The Vilayet of Scutari is left a relative narrow jurisdiction along the Adriatic. In other words, Üsküb is a garrison city assuming protection over the last Ottoman territories. Prishtinë and Prizren are under its command. After 1878, the main transport axis internal to the Empire uses the path of Kaçanik to reach Prištinë and Novi Pazar. Between Salonika and Belgrade, the old Vardar-Morava corridor loses some of its leadership: The new Ottoman–Serbian customs imply restrictions on international trade and people flows.

Attention must be given to the role played by Preševo between 1878 and 1918: Its *kaza* seems truncated after the last Serbian expansion. The new boundary invents a spatial barrier across the "integrated" Morava corridor. For the first time since the arrival of the Ottomans in the Balkans, this obscure village is cut from its surroundings and shoulders for a while the responsibility of managing international trafficking. The surveillance area covers the new trigon between the Ottoman Empire, the independent Serbia, and the Principality of Bulgaria. Its position is critical. Independence Wars produce predictable effects on frontier settlements: These become the target of spontaneous population movements and receive waves of refugees. The whole Vilayet of Kosovo enforces its Muslim presence through the departure of many "Ottoman Bosniaks" from Bosnia and Herzegovina and the Sandžak under Austrian dominion. Other Albanian-speaking Muslims of the former Ottoman areas follow the same exodus routes.

In 1918, the geographic framework of Fig. 7.19 changes again. The recent Serbian–Ottoman boundary is abolished, whereas Albania secures its national perimeter. The main spatial disruption has moved downstream and divides the first Kingdom of Yugoslavia from Greece. The new authorities adopt an administrative network that has nothing to do with the former historical and geographic divisions (Fig. 5.13). The *Vardarska Banovina* disjoins Metohija, leaving Peć and Mitrovica to the *Zetska Banovina*. It includes all Southern Yugoslavian lands—i.e., the entire present (FY)ROM area—and stretches until Leskovac to the north, thus wiping out the traces of the first Serbian national limits. These are political choices with serious consequences: They approve the internal non-recognition of past national boundaries settled by force and revolutions over the nineteenth century.

The Second Yugoslavia settles new internal frontiers that once again make a clean sweep of all past administrative maps. It is a well-known fact that Tito "reinvents" a nationality (*narodnost*) as a support for the creation of the *Republic of Macedonia*. Kosovo is also "reinvented": Its size and spatial imprint make reference to certain limits of Ottoman kazas. Precisely, the hinterland of Skopje expresses its singularity through the demarcation of international borders that were never more than internal limits until then.

Let us move to the last phase. In 1991, (FY)ROM declares its independence and manages to escape Yugoslavian/Serbian military reprisals. The internal border attains international status. This change tailors the fate of Preševo district. This community had become nominally "Serbian" with the choice of Tito to push southward the limits of the Republic of Serbia within socialist Yugoslavia. After Tito's decision, half a century later, a district largely populated by ethnic Albanians ends up split between (FY) ROM and Serbia.

Two other statements can help understand the one given in Fig. 7.19:

- The region formerly claimed by Bulgaria extends south of a line Kragujevac-Šar Planina. For a long time, Bulgaria refused to acknowledge any "Macedonian" nationality and considered the Slav-speaking populations of the area as an integral part of the Bulgarian nation. WWI and WWII temporarily provide the opportunity to implement these territorial ambitions.
- The other "unexpected" borders of this map are attributable to the spatial expansion of Montenegro during WWI and the temporary gains of Albania during WWII.

As already stated in 3.4.4, the status of Kosovan boundaries is still uncertain today. The administrative boundary line (ABL), which results from the principle of *uti possidetis* and determines the present shape of the *country*, is the convenient expression of a semi-legal territorial system inherited from the war. Behind the fine-sounding words "unsettled political and constitutional status of Kosovo," there is a raw, pragmatic reality that has taken precedence over legality for years, as is often the case when disagreements become embedded in the long term.

Present Kosovo has imposed its "autonomy" and territorial imprint against a background of forced migrations. To welcome the estimated one million Albanian refugees and Albanian-speaking Romani from Kosovo (Ashkalis-Evgjit) during the war (1998–1999), northern Macedonia temporarily changes into a land of asylum, at the same time as remote Albanian Alpine districts. Conversely, approximately 300 000 "internally displaced persons in Serbia," i.e., Serbs from Kosovo denied of refugee status, seek protection, and home in Central Serbia. Kukës in northern Albania, Brazda north of Skopje, and Kraljevo in Southern Serbia become famous camp sites. Such encampments get renowned because they rapidly reach the size of cities before their full eradication a few years later. When these displaced crowds overcome local reception capacities, they have to find shelter deeper in Serbia, (FY) Macedonia, and Albania—even in the outskirts of capitals. Dozens of thousands of refugees are thus forced to find temporary accommodation wherever possible as it always happens when military conflicts are raging.

Unlike other contested ex-Yugoslavian delineations, the new Serbian–Macedonian boundary may appear to escape any questioning. The events show a completely different reality. These land confines have experienced the impact of disorders related to the fights in Kosovo: They quickly turned into a shelter area, laying the ground for military and paramilitary operations, gun traffic, and local rebellions.

In February 2001, an agreement signed with the authorities of the former Federal Republic of Yugoslavia fixes the new demarcation line with Kosovo, west of

Skopje. The oncoming independent Kosovar representatives, however, reject the agreement as illegal. Tensions are rising in the neighborhood of Tetovo and later Kumanovo. With the support of the *Albanian National Liberation Army*, a few village communities refuse to comply. Stenkovec-Brazda, Tanuševci, Vaksince, Lipkovo, and Aracinovo are coming forward to demand their rights. They change into battlefields coveted by the regular Macedonian Army and Albanian rebels. After Sarajevo, Skopje has to deal with fights that take place in its immediate outskirts, even in front of its "walls." The situation is always critical when self-promoted personalities declare "free territories."

The crisis comes to an end with the Ohrid Agreement in June 2001. During the insurgency, 170 000 displaced people have joined the ranks of 360 000 Kosovo Albanians that (FY) Macedonia had already accepted in 2001 (Refugees 2001). Local selfgovernments are back. Along the border, the Albanian ethnic villages inherit a practical *semi-independence* as defined earlier. After the official recognition of the independence of Kosovo in 2008 by (FY) ROM, the authorities of Prištinë sign a second border agreement. The demarcation is adjusted in 2009 (Meha and Selimi 2010).

The valley of Preševo deals with a similar crisis in 1999–2001. An uprising throws the isolated district into a state of unrest for 17 months. The UÇPMB (*Liberation Army of Preševo, Medvedja and Bujanovac*) sets fire to a region populated by 70 000 Serbian citizens of Albanian origin. The Western press sees little value in the event, although this might have deserved greater attention. As in (FY) Macedonia, a bilateral agreement closes the political and military conflict (Covic Plan and Konculj Agreement) without addressing the irredentist/secessionist claims of some rebels. Local autonomy is reinvigorated. The decision is called "decentralizing power to local authorities." This euphemism reflects an already effective state withdrawal.

The conclusion is clear: To preserve peace and keep untouched the existing "boundaries," states have to relinquish some of their privileges in favor of local powers. This method makes it possible to avoid hazardous cascade effects. Of course, these agreements officially entail abandoning the collective use of weapons and transferring—at least partially—the police control capacity to international military contingents. Such a combination of decisions easily explains the subsequent explosion of cross-border traffics.

Land division pulls a plug on the organic links between former Yugoslav districts. A new mixed-border police is created. Serbian officers and UNMIK forces (*United Nations Interim Administration Mission in Kosovo*) take control of a 10-km-wide corridor where traffick is strictly surveyed. Six border crossings come into existence within the "window" of Fig. 7.19. The Depce-Muçibabë custom office opens in 2003; the Belanovce-Staçiq should open soon, if not already.

Of course, the non-recognition of legal boundaries slows down the creation of well-structured border posts. Road networks also have to adapt to the greater international movement of goods and people. Remodeling works take years and require heavy financial investments. By way of illustration, the highway from Albania to Kosovo—under construction—shall soon become the first entranceway to Kosovo as a replacement for Kaçanik. Gravity centers are moving. This Balkan area experiences considerable changes.

Another spatial discontinuity plays a crucial role. After Mitrovica district, the second largest Serbian enclave of Kosovo occupies the upper Lepenac-Lepenci valley, southwest of Ferizaj-Uroševac. UN checkpoints have long been controlling the Štrpce-Shtërpcë district, sometimes assimilated to the "Siriniqi" Valley. Their presence has contributed to the creation of a remarkable enclave before the police forces of Kosovo were to be given the responsibility for maintaining security.

The fate of Štrpce is singular: The city is a winter and summer tourist resort, just down the slopes of the Šar Planina and near the most developed ski station of socialist Kosovo. Local conditions have remained unsteady until recently. Separated Serbian, Albanian, and Romani communities have been severely impacted by the collapse of the tourist economy. During the 1998–1999 conflict, the valley registers little human victims, but most Albanian speakers leave their homes, fleeing to (FY) Macedonia and Albania, and a great many of their houses end up destroyed. Some of them shall come back once the dust settles. The valley is said to claim among the highest rates of return refugees in Kosovo. Simultaneously, many Serbs families originating from Uroševac and Prizren have settled in the area, just before the occupation of NATO. Thus, recent events would have produced limited negative effects on inter-ethnic relations, unlike what happened in other Kosovan places.

In spite of this, Štrpce has to face many problems today, among which administrative and land planning issues come first. Unlike Mitrovica county, which focuses most of the international attention, Štrpce has no territorial continuity with Serbia. It makes any material and economic support from Belgrade much more difficult to supply. The region is always described as "isolated." The question becomes acute when Kosovo declares its independence in 2008. Belgrade asks to create parallel local governments in all areas populated by Serbs. Thus, a series of multilayered competing bureaucracies governs Štrpce:

- Belgrade still has nominal authority on Kosovo; however, its application is currently limited to the Serbian enclaves.
- Supported by the UNMIK and NATO forces, Priština is supposed to exercise authority over any Kosovan area, including the Serbian enclaves. Located only 50 km away from Štrpce, Priština is the ultimate embodiment of a "close national" government.

At a local level, two parallel municipalities (mayors/city councils) oppose each other:

- Supported by the UNMIK forces, Priština considers the first assembly as the official one.
- Belgrade accredits the second one as a separate entity and regards any political opponent as an agent of "collaborationist" forces.

Formally speaking, none of these groups assumes any ethnic affiliation. Disputes are supposed to be fueled by legal issues exclusively. Such litigation shows once again how "local governments" emerge from "upper" administrative and political collapses.

The competition between both levels spills over into the economic field, in a context of fierce corruption. The valley faces the challenge of rethinking its economic future—as well as Kosovo at large. As already noticed, despite a poorly productive farming background, Štrpce has inherited few "heavy" infrastructures (International Crisis Group 2009):

- The *Brezovica resort* is located a few kilometers south of the city. Credited with the most rewarding position and attractive potential of all Balkan mountain centers, Brezovica is competing with Sarajevo and Bansko (Bulgaria). The integrated resort place is the result of a socialist project. Most facilities remain operational, although aging and needing maintenance. Unfortunately, limited tourist movements do not facilitate economic recovery. In order to comply with the vast privatization plan elaborated by NATO and conducted by the *Privatization Agency of Kosovo*, the socially owned enterprise (SOE) in charge of the resort has been put up for sale and waits for private buyers. The asset transfer is stopped for now because of disputes about the legal status of the SOE. In the grand tradition of post-socialist transitions, the question arises of who shall enjoy the purchase benefits.
- The *Weekend Zone* is a place of real estate investment just down the Brezovica resort. Although located within the Šar Planina National Park, the area has undergone notable development over the last decade. Many investors have built illegal villas and chalets, some of them luxurious, just next to small plots erected during the socialist era. The *nouveaux riches* of Kosovo, equally Serbs or Albanians, have had the good taste to establish their summer and winter residence in a preserved environment. In front of them, local governments have to solve serious tax collection problems. Managing cadastre and properties would be for them a sure way to stop drifting land practices.

A short distance away, Ferizaj-Uroševac has well earned its position of new Kosovan border city. The place is located 20 km off the Macedonian boundary. This is the first notable town to be crossed when entering Kosovo from Skopje. Impeded by the combination of harsh environmental conditions—rugged mountains and marshland basin, Ferizaj has grown lately, unlike its competitors. The building of the Prištinë-Skopje-Salonika railway offers the opportunity to gain momentum in the 1870s (Fig. 7.20). Trading shops and little industry spread out in the outskirts of the city. The population census of 1931 describes the former village as a mixed city of approximately 5 000 dwellers.

The second expansion phase begins in the 1950s. Ferizaj-Uroševac manages to drain the economic and demographic potential of the countryside and receives administrative responsibilities. The *naselje* reaches 12 000 residents in 1961 (Savezni Zavod za Statistiku 1965). The development policy of the socialist government changes the city into a small industrial center able to welcome a plentiful local workforce: Bricks and tools factories, construction units, agricultural

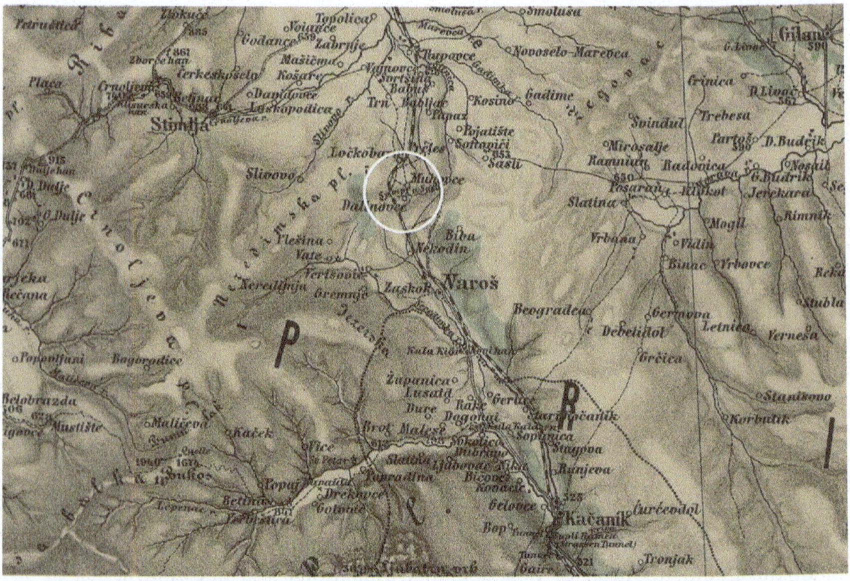

Fig. 7.20 Southern Kosovo in 1876. *Sources* French National Library (Königlisches Kartografisches Militärishes Institut 1876). *At this early date, there is no mention of Ferizaj/ Uroševac on the map*

companies, food manufactures, and other local implementations of purely state-owned enterprises allow the emergence of a modern city, with its clusters of buildings. In 1981, Uroševac reaches 38 000 residents and its district 114 000 (Statistical Office of the SFRY 1983). The most recent census (2011) found that the district gathers about 110 000 inhabitants (Statistical Office of Kosovo 2012). Half of the latter are attached to the city *stricto* sensu. With the war, almost all Serb residents have run away. This allows for the settlement of new incoming Kosovo Albanians. Ferizaj thus climbs to the third place among Kosovan cities, behind Prištinë and Prizren.

Figure 7.21 clearly illustrates the strategic location of Ferizaj/Uroševac. The city neighbors the latest slopes of Podguri i Prizreni and overhangs the southern part of Kosovo Polje. It fixes the hydrographic limit between the Sitnica River basin (running northward) and Lepenica River (flowing southward).

Old testimonies and maps describe Kosovo as a region reluctant to farming (ВемиЋ and Стругар 2007). In former times, a semi-permanent lake occupied the area north of Ferizaj and east of Shtime. It took the name of the most nearby village, Svrćin. The large swamps of Rahovica-Sazlija and Robovac used to stretch along its banks. The overall area covered by water reached an estimated 10 000 ha. No map postdating the 1850s mentions these swamplands any more. The southern Kosovo plain thus becomes agricultural land lately, thanks to an active drainage. The number of rural settlements increases. The 741 imaging of Fig. 7.21 makes it

Fig. 7.21 Ferizaj-Uroševac and its surroundings. *Sources* USGS, mosaic of two Landsat ETM images, 185/30, 08/22/2000, 185/31, 08/22/2000, WGS84/NUTM34, High Pass Filter merged 741-Panchromatic visualization

possible to imagine the extent of these former basins (dark areas): Cultivated with dry cereals and fodder plants or left to open pasture, most current arable plots concentrate here. As in Albania, the extreme parcel division is typical outcome of unachieved land reforms.

As reported by many observers (Beilock 2005), Kosovo has developed a high degree of specialization in agriculture, and the basin of Ferizaj is a perfect illustration of this. A complex world emerges from the conjunction of old demographic pressures and chronic industrial underdevelopment. The departure of Serb populations and the settlement of Albanian migrants have disrupted old geographic balances. Many dwellings have been destroyed during the war; others have been rebuilt, chiefly in large cities, therefore strengthening the urban expansion. Houses and fields have changed hands, but the process still needs to be studied. Such confusion necessarily affects agricultural productivity and performance. Obviously, a combination of factors place limits on the development of the countryside, among which:

- The extreme land division and the limited average size of agricultural holdings. In the district of Ferizaj, in 2008, about 80 % of all agricultural holdings had fewer than 1.5 ha to farm (Statistical Office of Kosovo 2010).

- The suburban development encroaching on arable lands.
- The wastelands and uncultivated plots abandoned by migrants working abroad.
- The disputed propriety claims arising from randomly referenced exchanges of real estates after the war: the contested uses of land.

All things being equal, Kosovo maintains its "leadership" in European agriculture —primary activity is still the main economic sector—although the dismantling of cooperative farms has raised major concerns. In response, the Ministry of Agriculture has decided to edit a *Rural Development Plan* initiating major reforms. Despite the absence of one-third of property owners and the lack of reserve lands, the aim is to achieve land consolidation. Significant problems stand up against the authorities: Checking property rights after 20 years of general breakup, planning the deregulation and reactivating the cadastre is no easy task. The question is almost the same as in Albania: The socialist collapse has wiped out the administrative services that were core institutions of the former regime. The process is further enhanced here because the central authorities are assimilated to *colonial* powers. The privatization of the SOE adds another difficulty: Is any land transfer from the "social sector" to private owners legal? Justice courts find here endless cases to discuss.

In this environment, the countryside offers diverse landscapes. Production areas and irrigated fields concentrate in the most populated places, i.e., at the lowland/mountain interface. The phenomenon is blatant, especially in the outskirts of cities. The demographic upheaval of recent decades ushers in a concentration of people in the few places likely to employ displaced households. Through this and other channels, the population of Kosovo manages its rural–urban drift. The development of Ferizaj is a characteristic of such a conversion (Boussauw 2012).

Figure 7.22 illustrates the different stages of Ferizaj's growth. The Ottoman "city" of the late Tanzimat and First Yugoslavian periods is gone, leaving a few

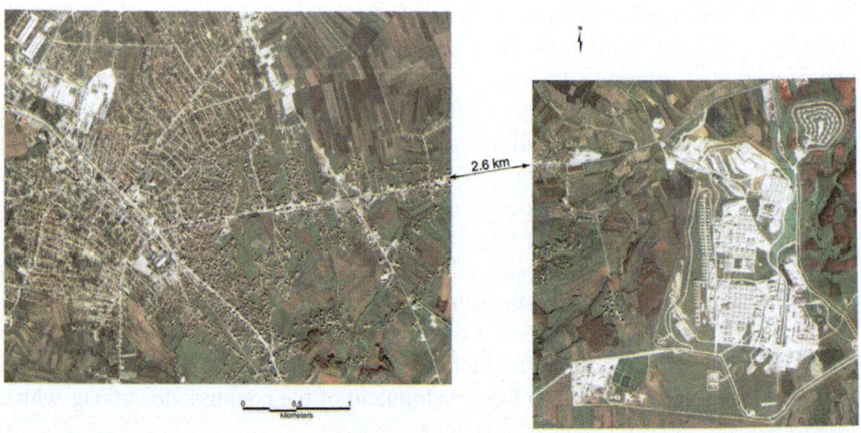

Fig. 7.22 The four faces of Ferizaj-Uroševac. *Source* Google Earth imagery, 11/18/2004

scattered buildings. Instead, the socialist town occupies large areas where administrative blocks, plants, and dwelling complexes gobble up the arable lands of the plain. Formal planning has imposed limits to the urban growth while increasing population densities. Housing blocks and factories make up an organized framework.

From the 1990s onward, another city rises from the ground. Spontaneous residential suburbs develop alongside the main eastern road axes. Despite the departure of the Serbian population, the place grows in line with the settlement of migrant flows. Taking profit of repatriated funds of people working abroad, "German"-style villas form clusters. The "new city" has already doubled the old one in 2004, when QuickBird provides its snapshot. Buildings scatter in the countryside as far as to integrate former hamlets.

A fourth city deserves special attention. As a symbol of its involvement in solving the Yugoslavian crisis, the US Army has built the Bondsteel Camp 5 km east of Ferizaj, whose dimensions are impressive. This military complex created in 1999 occupies a surface equal to the neighboring city. Since the early 2000s, it welcomes thousands of soldiers of the KFOR units. Ferizaj directly enjoys benefits from the presence of the American base. Despite rumors about a potential closure of the Camp, the US military complex is still described as the widest in Europe. With such a close presence, Ferizaj and its surrounding region, until Skopje, can feel relatively "safe." Apart from anything else, the choice of this Central Balkan area is symptomatic of a highly strategic position.

Concluding with this example, it should not be forgotten that, as a consequence of their respective "independence," Bosnia, Kosovo, and Montenegro—to a lesser extent—have become microstates whose global remoteness has never been more acute. At higher scales, this is evidenced by the concept of "uncertain centralities." Decades are needed to build new territorial balances and restore some economic capacity. Adversely affected by limited resources and "heavy" socialist pasts, these weakened bodies provide conditions for the ascent of local versus mobster powers in search of immediate economic efficiency.

7.6 Misleading Messages from Balkan Heartlands

Examples such as these abound. A large variety of cases would deserve appropriate investigations: old versus new Balkan frontiers, contested versus assumed border situations, areas affected by ethnic–confessional troubles versus districts supporting a "clean" integration, and poor and isolated mountains versus wealthy industrial plains. Blind spots and dead-end spaces demonstrate the relevance of a high-scale GIS analysis to illustrate and explain composite geographic frameworks.

Throughout this research, the combination of methods shows tangible results. Although field surveys are usually difficult to achieve, especially in border districts, remote sensing imagery constitutes an invaluable source of information. Image analysis is particularly efficient in helping people understand more about "secret" places. Lessons learned are all the more essential as boundaries are almost invisible from space—linear objects are seldom on earth—and studying frontiers required looking beyond appearances. The acknowledgment is even more valid when dealing with historical boundaries.

The post-1990s transition period seems almost complete (Herrschel 2007). Renewed public and private institutions are now trying to make strides in normalizing overcontrolled and poorly studied districts, whereas recent struggles and open conflicts slowly sink into memories. Many regions long maintained in "backyard" positions are ready to dust off the principle of mutual recognition. Industrial, agricultural, and population censuses are to be published soon that should allow measuring the extent of changes.

Borderlands reflect the great paradoxes of our times. Providing information is already a way of extracting them from their remoteness. This new "right of scrutiny" on border issues allowed by more flexible access rules to the relevant information lets us get over past clichés and discover complex systems influenced by geography and local conditions. Such a great variety of cases suggests that there are only exceptions to the rules governing the future of border areas. Let us, for instance, consider the Danubian demarcation between Bulgaria and Romania. The river is one of the least contentious Balkan boundaries. However, is it possible to conceive that one of the most strategic plants ever created has been established just next to the borderline? The nuclear power station of Kozloduy is located on the southern Danube riverbank, stuck in the middle of rural landscapes, dominating both Romanian and Bulgarian countryside areas. How has such an aberration—in strategic terms—come into being, while hundreds of kilometers of border areas have been cleaned from any human presence on the pretext that homeland needs to be secured?

Another singularity may convince us to measure the extent to which border issues are ridden with contradictions and paradoxes. Although he ordered to secure hundreds of kilometers of boundaries from civilian intruders, people usually ignore that Enver Hoxha did not hesitate to establish his summer villa in the village of Tushemisht, along the shore of Lake Ohrid, about 1 km away from the Macedonian borderline. In some cases, protecting the country from external threats with bunkers and heavy weaponry seems not to contradict certain recreational uses of land. The nearby city of Pogradec was already a summer escape for wealthy Albanian families in the 1920s and 1930s, but was converted into a popular resort place under Hoxha's rule (Starova 2003). Having a bath, fishing or sunbathing on the beach was allowed under the surveillance of border guards boarded on powerboats or perched on watchtowers. In this Balkan version of *Baywatch*, waves do not suit to great surfing, rescuers are less sexy, and defense systems replace beach huts—in case anyone would accidentally cross the invisible "blue" boundary of this magnificent lake.

Obviously, mapping frontiers reduces knowledge uncertainties; it also reveals unexpected situations that deeply alter our mental models. More detailed research is still possible in a limited number of cases benefiting from years-long surveys.

References

Backer FJ (1926) Protocole des conclusions de la commission de délimitation de la frontière gréco-turque. Macris, Athens
Beilock R (2005) Rethinking agriculture and rural development in Kosovo. South-Eastern Europe J Econ 2:221–248
Béthemont J, Sivignon M (2002) Effets de frontière en Thrace occidentale. Revue de géographie de Lyon 77(4):367–376
Blanc A (1963) L'évolution contemporaine de la vie pastorale en Albanie méridionale. Revue de géographie alpine 51(3):429–461
Boussauw K (2012) Challenges, threats and opportunities in post-conflict urban development in Kosovo. Habitat Int 36:143–151
Civici A (2010) Dé-collectivisation de l'agriculture albanaise (1989–2002). Une transition spécifique ou identique à celle des autres pays de l'Est? Editions Universitaires Européennes, Saarbrücken
Clayer N (2006) Aux origines du nationalisme albanais: la naissance d'une nation majoritairement musulmane en Europe. Karthala, Paris
Crampton J (1996) Bordering on Bosnia. GeoJournal: An Int J Hum Geogr Environ Sci 39 (4):353–361
Darques R (2004) L'Albanie en transition: mutations démographiques et recomposition territoriale (1989–2001). Espace, Populations, Sociétés 3:559–575
Doll B (2003) The relationship between the clan system and other institutions in Northern Albania. J SE Eur Black Sea Stud 3(2):147–162
Drejtoria e statistikës (1959) Anuari statistikor i RPSH. Drejtoria e statistikës, Tiranë
Drejtoria e statistikës (1973) Vjetari statistikor i RPSH 1971–1972. Drejtoria e statistikës, Tiranë
Drejtoria e statistikës (1991) Vjetari statistikor i Shqipërisë. Drejtoria e statistikës, Tiranë
Durham E (1909) High Albania. Arnold, London
Erkânıharbiye-yi Umumiye (1902) Rumeli-i şahane haritası
Farrand RW (2011) Reconstruction and peace building in the Balkans: The Brčko experience. Rowman & Littlefield, Lanham
Grandits H, Taylor K (eds) (2010) Yugoslavia's sunny side: A history of tourism in socialism (1950s–1980s). Central European University Press, Budapest, New York
Gruber S, Pichler R (2002) Household structures in Albania in the early 20th century. Hist Fam 7 (3):351–374
Herrschel T (ed) (2007) Global geographies of post-socialist transition: geographies, societies, policies. Routledge, London: New York
INSTAT (2000) Regjistrimi i pergjithshem i njesive ekonomike bujqesore, Qershor 1998. INSTAT, Tiranë
INSTAT (2004) Regjistrimi i popullsisë e banesave, Shkodër, 2001. INSTAT, Tiranë
International Crisis Group (2009) Kosovo: Štrpce, a Model Serb Enclave? Europe Briefing 56
Jeffrey A (2006) Building state capacity in post-conflict Bosnia and Herzegovina: the case of Brčko District. Political Geogr 25:203–227
Karnavas MG (2003) Creating the legal framework of the Brčko District of Bosnia and Herzegovina: a model for the region and other postconflict countries. Am J Int Law 97(1): 111–131

Kaser K (2012) Household and family in the Balkans: two decades of historical family research at University of Graz. Lit Verlag, Berlin
Kiepert H (1862) Das Fürstenthum Zrnagora oder Montenegro. Zeitschrift für allgemeine Erdkunde, 13-1862 edn. Reimer, Berlin
Königlisches Kartografisches Militärishes Institut (1876) General Karte von Bosnien, der Hercegovina. von Serbien und Montenegro, Vienna
Lampe JR, Jackson MR (1982) Balkan economic history, 1550–1950. Indiana University Press, Bloomington
Lane RW (1923) Peaks of Shala. Harper & Brothers, London-New York
Meha M, Selimi B (2010) The challenges of the border demarcation Kosovo-Macedonia. Paper presented at the FIG Congress 2010, Sydney, Australia
Peyre M (1935) La réforme agraire en Yougoslavie et ses effets dans les provinces de Bachka et de Srem. Annales de géographie 44(247):58–64
Refugees USCf (2001) World refugee survey 2001: An annual assessment of conditions affecting refugees, asylum seekers, and internally displaced people. U.S. Committee for Refugees, Washington D.C
Reinkowski M (2003) Double struggle, no income: Ottoman borderlands in Northern Albania. Int J Turkish Stud 9:239–253
Roux M (1979) Problèmes agraires yougoslaves. Etudes rurales 73:97–124
Roux M (1995) Les Albanais en Yougoslavie. Maison des Sciences de l'Homme, Paris
Rugg DS (1994) Communist legacies in the Albanian landscape. Geogr Rev 84(1):59–73
Savezni Zavod za Statistiku (1965) Popis stanovništva 1961. Rezultati za naselja. Stanovništvo i domaćinstva u 1948, 1953 i 1961. Beograd
Scheda J (1856) General Karte des Oesterreichischen Kaiserstaates. Expeditions-Bureau, Vienna
Schon R, Galaty ML (2006) Diachronic frontiers: landscape archaeology in Highland Albania. J World-Syst Res 12(2):230–262
Selenica T (1928) Shqipria më 1927. Sthypshkronja Tirana, Tirana
Sivignon M (1977) Itinéraire de géographie rurale en Albanie. Méditerranée 28(1):3–17
Starova L (2003) Le rivage de l'exil. Editions de l'Aube, La Tour d'Aigues
Statistical Office of Kosovo (2010) Agricultural Household Survey 2008. Pristina
Statistical Office of Kosovo (ed) (2012) Population and housing census in Kosovo. Preliminary results. Pristina
Statistical Office of the Republic of Serbia (2011) Municipalities and regions in the Republic of Serbia. Belgrade
Statistical Office of the SFRY (1983) Popis stanovništva, domaćinstava i stanova 1981. Savezni zavod za statistiku, Belgrade
Uvalic M (2010) Serbia's transition: towards a better future. Studies in Economic Transition, Palgrave Macmillan, New-York
Вемић М, Стругар М (2007) Атлас старе Србије: европске карте Косова и Метохије. Svetigora, Cetinje-Belgrade

Chapter 8
Broken Lands and Hurt Populations: Eastern Macedonia

Abstract This last chapter focuses on the Greek–Bulgarian frontier as a significant example of land transformation over time. The Rhodope region is examined through the perspective of historical demography from 1912 to present day. Thematic mappings show how Bulgarian and Greek border districts have managed to overcome a succession of major territorial changes and crises after the collapse of the Ottoman Empire.

Keywords Population dynamics · Thematic mapping · Ethnicity · Confession · Land partitioning · Iron Curtain · HGIS methodology · Data comparability · Refugees · Migrations · Border communities

Spatial analyses need to be completed with an alternative approach directly involving the *human variable*. What is the price to pay for these countless frontier fluctuations? Putting a human face to the issue is the trickiest part. Even dozens of years after the events, the subject still arouses intense passion and emotion. Bitter experiences echo painful memories. Countless "passively or actively" destroyed places pervade the environment, in such a way that few Balkan confines eventually deserve the name "soft borders." Most boundaries have been cut with a scalpel—the most appropriate tool for creating linear artifacts. With so many victims and those left behind, Balkan borders are "hard" demarcations, doubtlessly.

Population dynamics are a complex combination of opposing forces. One cannot understand Balkan demography without considering the relentless political cycles of local uprisings-fierce military repression. Wars have terrible outcomes on local communities. This "destroying machinery" has been running for decades (Prévélakis 1994; Sivignon 2009; Glenny 2012; Mazower 2000; Hall 2011). In old times, rural residents were among the first victims of the insurrections run by Andartes, Komitadji, Chetniks, and all other "resistant fighters" against the "Turkish occupation." The underdevelopment of the countryside is not only the result of a decaying Ottoman Empire unable to catch up with the vigorous economic growth of Western Europe (Chirot 1989; Lampe and Jackson 1982).

© Springer International Publishing Switzerland 2017
R. Darques, *Mapping Versatile Boundaries*,
Springer Geography, DOI 10.1007/978-3-319-40925-2_8

The creation of new borders and the emergence of fringe territories cause population movements merely studied, owing to the lack of sources (Kasaba 2010). Some observers recount here and there the departure of thousands of families forced to leave their home and to settle in the Balkan countryside or in Anatolia. A large number of multilateral transfers between Greece, Bulgaria, Serbia, Bosnia, Turkey, and Albania are reported, before and after each boundary shift. Historical and demographic sources are incomplete and often confused. When necessary, entire village or city communities have been urged to temporary or permanent exile, whereas other semi-nomadic groups—Vlachs, Gypsies, Sarakatsani—were gradually exhorted to hang their hat and to choose reliable homelands. Spontaneous movements occur side by side with rigorously planned transfers (Treaty of Lausanne, for example) (Andréadès 1928).

This "chaos" results directly from what has been called "uncertainty." Measuring the human impact of such disorders: Is it even within reach? The challenge of this chapter is to build a powerful GIS tool capable of giving an objective basis to stormy debates. A thematic mapping based on reputable sources shall complement the cartographic work already achieved. But how could we distinguish "reliable" sources from other specious data?

A map, even old, is somehow "lying" because any graphic representation emerges from a selection of information. Statistical analyses deserve the same defaults. Statistics even reach higher suspicion levels. Neither impartial nor systematic, they are outcomes of deliberate biases based on ideological grounds.

Since their establishment as nation-states, most Balkan countries have been developing a tendency to self-examination and mutual confinement. Soon after their independence, despite persistent skirmishes, the new authorities begin to produce official statistics aimed at showing how they can master their own destinies. Over the spoils of Ottoman Rumelia, statistics become a playground where states can engage in fierce competition.

Against this background, the existing data are of contentious and varying qualities. Over the nineteenth and twentieth centuries, official statistics never stopped being propaganda and war instruments, especially in the hands of the new *kingdoms*. In Rumelia, the main source of information was the Ottoman government itself. Unfortunately, little analytical data of the Ottoman era have survived until today, despite the efforts of public officials to produce financial, administrative, cadastral, demographic, and economic information.

There are many methodological flaws between national institutes that may explain why international statistical comparisons are so difficult to conduct. The most famous example relates to the ethnic-confessional attributes of populations, as they were defined at the end of the nineteenth century. The *millet* confessional system governs the late Ottoman *counts*: People are categorized according to their declared religion (Behar 1998a, b; Mentzel 2000; Shaw 1978). For the Sultan and his civil services, confession is the only formal way of identifying individuals, regardless of any language, origin, or political allegiance. In response, independent state authorities prefer to separate nation from confession. Making such antagonistic perspectives compatible seems an impossible objective. How could these *sensitive* data be decrypted and mapped?

Studying and assessing the impact of versatile boundaries in human terms is a high-risk venture. To this end, special attention shall be given to all known sensitivities, as far as possible. The forthcoming statistical analyses result from a lengthy information gathering. It should not be forgotten that past censuses, counts, and surveys are irreplaceable tools for researchers, despite their actual or perceived inadequacies. Rejecting their use would lead us down a blind alley. The best way to solve data incompatibility is to gather as much detailed statistical information as possible. This is a basic principle: The higher the aggregation level, the stronger the likelihood of error. With very detailed statistical information, further data treatments become possible.

With such rules in mind, the Greek–Bulgarian *interface* has been identified as the most appropriate area to study. As we have already mentioned, the *new* boundary between Greece and Bulgaria is settled in 1913. Any existing data over the past century have been gathered without noticeable restriction, leaving apart access problems related to rare materials preserved in secured archives. A large GIS database has been designed. This tool covers a rather small area at the crossroads of Eastern Macedonia and Rhodope. The most challenging part of the process consists in developing an appropriate methodology.

In this part of Rumelia lately "liberated" from the Ottoman rule, *new* national frontiers have created the conditions for the development of diverging evolutionary patterns among places formerly linked by a common destiny. Although today one can pass through a series of empty places surrounded by mountains and forests, the situation was reversed a few decades ago. Rhodope and Eastern Macedonia appeared in a different light. They were living countryside areas, whose horizons were not limited by oppressive boundaries.

The big breakthrough comes with the Balkan Wars and WWI. Bulgarians, Macedonians, Vlachs, Greeks, Muslims, Sarakatsani, and Pomaks lose control of the situation and find themselves caught up in the upheaval of a global crisis. A second phase starts in the 1940s, when the communists take power. The remaining rural and urban communities are disempowered by global militarization. The process of relegation grows as fast as the boundary becomes more supervised and controlled (Schindler et al. 2011). The *Iron Curtain* period exemplifies the climax of antagonism. Dividing walls lead to strictly opposed planning models.

As a result, it does not seem farfetched to consider the year 2007 as a historic moment. With the European integration of Bulgaria and Romania, the Balkans discreetly enter a new era. Caught up into *globalization*, few people identify the historical dimension of the "unification" scenario promoted by Brussels.

The Rhodope area had been addressed previously at a macroscale level (Sect. 5.1). The approach, however, had not emphasized the various facets of the geographic divide between Greece and Bulgaria. What has happened to the old Ottoman Vilayet? Will the socialist collapse and the European integration process create the conditions for a new era of *convergence*?

8.1 Why Eastern Macedonia? The Choice of a Research Field

Having in mind the sticky problem of data sources—accessibility, reliability, and comparability across space and time—a section of the former Iron Curtain has drawn our attention. The area extends from Lake Doiran at the Greek–Macedonian border to Central Rhodope (Figs. 8.1 and 8.2).

Three main reasons have led us to focus on this territory composed of five adjacent *districts* (Serres, Drama, Xanthi, Blagoevgrad, and Smolyan):

1. Running a high-scale analysis supposes an in-depth knowledge of the field. Northern Greece and Southern Bulgaria are spaces that I have been visiting many times over the last decades. Exploring as many places as possible was a sure way to establish and develop contacts with local administrations.
2. Data collection has been achieved in Salonika, Sofia, and Athens, mainly. Through years of work and literature search—in both Greek and Bulgarian—a consistent body of statistics has been gathered. Despite all efforts to access the original Ottoman sources, the search for Turkish archives ended in stalemate, unfortunately. Months were sometimes necessary to negotiate access rights: Many documents remain "sensitive" from a historical and political point of view. Other unpublished data require on-demand compilation by the competent statistical departments. In other words, to some extent, a "where data exist" precondition has guided the research work, and these borderlands have been found to satisfy eligibility criteria.
3. The informed administrative units of Rumelia determine the size and shape of the study area. Integrating the late imperial surveys into the work was simply not an option. This reference is precisely what makes any further geographic development understandable. It brings light on a primary *transition* stage that has far-reaching implications. Looking back over a century, the approach does not imply that the Ottoman *Vilayets* formed coherent geographic "units"—the *Tanzimat* reforms provide evidence they did not. In brief, the latest Ottoman data and maps provide the starting and focal point for further research developments.

The study area is placed today under the governance of two chief cities, Salonika and Sofia. At the end of the nineteenth century, the Macedonian metropolis was the second or third Ottoman center in the Balkans, behind Istanbul and side by side with Edirne. Sofia was no longer the provincial town it used to be under the Ottoman rule. As the new capital of the young Bulgarian state, the city had entered an active growth phase.

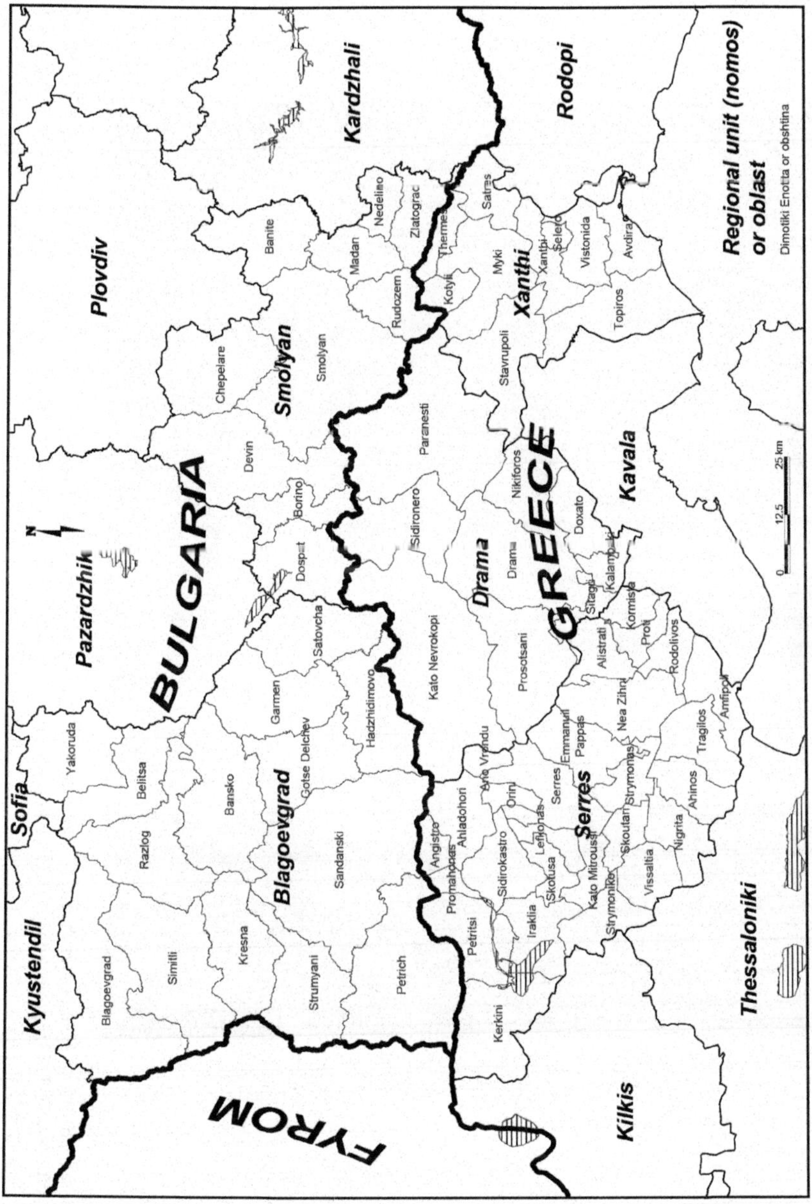

Fig. 8.1 Administrative divisions in 2015. Five districts. *Source* GFK Geomarketing

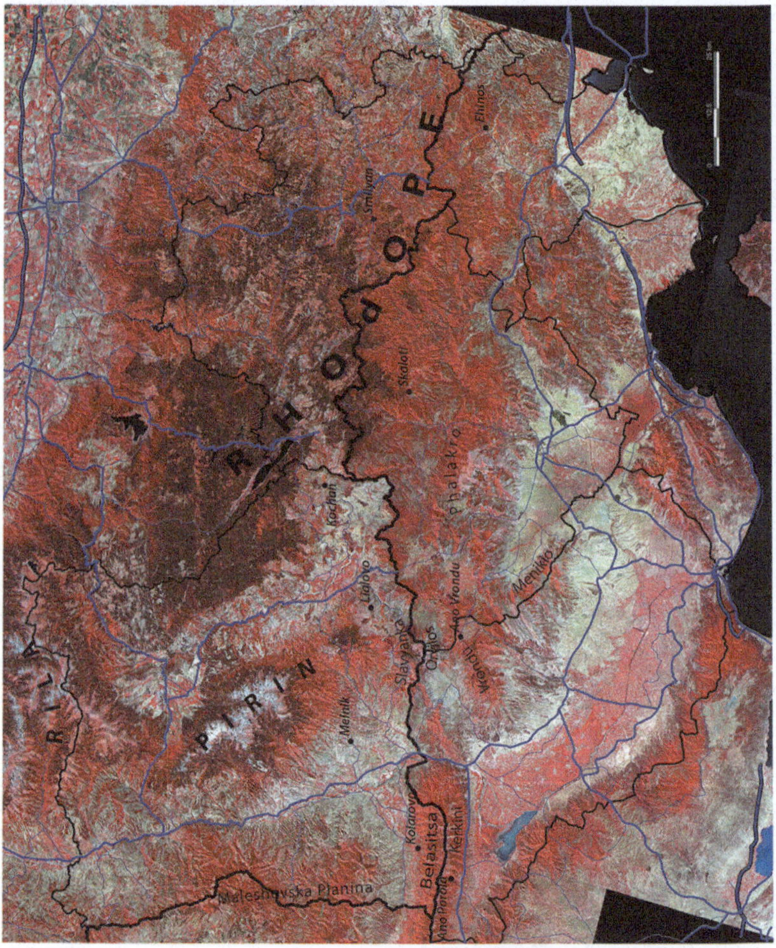

Fig. 8.2 Landsat MSS view. Five border districts. *Source* NASA Landsat program, USGS. Landsat 2 MSS orthorectified imagery (08/22/1977, 08/21/1977, 07/16/1977), WGS84/NUTM35 projection, ATCOR3 atmospheric correction, NIR visualization. *The nine villages studied in Sect.* 8.6 *are to be found alongside the boundary*

Passing through Eastern Macedonia[1] and Western Rhodope (the latter belonging to Thrace), two main rivers establish a connection between the Aegean Sea coast and the Balkan hinterland: the valleys of Strymon–Struma and Nestos–Mesta. The underdevelopment of these geographic confines has long been recognized, with a countryside entirely dedicated to forestry and extensive livestock farming.

8.2 Documenting the Impossible: The Methodological Challenge

Since the beginning of the twentieth century, as a result of an advanced land partitioning, few mappings cover more than one Balkan state. When macroscale works exist, they fall in the area of economy and political science and suffer from consistent methodological biases (reliability and statistical processing). Geographers commonly consider these global analyses as highly questionable, but the criticism also refers to the poor conclusions such studies eventually produce in the eyes of field practitioners.

Shall we assume that any "unified" cartography is ineffective (Gartner and Ortag 2010)? Is this transnational project simply achievable in methodological terms as it applies to countries that think and act as independent entities?

8.2.1 Scale Level Mapping and Data Collection

The decision was made to build the most accurate spatial and statistical database. In fact, many attempts of transboundary mapping adopt the district level as a reference framework. These small-scale studies produce results of no great significance because of aggregation problems. Avoiding this pitfall implies to create a fully functional geodatabase covering Greece and Bulgaria at different scale levels (St-Hilaire et al. 2007):

(a) Greece:

- district (*nomos-νομός, regional unit-περιφεριακή ενότητα* after 2010),
- district sector (eparhia-επαρχία, *dimos-δήμος* after 2010),
- municipality and commune (these labels evolve in step with reforms, *dimos-δήμος* and *kinotita-κοινότητα* until 1998, *dimotiko diamerisma-*

[1]Formerly known as *Pirin Macedonia*.

δημοτικό διαμέρισμα and *kinotita-κοινότητα* from 1998 to 2010, *τοπική & δημοτική κοινότητα-local & municipal communities* after 2010),
- and hamlet (*ikismos-οικισμός*)[2]

(b) Bulgaria: district (*oblast-област*, formerly called *okrug-окръг*), commune (*obshtina-община*), "populated places" or localities (*naseleni mesta-населени места*) divided into cities (*gradove-градове*), and villages (*sela-села*)[3]

(c) Ottoman Empire: province (*vilayet*), district (*sandjak*), district sector (*kaza*), and commune-village (*nahija*).

Reliable mappings are based on units of comparable surface on both sides of the boundary. Until 1998, Greece's municipalities-communes were much smaller than the Bulgarian ones. Significant surface variations prevented any comparison between "municipal" levels. The administrative reforms of Kapodistria (1998) and Kallikrati (2010) have solved the problem. *Dimi* and *obshtini* (or *dimotikes/topikes kinotites-obshtini*) are now relevant inputs.

About **800 places** (the most detailed units) are reported in 2011 across the five districts under focus. Experienced observers would denote a slight bias in the spatial distribution of these places. This phenomenon arises from a difference in scale: The *naseleni mesta* are the smallest Bulgarian census units, but they overcome the size of the Greek *ikismi*. As a result, maintaining some equivalence between countries has led us to ignore the *ikismi* and adopt the *topikes* and *dimotikes kinotites* of 2011 (*dimotika diamerismata* of 2001) as a basic background map (Sect. 8.5).

This GIS database has been prepared with a holistic perspective in terms of population change. Precise data allow any further combination, aggregation, and subtraction imposed by repeated changes of administrative boundaries over the years. Data accuracy is essential to our goal. This is the only way of connecting fieldworks with statistical analyses. At lower scales, observing the complexity of the region, its high fragmentation, remarkable ethnic mix, and fluctuating urban–rural relations is almost impossible. Accuracy and reliability are two faces of the same coin.

---

[2]The so-called *Kapodistria* administrative reform (1998) has replaced the 6 000 urban and rural Greek communities (*dimi* and *kinotites*) with approximately 1 000 entities. Between 1998 and 2010, the ELSTAT (Hellenic Statistical Authority, formerly NSSG) has maintained two infra-territorial levels that are neither administrative nor political authorities: the *dimotiko diamerisma* (municipal sector) and the *ikismos* (hamlet). Officially, apart from the ELSTAT needs, these two subdivisions have no legal existence.

The *Kallikrati* plan (2010) reshuffles the cards once again. The number of *dimi* (first level units) is reduced to 325.

A last level (the *eparhia-επαρχία*), which was eliminated in 1998, has not been informed. *Eparhies* might be seen as by-products of the former Ottoman *kazas*. This territorial level has been rehabilitated by the Kallikrati plan: The "new-new" municipalities reach the same size as the pre-1998 *eparhies*.

Anyway, reform after reform, the administrative framework loses its consistency.

[3]There was an intermediate level between *oblasti* and *obshtini*: the *okolya-околия*. See below.

8.2.2 Mastering the Administrative Variability

Over the course of the past century, constant administrative reforms have been a major hindrance to historical mapping. Once minor, at other times systematic, hundreds of changes have affected the living environment of village communities and provinces, to such a point that any inter-census mapping implies a painful work of data recombination. Seen with an external eye distance, municipalities and communes suffer from significant size and shape distortions. There is a widespread yo-yo effect, which makes local communities assemble, divorce, or dissolve in larger groupings. Growing and shrinking accordingly, local governing units play the role of pawns in a political chess game. *In fine*, such enhanced variability gives currency to the idea that central authorities are engaged in low regulation. "Independent" local powers would need no specific, i.e., "geometric," definition of their territories (Dimitropoulos 2008).

These systematic changes originate in the national integration process (Péchoux 1977; Koulov 1992). In the months or years following their "liberation" from the "Turkish occupying forces," local communities find themselves free of any effective government. During the transition period, they continue to function with inherited, yet obsolete, imperial rules. Early decision makers are at first military officers. As a second step, civilians come into power. The transfer process has not always been a smooth one.

An alternative view would be that the political instability of the interwar period leads to constant internal restructuring because of confused planning policies. The main historic-political events produce an in-depth geographic reordering, as in the 1910s–1920s (Balkan Wars, WWI, Asia Minor disaster, and population exchanges) and in the 1940s–1950s (WWII, Civil War, first communist period).

Detailed population censuses[4] are great tools for studying these changes. Our statistical database accounts for hundreds of administrative transfers.[5] This operation has been fulfilled thanks to the relational database capacities of MS Access.

8.2.3 Versatile Place Names

The constant revision of place names is another issue to address. Soon after the liberation and as a result of population transfers, most settlements give up their

[4]Many publications covering the administrative changes are proved to be inaccurate (Χουλιαράκης 1973–1976).

[5]The village (*kinotita*) of Anatoli (district of Serres) exemplifies the upheaval experienced by many communities: In 1905, within the limits of the present municipality, almost ten localities are reported, most of them being populated by Muslims. Only four of them remain in 1928. Meanwhile, four other settlements are built to meet the needs of Asia Minor refugees. From this total, only three hamlets had survived in 1998. Such a "simplification" process is frequent.

Fig. 8.3 Public ceremony for the change of Gorna Dzhumaya's name (1950). *Source* www.lostbulgaria.com

Ottoman names (Fig. 8.3). The change may be total or partial, and its pace is determined by variable local conditions and political contingencies. For instance, Turkish designations are preserved for decades in the main Muslim villages. Instead, "ethnic" Greek and Bulgarian places spontaneously vote in favor of the adoption of "new" names. Of course, the conversion often poses a challenge. Difficulties have been identified, for example, when resettled communities keep the same village names despite relocations—thus replacing other abandoned toponyms. And the same is true when local place names are simply transliterated from their "archaic" Ottoman into a first Greek or Bulgarian form.[6] Many other spelling revisions follow afterward .[7]

[6]Transliteration is a complicated problem: it implies dealing with three different alphabets and relatively "young" national languages whose linguistic codification is not fully achieved.

[7]For example, the mixed Christian-Muslim locality of *Boblic-Боблич-Μπομπλίτς* (*kaza* of Drama) is renamed *Vuvlits-Βουβλίτς* in 1913; it temporarily becomes *Vuvlitsion-Βουβλίτσιον* before gaining its present name of *Pyrgi* in 1928.

In Bulgaria, because most settlements were inhabited by Slavs and Pomaks, toponymic changes are more seldom. Despite this, the communist regime has replaced many Ottoman names that had outlived half a century. For instance, the village of *Damyanitsa-Дамяница*, located south of Sandanski, used to be called *Orman Chiflik-Орманъ чифликъ* until 1946. In the same way, the Pomak areas keep Ottoman toponyms until the first years of the communist period, when most of them are changed.

In such conditions, research quickly turns into a "genealogical" quest. Practical difficulties are considerable, all the more so since expectations are high: Hundreds of places have been abandoned by their former inhabitants and driven into ruin after war events, population exchanges or, more simply, rural depopulation. This is obvious when considering the spontaneous migration flows that follow the Turkish defeat of the first Balkan War. These years of uninterrupted state of war, movements of armed troops, occupations, and liberations have pressed hundreds of thousands of people to run away outside any organized policy and protection. In the absence of adequate sources, these departures have never been studied in detail (Peace 1914). The late Ottoman and early national censuses that report "unidentified" place names are the only sources providing evidence of the existence of those lost communities.[8] Besides, when a village name adopts three or four different forms according to the national source, retrieving reliable data proves a difficult task.

8.2.4 Thematic Mapping Without Topographic Maps

Until recently, Macedonian and Thracian confines were military fields on which the army extended its exclusive power domain. Not to bring back painful memories, the bilateral Greek and Bulgarian claims over their respective *lost territories* shall remain outside the scope of our research.[9] The two periods of Bulgarian occupation/liberation in Eastern Macedonia are present in everyone's mind.[10] Obviously, since its creation, this borderline has always been classified as "sensitive."

When confronted with such frontier districts, geographers in search of maps have to cope with severe shortages. Border maps are sparse. When available, these are old topographic sheets produced by the French, German, and British armies in the 1910s and 1940s, as support to military goals (Δημοτική Πινακοθήκη Θεσσαλονίκης 2004). With the establishment of the Iron Curtain, the Greek and Bulgarian armed forces manage to edit their own map covers. National geographic departments develop programs in the 1960s and 1970s. Since then, no document has been updated. Any unauthorized person was absolutely denied access to any map until the 1990s. Overcoming the lack of mapping support implies to use alternative providers in the last instance. A "black market" provides most civil administrations with such materials for years. In their defense, it must be said that these documents are vital for local and regional authorities in running their planning projects. The interdiction has been lightened recently (Sect. 3.3).

Incomplete and outdated maps were the only tools available for identifying and locating the hundreds of places, villages, hamlets, and cities mentioned in past

[8]Such a reconstruction is not a dematerialized form of "archeological investigation," however. Until recently, older people had kept personal memories of this *past world*.

[9]See the abundant historical literature on this matter, especially about Pirin and Aegean Macedonia (in Greek and Bulgarian).

[10]1913–1919 and 1941–1947.

censuses. Each single case was another puzzle to solve. When maps were completely lacking, further information was to be found in historical records or through field investigations (Darques and Deslondes 2008). One thing was to geocode old maps through a process of picking up 50–150 pair-points (Sect. 4.2). The exercise authorized a limited freedom: Most places could remain unidentified while focusing on a limited number of ascertained landmarks. Another thing is to manage the challenge of thoroughness. In the long term, statistical comparisons entail providing *systematic* information on 800 settlements. Once again, this endeavor took several months of full-time work to be completed.

8.2.5 Error Estimation and Data Interpolation

The database structure is based on a reward information retrieval, beginning with the latest administrative and statistical data. Many historical works use simple mapping techniques and settle for minimalist approaches to these matters. Even when research is achieved at mid-scale level, statistical adjustments and border changes are often scrapped. The effect is obvious: Elaborated through such unlikely methods, most thematic mappings are at best inaccurate, at worst ineffective and error producing (Foody 2007).

Going back in time, the number of missing data increases. Within the five districts under scrutiny, 776 populated places are reported in 2011 (*dimotikes & topikes kinotites* and *sela-gradove*) with an overall population of 831 000 people. All settlements have been processed. On the other hand (Φιλιππίδης 1912; Yosmaoğlu 2006), the eight *kaza* of 1905 accounted for 514 localities, plus eighteen Muslim villages in the district of Demir Hisar (Sidirokastro) and hundred settlements recorded as groupings around Nevrokop (Gotse Delchev). Only 374 of these 514 places have been clearly identified and located (approximately 75 %). 34 000 inhabitants remain out of reach, which account for 10 % of the 338 000 reported souls. These used to live in sparsely populated places, most of which have disappeared soon afterward.

Left blanks usually require specific processing. However, given the advanced variability of spatial and temporal data (ethnic structure, main migrations, immense destruction caused by war,[11] etc.), the suggestion to fill in missing data has been abandoned. This would have introduced many more errors than proper corrections. Smooth areas on maps would have replaced the municipalities where high spatial and temporal *frequencies* prevail.[12] As a consequence, many interpolation techniques developed in other research projects had to be rejected (Gregory 2002; Lixin

[11]Just consider that the WWII and Civil War periods have led to the complete destruction of dozens of settlements in Northern Greece, where fighting was most intense.

[12]These terms are taken from remote sensing techniques. High or low frequencies express the spatial heterogeneity of neighboring pixels within a raster image.

et al. 2002; Gregory and Ell 2006). Spatial interpolation methods that refer to neighborhood analyses do not make sense here for the above-mentioned reasons. In addition, temporal modeling techniques are defective when data are subject to frequent and unexpected accidents, or when too many details are missing.

8.2.6 Bringing Solutions to Temporal Issues

A last methodological challenge to take up concerns temporality issues. As one might expect, Greek and Bulgarian censuses are absolutely not coordinated in terms of dates and methods (Fig. 8.4).

Over the course of the twentieth century, four censuses have been carried out simultaneously:

- 1905: last Ottoman counting (Hilmi Pasha).
- 1920: first regular census on both sides of the border after the Balkan Wars and WWI.
- 2001 and 2011: post-communist censuses achieved according to the international agenda and method.

Between those given dates, especially during the communist period, population censuses adopt a systematic time lag strengthened by diverging account methods (Cook 2004). This is no coincidence. The data are made not to match, on purpose. Most experts consider that the pre-WWII censuses are reliable and fairly well documented. After WWII, political control is getting stronger. In both countries, population characteristics are no longer collected (religious–linguistic–national identity, for instance).

In Greece, the surveys are usually carried out in spring. All "sensitive" data are gathered at a broad scale: Little information is available locally (*dimos-kinotita* level). In Bulgaria, these are conducted at the end of the calendar year. Besides, the communist authorities believed the censuses to be internal political tools submitted to planning purposes. They were not meant to be published.

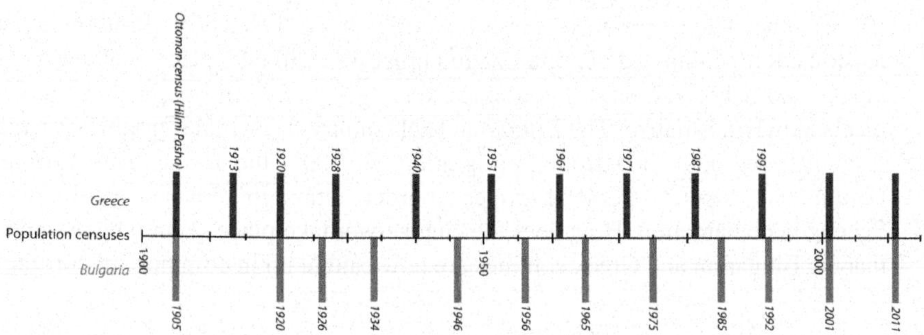

Fig. 8.4 Chronology of Greek and Bulgarian population censuses after 1900. *Source* ELSTAT, NSIB. *Reference is only made to the censuses that concern the Greek–Bulgarian border area*

The subject could be debated at greater length. When civil authorities take the decision to alter their collection methods on a regular basis, experts already have to create suitable algorithms able to restore some degree of internal data comparability (Desrosières 2009). The problem is much more intricate with data arising from different sources/countries. Only accredited statisticians can overcome such difficulties with full access rights and detailed information at their disposal. Therefore, we shall keep in mind the limits of the mapping exercise proposed hereafter.

Using mean annual data is a common way of addressing the intercensal variation issue. But time-sequencing constraints also had to be managed. Because the Balkan area has been submitted to a long-term political–territorial instability, defining a periodization was full of risks. Although other temporalities could have been accepted, it was decided to adopt quite conventional historical benchmarks:

- 1905–1934 (BG) and 1905–1940 (GR). The prewar period covers major events: the last years of the Ottoman Empire, the Liberation, the WWI disorders, and the first two decades of national integration. Migration patterns and war destruction make up a complex demographic background. The five districts are not fully informed: Detailed results of the latest Ottoman counting are missing for the Vilayet of Edirne. Moreover, because research focuses on borderlands, it was decided to ignore the *kaza* of Kavala—badly informed in the Ottoman survey (Karpat 1978, 1985; Panzac 1993).
- 1940–1951 (GR) and 1934–1946 (BG). The period roughly covers WWII, the Greek Civil War, and the establishment of the communist regime. These are tough times too. The Greek–Bulgarian frontier is called to move for the last time. Through army abuses and occupation, the countryside becomes fighting land with all related results.
- 1951–1991 (GR) and 1946–1992 (BG). With the Iron Curtain, the degree of militarization increases. On both sides, traffics are strictly regulated within the "control zone." Cross-border exchanges slow down to a minimum. The communist state applies its planning strategy with stunning outcomes: rural development and modernization, industrialization of the countryside, reinforcement of minor administrative centers and cities. By contrast, the Greek urban growth results in a vigorous decline of all remote places.
- 1991–2011 (GR) and 1992–2011 (BG). This is "reconciliation" and "convergence" time. The communist collapse contributes to the easing of international tensions. Reopening and demilitarization processes will take place over a span of years—old habits die hard. Economic exchanges, trade, and population movements slowly normalize. The Bulgarian backcountry faces major troubles arising from massive and uncontrolled emigration. In 2007, the former Iron Curtain becomes an "open" internal European border, although replaced by the new Schengen demarcation. How does the changeover take place? Can it be assumed that the Bulgarian and Greek confines are now caught up in common misfortune?

8.3 Lands and People at a Crossroads: Mapping a Historical Divorce

Figures 8.5 and 8.6 refer to two separate periods of history when space can be regarded as "open": the late Ottoman Empire and the supposed "united territories"—although still much fragmented—of the early integration of Bulgaria into the European Union. Meanwhile, a hundred years of major revolutions have elapsed. This century-long "detour" does not deserve being called "transition," obviously.

In 1905, the Mesta-Nestos and Struma-Strymon Valleys adopt an old scheme: two parallel river basins separated by the Pirin heights and their southern extension of Orvilos and Phalakro. Few passes cross this mountain chain. The main one runs through Razlog (Predel). Another connection links Gotse Delchev with Sandanski. Further south, a narrow road connects Serres with the basin of Kato Nevrokopi via Vrondu.

The Struma-Strymon Valley seems more populated than the one of Mesta-Nestos at the beginning of the twentieth century. However, this assessment has to be weighted because the Ottoman authorities have encountered problems in surveying mountain communities (Pomaks, Vlachs). The semi-nomadism of these pastoral groups is one among many factors that drive the overall underestimation of their size and the difficulty in locating their settlements. "Mobile" communities are not named in reference to villages; their existence is made obvious through the use of generic toponyms. Furthermore, many Muslim settlements could not be identified because most of them have vanished in the early years of the twentieth century without leaving a trace.

Figure 8.5 demonstrates that the main imbalance is north–south. The largest communities concentrate in the Mediterranean area. With its 43 000 people, Serres is the leading city of this eastern part of the *Vilayet*. Its district gathers the bulk of the population. Drama reaches 14 000 souls—without considering the 18 000 inhabitants of Kavala. The other *kaza* "centers" are head villages[13]: Petrich, Nevrokop, Gorna Dzhumaya, Razlog, and Bansko do not exceed six to seven thousand people. Admittedly, the last three localities were only a few kilometers away from the Ottoman–Bulgarian borderline at that time. The mountainous hinterland—today Bulgarian—is almost exclusively rural. The Rhodope is sparsely populated by scattered agricultural/pastoral communities of different religious–national–linguistic affiliations.

Significantly, the entire urban system is made of foothills cities. The amount of arable land is limited, especially in the uplands of the Mesta and Struma rivers. Lower lands suffer poor environmental conditions: Permanent or temporary lakes surrounded by swamplands and wastelands are ordinary. The basin of Serres is the most significant example of that kind (Fig. 8.7).

[13]The Greeks would say *komopoli-κωμόπολη* or *kophalohori κεφαλοχώρι*.

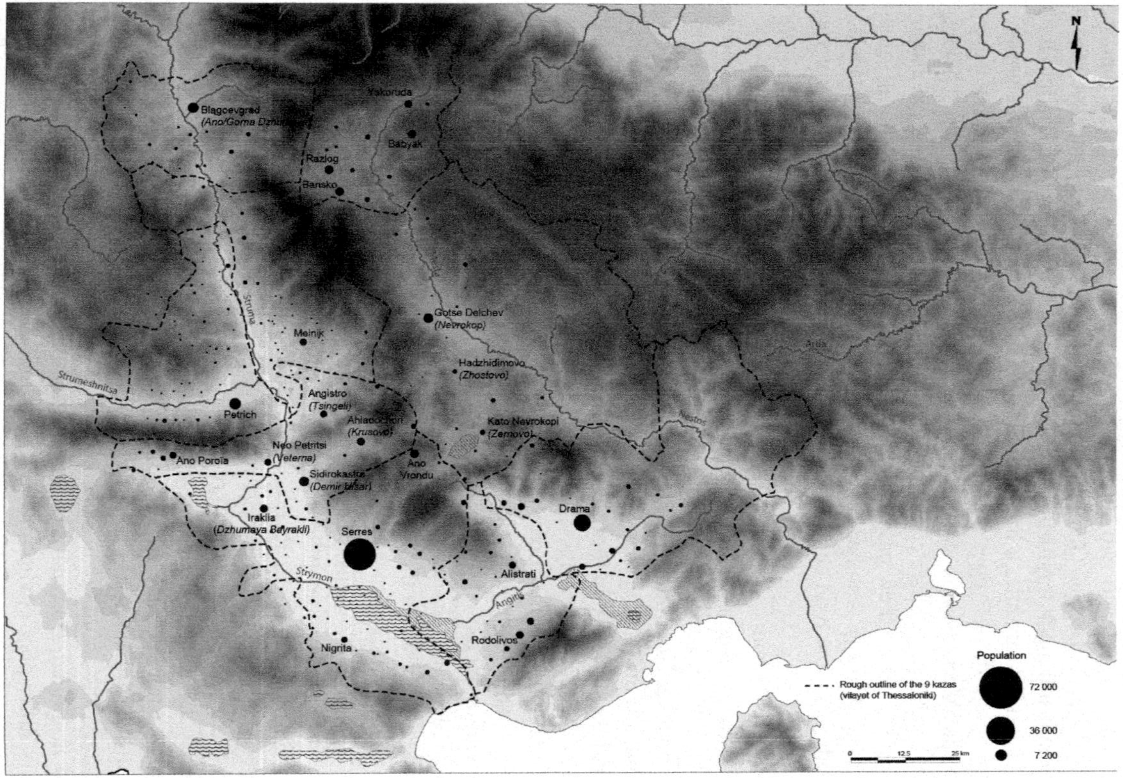

Fig. 8.5 Numbered population in 1905, eastern Vilayet of Salonika. *Source* Ottoman counting 1905 (Φιλιππίδης 1912). Administrative network: *Carte des écoles chrétiennes de la Macédoine*, (Carte des écoles chrétiennes de la Macédoine 1905). Data by cities and villages. DTM: NASA SRTM program

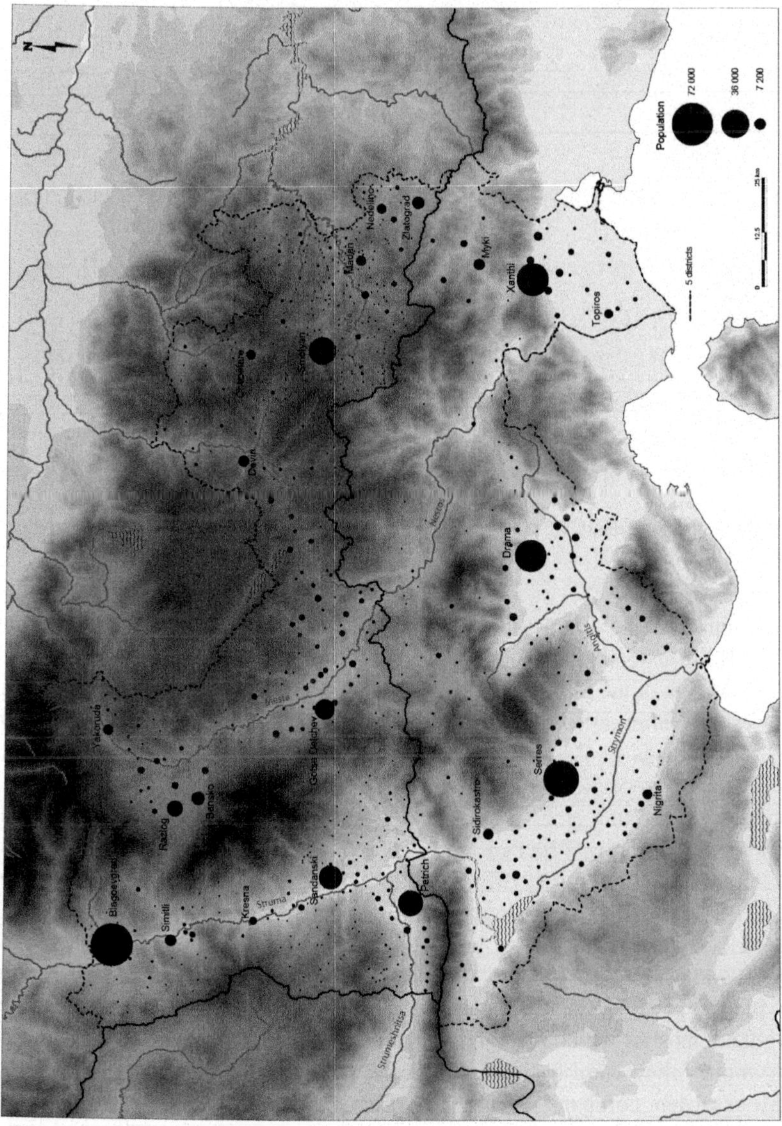

Fig. 8.6 Present population in 2011, five districts. *Source* Population censuses, ELSTAT and NISB, 2011. Data by cities and villages (dimotikes-topikes kinotites and naseleni mesta). DTM: NASA SRTM program

Fig. 8.7 Swamp and wastelands of the Serres-Drama region in 1908. *Source* British War Office 1908 (British War Office 1908)

At the beginning of the century, the so-called *Ehinos* Lake occupies the southern part of the Serres Basin, extending its quiet waters and riparian vegetation from the edges of the city to the mouth of the Strymon at Amphipoli. Another semi-permanent water area stretches west of Sidirokastro, the Butkovo-Kerkini Lake. Some marshy fields also cover the surroundings of Zernovo-Kato Nevrokopi, the bowl of Philippi (south of Drama), the floodplain of Nevrokop-Gotse Delchev, and the outskirts of Petrich.

Drainage systems and other major infrastructures shall make these acreages farmable later on, in the first half of the twentieth century. In 1929, the American Monks-Ulen company begins to drive a wide-ranging dewatering program of the Ehinos Lake. Before the works were commenced, the enterprise claimed that, across the 1 100 km^2 of the Serres basin, 290 (25 %) were devoted to lakes and permanent swamps, 282 (24 %) to floodable areas, and 168 (14 %) to marshy spaces (Μπαρτζούδης 1995). One can easily imagine the human pressure on the remaining land, most of which received dry cereal and tree crops. Lower-lying areas of the Strymon Valley were frequently exposed to devastating floods.

The Monks-Ulen company carries out drainage, antiflooding, and irrigation works during the years 1928–1938, as in other districts of Greece. The project is conducted under the supervision of the state (Ministry of Agriculture) in the frame of a great national program for rural development, the aim of which is to settle the Asia Minor and Thrace refugees (Treaty of Lausanne, 1923). The *nomos* of Serres welcomes 70 000 refugees according to the census of 1928. Resettled populations inherit improved fields. Dozens of village communities are established in the middle of formerly flooded lands. The first hydro-agricultural dam across the Strymon is ready in 1932, which results in stabilizing the Kerkini Lake. Downstream, 345 km^2 of land receive irrigation. Other improving works are achieved in the 1950s and a higher dam is built in 1982. The acreage of irrigated farmland finally reaches 850 km^2.

Assuming that one century of historical demography cannot be summarized briefly, it should be added that, indirectly, the urban network has made the most of the opportunities provided. On the one hand, the general strengthening of the Aegean districts is obvious, as a result of rural–urban migrations, longstanding support to agricultural development, and service sector expansion in the post-WWII period. On the other hand, the Bulgarian planning system should not be underestimated. In Bulgaria, no such geographic "revolution" took place, i.e., land use changes are not thorough. The map dated 2011 is clear on that point. This fact does not mean that the Bulgarian Rhodope has been maintained asleep for ages. To put it simply, the areas concerned are much smaller. In the years following the annexation (1913), priority has been given by Sofia to maintaining and entrenching control over borderlands, which implied creating a new urban network. This was meant to compensate for the influence of the Greek southern cities, thus limiting their attractiveness to the north (until the Kerkini–Orvilos barrier).

Over the century, Nevrokop-Gotse Delchev and Petrich have succeeded in maintaining their position in the urban hierarchy, although they had to face the emergence of rival cities promoted by the central authorities. The examples of Sandanski, Smolyan, and Blagoevgrad are particularly significant.

Figure 8.8 focuses on the fate of eight cities. Two tendencies are readily identifiable. At first, the Bulgarian municipalities witness an exponential increase. Although the first half of the century is a period of slow expansion, a soaring growth initiated by the communist regime is registered in the 1950s and 1960s. Until that date, the Rhodope used to be ruled by outside urban centers—Plovdiv, Edirne, Serres, Salonika, and Sofia. This spatial organization inherited from the Ottoman era was no longer sustainable. The communist authorities organize the development of remote districts. One of the stated goals includes securing the areas inhabited by Pomak populations, who suffered from their isolated positions against the Greek and Turkish boundaries.

The Bulgarian government decides to promote two urban centers: Smolyan and Kardzhali. After a comprehensive reform of the administrative network, these cities take the lead in two new *oblasts*. Kardzhali's case will not be discussed, as the city stands outside of the research area.

At the beginning of the century, Smolyan was nothing like the powerful city we know today. According to the *Committee for Western Thrace*, 1 912 "Turks," 193 "Greeks," and 1 003 "Bulgarians"[14] populated *Pashmaklı* before the Balkan Wars

---

[14]The "Turks" are in fact Muslims, the "Greeks" are probably Patriarchist Bulgarians, and the "Bulgarians" are certainly the Exarchist Bulgarians mentioned in the Ottoman census of 1905.

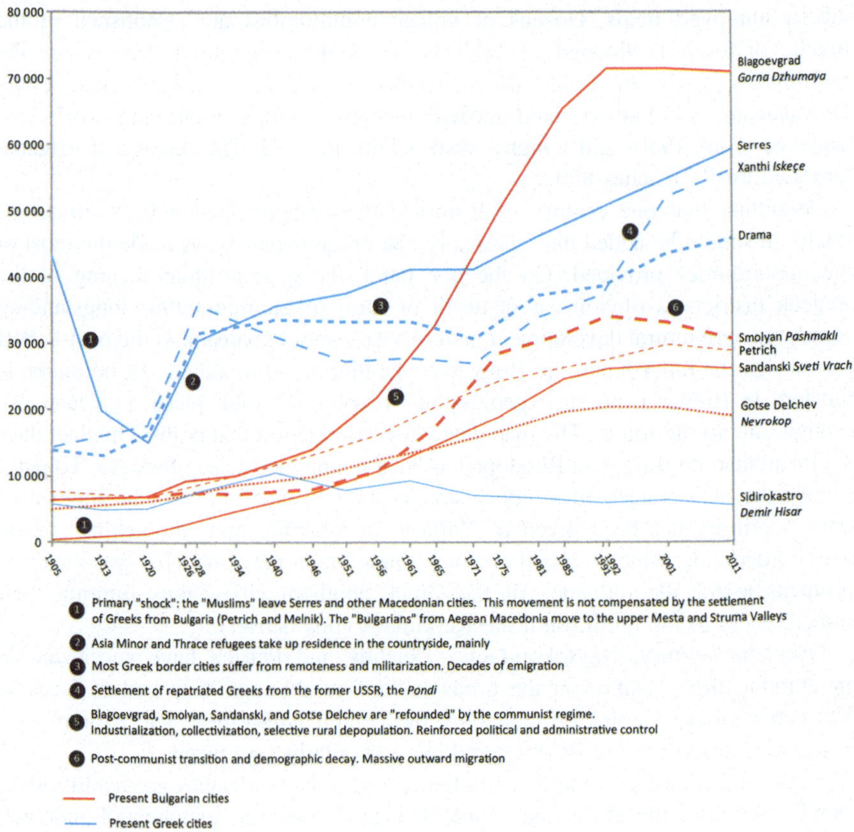

Fig. 8.8 Population growth in the main cities, 1905–2011. *Source* Greek and Bulgarian censuses, Ottoman counting

(Comité de la Thrace Occidentale 1922). Surprisingly, this balance of power has survived the events of the years 1900–1930. Because few people had left Pashmaklı during the Balkan Wars, few incoming refugees have been tempted to settle down. In 1934, the city adopts a new name: *Smolyan*. The same year, the population census registers 2 903 residents, of whom only 39 are born outside Bulgaria. The Muslim community is composed of 1 518 souls (52 % of the population). This stable environment is exceptional when considering the huge changes taking place in most border areas at the same time. However, the real "awakening" will happen later. In 1960, Smolyan links its fate with that of two neighboring villages further down the Cherna Valley, Raikovo and Ustovo. The new conurbation climbs instantly to 11 000 people. Helped later by massive investments, the city reaches 35 000 souls at its acme, in 1992 (Figs. 8.9 and 8.10).

Fig. 8.9 The market of Ustovo (Smolyan) in the 1930s. *Source* www.lostbulgaria.com

Fig. 8.10 Smolyan today. Main road along the Cherna River. *Source* R. Darques, 2006. *The urban agglomeration has developed along a large road where administrative buildings, dwelling blocks, industrial units and, more recently, resort centers gather. From the old settlements, few buildings have survived to the present day*

The case of Blagoevgrad/Gorna Dzhumaya might also be considered exceptional with regard to its spectacular growth. In 1905, the city has no more than 6 000 residents, of whom 4 500 (approximately 75 %) are Muslims. With the first Balkan War, most of the "Turks" move abroad and settle in Thrace or Anatolia. The following years, Gorna Dzhumaya receives 1 330 refugees from the Aegean Macedonia,[15] plus 680 from the "former Western Bulgaria" and Vardar Macedonia.[16] In 1934, the remaining Muslims are only 733 (7 % of the population). Only hundred kilometers away, Sofia considers the city as a strategic point. The control of the Struma Valley, surrounded by highly sensitive boundaries, is thus delegated to this emerging urban center. The official support of Sofia to Gorna Dzhumaya's development takes place shortly after the "liberation," thirty years before Smolyan's refoundation. On the eve of WWII, its population had already increased twofold. In 1950, the city changes its name to become Blagoevgrad; an intense construction and industrialization phase begins. Meanwhile, the countryside becomes involved in the process of collectivization. A series of crucial infrastructures are created: huge plants, hospital, administrative buildings, university, modern housing blocks, and transport means. From the 1990s onward, there are close to 70 000 inhabitants in the city (Figs. 8.11 and 8.12).

Gotse Delchev, Sandanski, and Petrich are intermediate cases. As cities close to the Greek border, they directly receive flows of "Bulgarian" refugees at the beginning of the century. In 1934, according to the Bulgarian census, 36 % of Gotse Deltchev's population is born abroad (3 195 of 8 767 people). The rate reaches 34 % in Sandanski and 35 % in Petrich. Meanwhile, the proportion of Muslims falls down abruptly. After the departure of many "Turkish" elements and as a consequence of the settlement of many immigrants/refugees, the population becomes predominantly Christian. In Nevrokop, the number of Orthodox Christians soars from 21 % in 1905 to 87 % in 1934. Petrich also moves from 29 to 97 %, whereas Sveti Vrach remains almost exclusively the Christian place it used to be. Such rates indicate the order of magnitude of changes. This first expansion has further outcomes: all three cities develop as administrative centers under the communist rule, helped by many agro-industrial units serving the surrounding collective farms (Figs. 8.13, 8.14, 8.15 and 8.16).

The upper graph (Fig. 8.8) discloses a second scheme. Greek cities follow trends that have nothing in common with their Bulgarian counterparts. Instead of increase, they maintain a steady profile. Had the *Great Catastrophe* not happened, the Greek Macedonian and Western Thracian cities would have faced an abrupt decline. Historical and geographic backgrounds are very different. Serres, Drama, and

---

[15]The Bulgarian early sources use the term "White Sea" (*Bialo more-Бяло море*) instead of "Aegean Sea". *Bialomorska Makedonia i Trakia –Бяломорска Македония и Тракия-* thus make reference to the Aegean Macedonia and Thrace.

[16]Bulgarian population census of 1934. These areas roughly correspond to present (FY)ROM.

Fig. 8.11 Gorna Dzhumaya-Blagoevgrad in 1919. A farming community. *Source* земи (1919)

Fig. 8.12 Blagoevgrad today. The modern industrialized metropolis. *Sources* R. Darques, 2005

Fig. 8.13 Nevrokop-Gotse Delchev in 1933. *Source* www.lostbulgaria.com. *Two decades after its "liberation," the city experiences a period of urban renewal, although still characterized by its Ottoman aspect and its limited spatial extent. The strongest growth phase will happen 20 years later*

Fig. 8.14 Gotse Delchev today. *Source* R. Darques, 2005. *This is not the urban revolution undergone by Blagoevgrad, but the hillocks of the foreground hide the northern and southern suburbs. Gotse Delchev has given birth to a second modern town next to the old city. Its influence zone trespasses the river basin visible in the background*

Fig. 8.15 Sveti Vrach-Sandanski in the 1930s. *Source* www.lostbulgaria.com. *The population of Sveti Vrach was only 500 in 1905. In 1934, a tenfold increase had been registered. The place has become a boomtown. The city's success is partly caused by the decay of Melnik. In the 1930s, Macedonia Street brings a "modern touch" to the surrounding Ottoman dwellings*

Fig. 8.16 Sandanski today. *Source* R. Darques, 2006. *The trees on Macedonia Street, almost unchanged, extend to the right*

Xanthi were well-established cities at the end of the Ottoman era. Despite a short distance from Salonika, Serres had no fewer than 40 000 people, a level that made it a weighty place within the Empire (Καφταντζής 1996). In 1905, Serres is home to 18 800 Muslims—44 % of the population. In Drama, they are 9 000 (65 %), in Sidirokastro 3 600 (65 %), and in Xanthi 8 200 (54 %). The main reason for the departure of Muslim communities is the disorders of the Balkan Wars and WWI. The remaining families shall be exchanged as required by the Treaty of Lausanne. Thus, unlike what has happened in Bulgaria where some "Turkish presence" has been maintained, Muslim populations leave Greek Macedonia never to return. This past episode is well known. The settlement of Greek refugees is the lifeline and safety net of the Aegean cities. No rural outflow could have compensated for such brutal depopulation (Figs. 8.17 and 8.18).

From the 1930s to the 1970s, the Greek cities undergo a period of stagnation. The border "sterilization" effect reaches its limits after WWII and the Civil War. As a result of land militarization, thousands of young workers decide to settle in the main Greek cities—most often Salonika and Athens—or Germany; Serres receives limited incoming flows. Some of these "migrants" shall return home in the 1970s–

Fig. 8.17 Serres in 1916–1918. Bulgarian officers. *Source* www.losbulgaria.com, Colonel A. Kolarov. *Although severely destroyed by bombings, Serres stands as a significant Ottoman city*

Fig. 8.18 Serres today. *Source* R. Darques, 2005. *From the Ottoman city, few urban districts have survived. The building fever that has seized the Greek provinces after WWII has played an important role in the urban renewal. The city structure compares poorly with that in Bulgaria: Here, no "block," no industrial building, or large factory is to be found*

1980s. Finally, a limited renewal takes place in the 1990s, after the "repatriation" of thousands of "Pondi."[17] Athens resettles many of them in Thrace, promptly.

Focusing two Bulgarian and Greek border cities might be a simple way of illustrating short-distance discrepancies. Petrich and Sidirokastro are symbols of two different government policies: colonization and promotion in the north, and relative decline and neglect in the south. In 1905, the two places are comparable to each other in size. One century later, Petrich far exceeds Sidirokastro—even though most Greeks view this border region today, and the whole of Bulgaria incidentally, as an underdeveloped marketplace (Darques and Deslondes 2008). After 1912, Petrich receives the repatriated "Bulgarians" of the upper Strumeshnitsa Valley and those originating from Western and Southern Macedonia. Through an opposite process, Sidirokastro receives part of the Greek population originating from the

[17]Greeks from the Black Sea and the USSR.

northern face of the Belasitsa-Slavyanka/Kerkini-Orvilos mouth. Later, the communist regime shall never stop favoring the development of far-southern lands (especially paying attention to vineyards, tobacco growing, and fruit crops) (Figs. 8.19 and 8.20).

Fig. 8.19 Demir Hisar-Sidirokastro in 1919. *Source* земи (1919)

Fig. 8.20 Sidirokastro today. *Source* R. Darques, 2005. *A few kilometers away from the boundary, Sidirokastro is one of those cities—head of kaza, afterward of eparhia—that have greatly suffered from a remote position. With a stable population across the century, the built surface does not far exceed the old site, but the urban renewal is almost complete*

In contrast, the Hellenic approach has allowed an advanced depopulation. Across the Aegean coastal plain, any development opportunity has long been considered unsafe. The Greek military officers and diplomats in charge of negotiating the peace treaties had early identified the problem in the 1910s (Peace 1914; Dalègre 1997). The Bulgarian authorities have been keeping an eye on the Aegean Macedonia until a late date, as well as other Thracian districts. For the Hellenic part, the seaside corridor is a strategic area whose protection from potential external attacks meets considerable enforcement difficulties: Reaching Aegean cities located 15–30 km away from the boundary takes just a few minutes. The interwar period is full of Komitadji–Andartes raids and military incidents along the Greek–Bulgarian frontier. This easily explains why the Pomak area and the hinterland of Xanthi have been classified as "special zones" until recently (Sect. 6.1.2).

8.4 Disentangling the Ethnographic Condition

The ethnic-confessional data of 1905 are key instruments for studying the region as it used to be before the settlement of the boundary (Φιλιππίδης 1912; Brancoff 1905; Кънчов 1900; Comité de la Thrace Occidentale 1922). Some Western observers may consider the information as belonging to times that are no more. However, underestimating the importance of this "background" would be a mistake. People in the Balkans have a keen sense of time and history: In their view, the beginning of the twentieth century is a recent past whose memories could be accounted by direct witnesses only a few years ago.

Most specialists in Ottoman history address the problem through "easy" solutions. They focus on *kaza* scale levels and avoid raising the issue of population dynamics while concentrating on single surveys (Panzac 1992; Karpat 1978, 1985). In fact, archival data on *kaza* units are relevant and comprehensive. At this scale, though, aggregation biases reduce the scope of research.

The population distribution (Figs. 8.21, 8.22 and 8.23) observed through the latest Ottoman counting[18] reveals the advanced mutual interpenetration of ethnic-confessional "communities" (Cowan 2008; Bieber 2000; Michailidis 1998). Muslims concentrate within major imperial cities where they make up the majority of households. Only a few towns are exceptions to this trend: Melnik and Bansko

[18]Coming back on the issue of the *millet* system and its many national interpretations is useless. In mixed areas, the term "community" is often misused as people interact with various other linguistic, religious, and cultural representatives in the everyday life—a situation that helps blur the identity of individuals. This point is much discussed by specialists. Some of them argue that, instead, such mixed contexts create powerful "community" versus national feelings and consciousness among people.

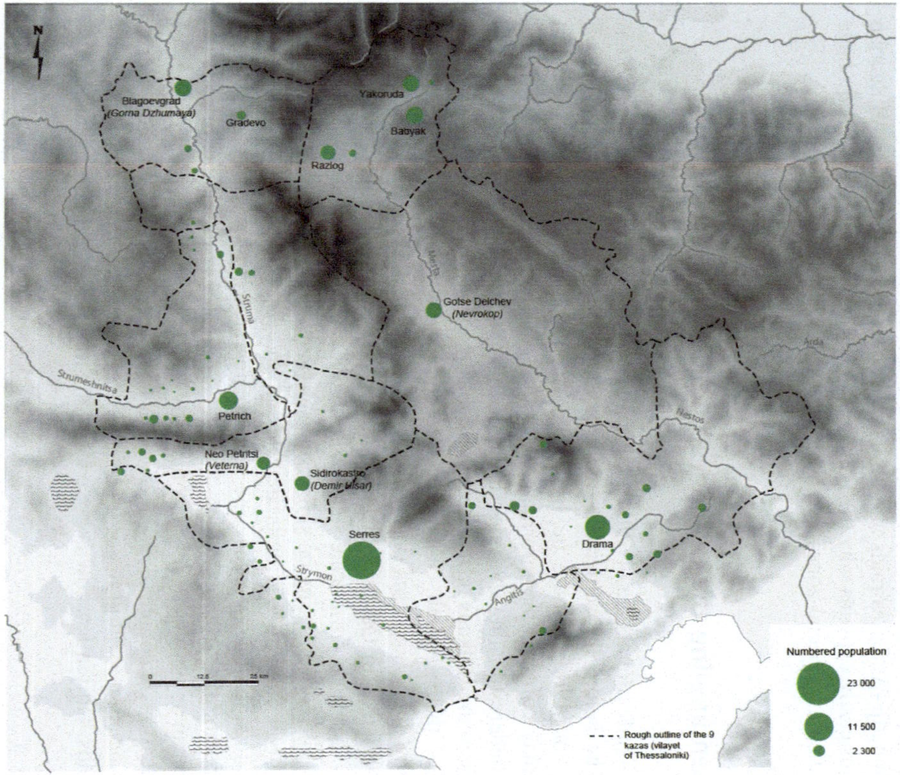

Fig. 8.21 Muslim population in 1905. *Source* Ottoman counting, 1905 (Φιλιππίδης 1912). Administrative network: *Carte des écoles chrétiennes de la Macédoine*, (Carte des écoles chrétiennes de la Macédoine 1905). Data by cities and villages. DTM: NASA SRTM program

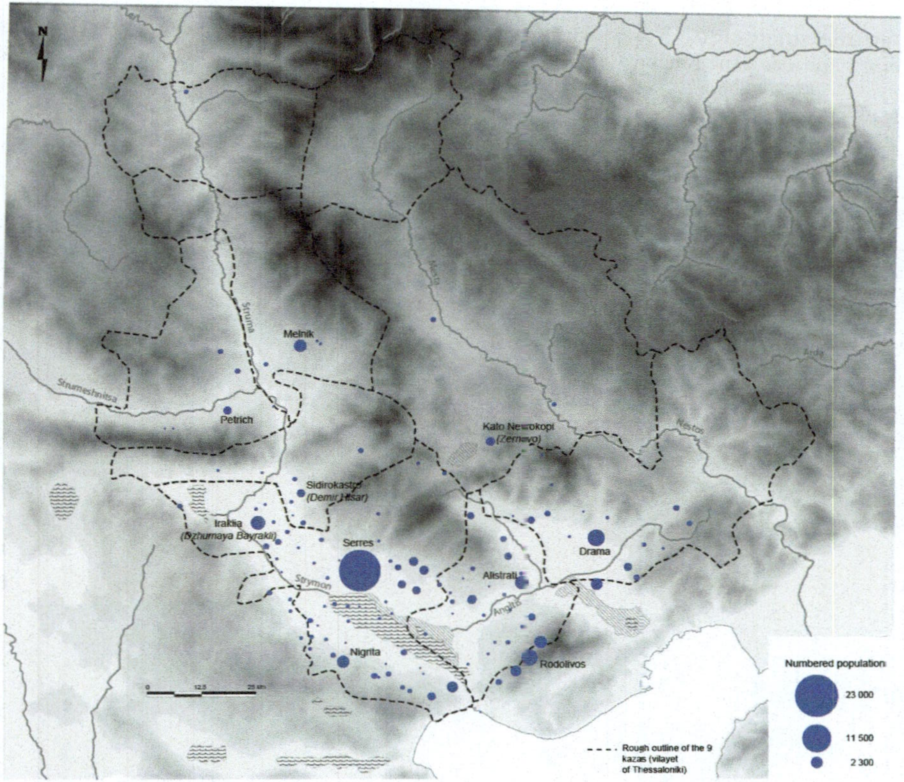

Fig. 8.22 Greek population in 1905. *Source* Ottoman counting, 1905 (Φιλιππίδης 1912). Administrative network: *Carte des écoles chrétiennes de la Macédoine*, (Carte des écoles chrétiennes de la Macédoine 1905). Data by cities and villages. DTM: NASA SRTM program

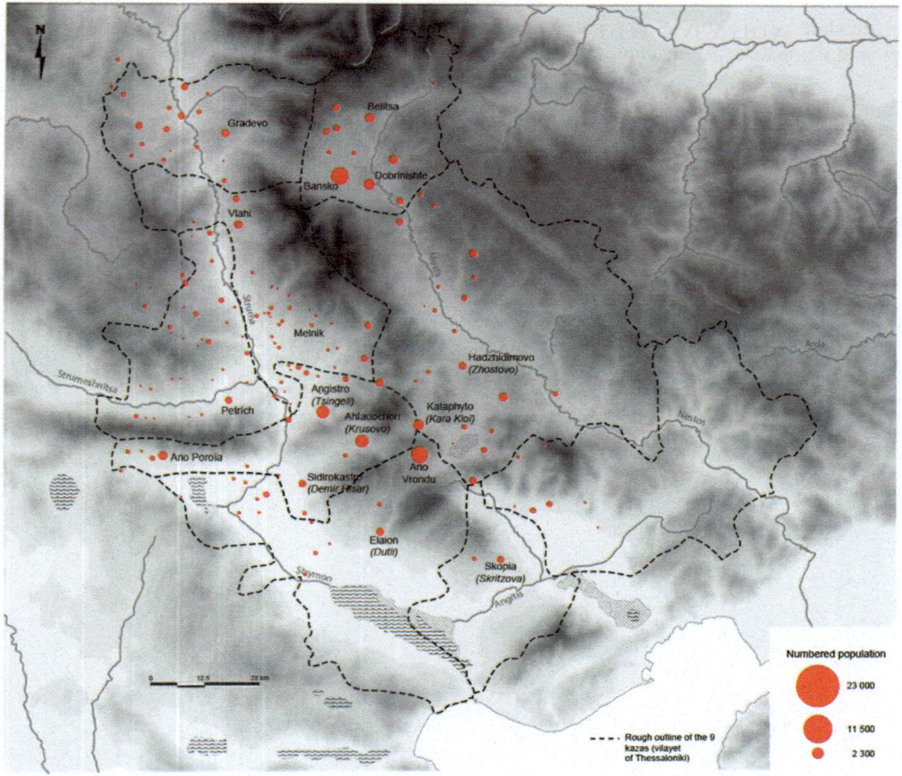

Fig. 8.23 Bulgarian population in 1905. *Source* Ottoman counting, 1905 (Φιλιππίδης 1912). Administrative network: *Carte des écoles chrétiennes de la Macédoine*, (Carte des écoles chrétiennes de la Macédoine 1905). Data by cities and villages. DTM: NASA SRTM program

belong to this category. Muslims seldom live in rural areas.[19] Some of them have spread over the lower Strymon basin and in the outskirts of Drama.

The Bulgarian population adopts a reverse profile. Bulgarian households prefer dispersion and willingly adopt rural housing (Mouzelis 1976). Their settlements are made of small units scattered across the upper Struma and Mesta valleys where they must cope with limited agricultural acreages. Some hamlets come close to the Aegean shore in the "far south." The biggest communities are to be found in what shall become Greece, around the Orvilos-Slavyanka Mount: Ano Vrondu, Krusovo, and Tsingeli gather between three and five thousand souls. Bansko is at the center of another cluster.

There are virtually no Pomak communities. Pomaks are not associated with other "Muslim" elements in the census—an awkward fact, indeed. They more likely merge into the "Bulgarian" sample (Neuburger 2000). As already noticed, the semi-pastoral groups living in mountain areas have not been surveyed properly. The Ottoman officials have focused their efforts on sedentary and established groups.[20]

The Greek population is some in-between case. Most Greeks live in the sedimentary plains of the Aegean area. Their communities are evenly distributed between cities and countryside. Melnik and Petrich are exceptions, at the extreme north of the "Hellenic lands."

After 1905, there will be no way left to address the ethnographic condition with comparable accuracy. Breaking with the *millet* system, Bulgarian and Greek states strictly limit the scope of the next censuses and impose their own perspective to the ethnic–linguistic–confessional problem. Minority issues are closely supervised, to such a point that scientists will be kept away from these controversial topics as soon as 1912. The "new" statistical institutes will never publish data at a local scale. In Greece, the 1928 and 1951 censuses are the only ones to get into the matter. Data are available at an *eparhia*-province level. The same phenomenon occurs in Bulgaria—the *okolyi* and *oblasti* are informed, not the *obshtini*. The census of 1934 is the last one to collect relevant information about religion, spoken language, nationality, and birthplace. Under the communist regime, these mentions "disappear" from the official questionnaires. They will eventually return in 2001.

Of course, new concepts and terminology bias the existing data, whenever available. Given the political context, people have early learned to hold any official questionnaire with an extreme suspicion, especially as regards identity issues.

---

[19]It should be remembered that Fig. 8.21 shows almost nothing about rural Muslim populations. The Ottoman data have been converted and interpreted through national prisms. As already mentioned, many Muslim village communities—especially small dwelling groups of the southern Struma and Drama region—could not be located, although mentioned in the sources. The same remarks apply to the Vlach, Sarakatsani, Gypsy, and Pomak settlements. In some cases, the population is recorded as a whole. For example, 90 Pomak and 10 Muslim villages of the Rhodope are recorded together: 35 000 people could not be represented on the map.

[20]For this reason, the operation led by Hilmi Pasha is called "counting" and not "census." It is based on established registers. Another explanation is that the authors of the publication intended to turn a blind eye to any "disturbing" data.

A dual-converging initiative from the top and the bottom thus works to restrict the value and accuracy of the information. Following the fate of minorities is, therefore, a hard challenge, with scarce statistical data enabling the analysis of any change. With a view to the internal logic of sources, research will be limited to the interwar period.

The Greek census of 1928 and the Bulgarian one of 1934 let us settle for a minimalist evaluation of the impact of the various wars. Both operations take place in a relatively quiet period, after decades of fights and large population transfers. These are among the most serious and complete censuses conducted over the century.

At first, it would appear that the spontaneous and/or organized population movements have reached their goal. In 1934, the Bulgarian area under study is empty of Greek speakers, a fact which lends weight to the idea of a closed "Greek question" in Pirin Macedonia. At this date, almost 10 000 Greek-speaking people live throughout Bulgaria, scattered around Burgas or established in the main cities. Only 2 500 of them declare their nationality as being Greek.

The Greek situation is more complicated in 1928. Despite war events and the violence of past decades, 12 000 "native" *Slavic Macedonian* speakers remain in the neighborhood of Serres and Drama. In the *eparhies* of Zyrnovo and Sintiki, they are 5–10 % of the population. Of course, the adopted "labels" prevent any cross-border comparison. There is doubt as to whether the informants declare true identities. At this date, 82 000 Slavic Macedonian-speaking people are referenced in the Hellenic Kingdom. The concerned communities typically concentrate in Western Macedonia, around Florina and Pella (Deslondes 1997; Ancel 1930). Beside the much discussed issue of the oncoming fate of these *Slavs*, it must be noted that the term *Bulgarian* is reserved for the Pomak populations of Western Thrace, in reference to their language. Surprisingly, at least from an external eye, the second most widely spoken language in the Greek Macedonian countryside is Turkish—a singularity inherited from the settlement of the Asia Minor and Thrace refugees (Table 8.1).

Going on further on this issue would be useless. The interest generated by declared nationalities, religions, and spoken or native languages—all data made up not to match on both sides of the boundary—leads to a deadlock. This is the reason why we decided to focus on birthplaces. These are the least risky of all available information as birthplaces do not determine the citizenship or "ethnic" membership of the registrants directly.

Figures 8.24, 8.25, 8.26, and 8.27 demonstrate that the boundary is more than a fictitious geometric outline: It decides on the fate of people. These maps illustrate the post-1912 reconfiguration of land. In a few years, a whole world has gone: The inherited Ottoman framework is swept away. The new boundary brings greater "clarity" through a change of scale: An "external" and "monolithic" segregating line far away from the main populated places replaces the many "internal" and "soft" Ottoman borders that used to crisscross each city, neighborhood, village, hamlet, and the whole countryside.

Table 8.1 Orthodox Christians of Greek and Slavic-Macedonian mother tongue in 1928

	Greek language	%	Slavic Macedonian language	%	Others (mainly Turkish)	%	Total population
Drama	82 593	83.9	3 148	3.2	12 667	12.9	98 408
Zyrnovo	9 244	70.2	966	7.3	2 954	22.4	13 164
Sintiki	29 690	76.4	3 843	9.9	5 307	13.7	38 840
Serres	61 930	84.4	2 747	3.7	8 726	11.9	73 403
Nigrita	29 834	91.6	558	1.7	2 141	6.7	32 074
Zihni	33 834	88.1	512	1.3	4 047	10.5	38 393
Xanthi	44 245	49.6	294	0.3	30 470	34.1	89 266
			And 14 257 Muslims of Bulgarian mother tongue	16.0			

Greek Eastern Macedonia
Source Population census, ELSTAT, 1928. Data by *eparhies*. Many prominent linguists consider the *Macedonoslav* as a variant of the Bulgarian language. The adopted name reveals the political position of the institution in charge of the census. In 1934, the Bulgarian census refers to the spoken language and not to the mother tongue, as in Greece. Any comparison is made impossible. At that time, hardly anyone declared speaking a Slavic language other than Bulgarian in southwestern Bulgaria and Rhodope—even the refugees originating from "Southern Serbia"

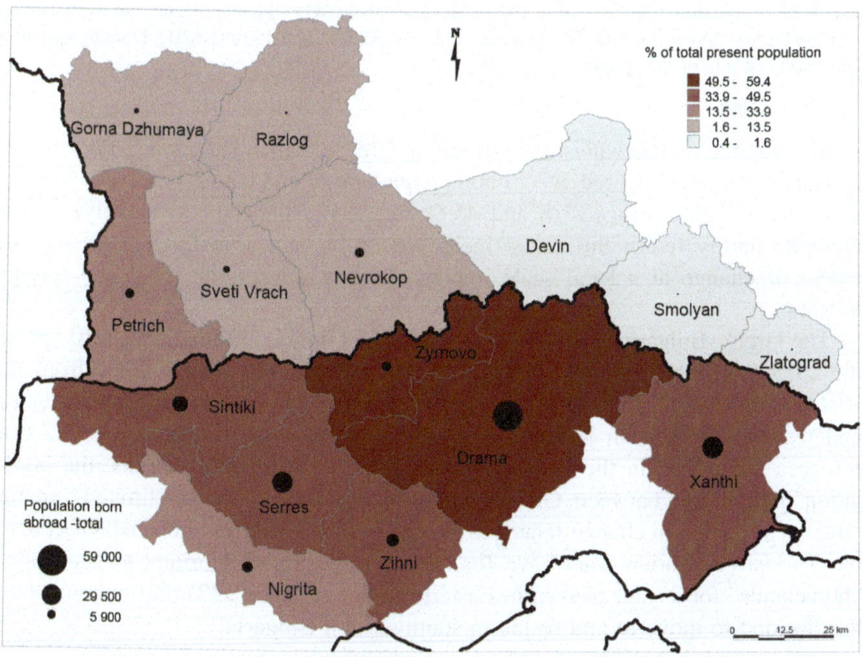

Fig. 8.24 Population born abroad in 1928 (GR) and 1934 (BG)-Total. *Source* Population censuses, 1928 (ELSTAT) and 1934 (NSIB). Data by *eparhies* and *okolyi* (limits of the time)

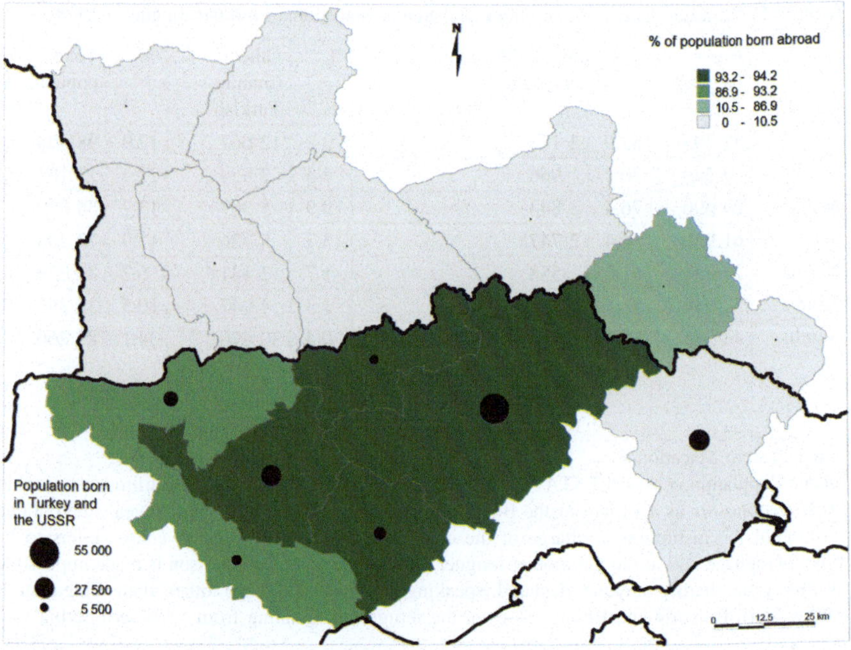

Fig. 8.25 Population born abroad in 1928 (GR) and 1934 (BG)—People originating from Turkey and the USSR. *Source* Population censuses, 1928 (ELSTAT) and 1934 (NSIB). Data by *eparhies* and *okolyi* (limits of the time)

The population renewal is more salient in Greece than in Bulgaria (Dragostinova 2011). The eight *okolyi* receive 23 500 people born abroad, and the seven *eparhies* welcome 172 000 refugees (8 and 45 % of the population, respectively). These amounts hardly return the actual incidence of massive transfers. Assessing the extent of change at a local scale implies investigating on the origin of foreign inflows.

The Greek–Bulgarian cross-boundary transfers are quite unbalanced: 20 800 of the 23 500 foreign-born individuals resettled in Pirin Macedonia come from the *Bielomorska Makedonia* (89 %). On the other hand, the refugees born in Bulgaria and resettled in Aegean Macedonia are only 12 400—i.e., 7 % of the 172 000 refugees mentioned in the Greek census. Despite painful destructions, the population interchange between Greece and Bulgaria has produced "limited" spatial outcomes. The main change relates to population inflows from the *East*. Of course, the Turkish–Bulgarian wars have forced a great many Bulgarians to leave their "homelands" for a safer heaven in Eastern Thrace (Ladas 1932), but none of them has decided to move to and reside in southwestern Bulgaria.

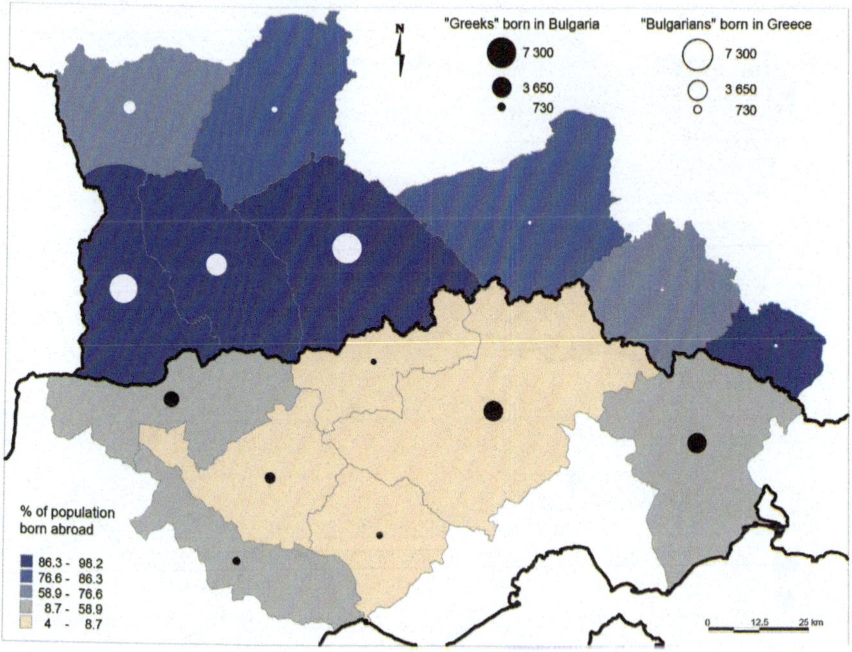

Fig. 8.26 Population born abroad in 1928 (GR) and 1934 (BG)—“Greeks” from Bulgaria and “Bulgarians” from Greece. *Source* Population censuses, 1928 (ELSTAT) and 1934 (NSIB). Data by *eparhies* and *okolyi* (limits of the time)

Surprisingly, these stupendous events have set up long-term geographic patterns: New regional imbalances that had come into being at that time have survived to this day. One of these “internal” limits—between Greek Macedonia and Western Thrace—is well known.[21] In Bulgaria, a similar “barrier” opposes Pirin Macedonia and Western Rhodope. The dividing line roughly follows the Dospat Valley. Such internal divisions are direct outcomes of legal provisions laid down by the international treaties regulating the settlement conditions of spontaneous/forced migrants.

When focusing on urban–rural relations, contrasts become sharper. The Bulgarian cities are a safe haven for half of the “repatriated” Slavic populations. The settlement of approximately 10 000 “foreign” people enhances understanding of the early urban expansion. Almost a third of Bulgarian city dwellers are foreign

[21]This geographic limit has been described extensively (Dalègre 1997; Péchoux 1968; Darques 2000).

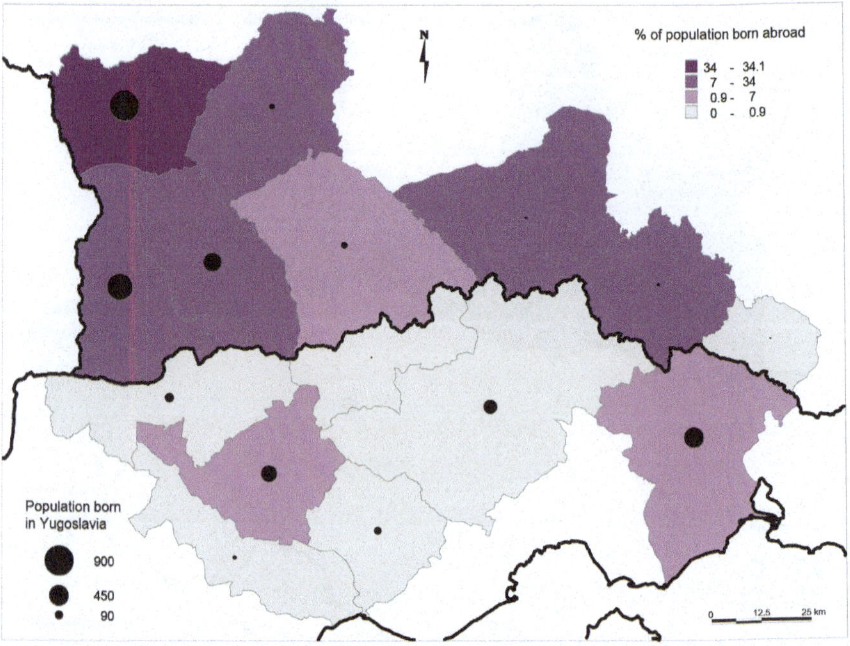

Fig. 8.27 Population born abroad in 1928 (GR) and 1934 (BG)—People originating from Yugoslavia. *Source* Population censuses, 1928 (ELSTAT) and 1934 (NSIB). Data by *eparhies* and *okolyi* (limits of the time)

natives in 1934. The outflow from Greece turns into rural–urban migration. The arrival of a few hundred households originating from Yugoslavia reinforces the overall trend (Table 8.2).

Bulgarian and Greek situations do not withstand comparison. The three Greek head cities receive approximately 50 000 refugees, of whom only 4 000 come from Bulgaria. The share of "immigrants" is almost twice higher than that of the Bulgarian cities: It attains 50–60 %. Drama is the place where the "substitution" reaches a peak. However,—and this is the main difference between Greek and Bulgarian development models—the inflow changes into a rural–urban migration for only 13 % of the refugees. In other words, the settlement of refugees produces a momentous urban renewal in Bulgaria, at the very time when the Greek countryside is overwhelmed by incoming flows. The Aegean coastal plains and interior basins undergo a "revolution" that the Bulgarian rural areas shall wait for half a century (Table 8.3).

Table 8.2 Population of the main Bulgarian cities by birthplace in 1934

Cites	Total population	Abroad total	%	Greek Macedonia	%	Yugoslavian Macedonia	%	Greek Thrace	Yugoslavian "Former Western Bulgaria"	European Turkey	Others
Petich	10,120	3,592	35.5	3,116	86.7	422	11.7	4	1	8	41
Nevrokop	8,767	3,195	36.4	3,085	96.6	37	1.2	10	3	16	44
Gorna Dzhumaya	9,977	2,130	21.3	1,333	62.6	660	31.0	19	19	8	91
Sveti Vrach	4,7.3	1,594	33.9	1,376	86.3	176	11.0	10	5	2	25
Melnik	552	220	39.9	211	95.9	8	3.6	0	0	0	1

Source Population census, NSIB. The definition of the "urban" class is not based on a demographic criterion. "City" refers to an administrative status in Bulgaria. In 1934, Melnik is the country's smallest city with 552 inhabitants. Ratios refer to the total population (left side of the chart), then to the total foreign natives (right side). Only places with more than 5 % people born abroad are mentioned

Table 8.3 Population of the main Greek cities by birthplace in 1928

Cites	Total population	Abroad total	%	Thrace	Bulgaria	Constantinople	Yugoslavia	Asia Minor	Pontos	Caucasus	Russia	Others
Drama	32,186	20,013	62.2	5,916	1,346	323	205	6,668	4,640	565	175	175
				29,6 %	6.7 %	1.6 %	1.0 %	33.3 %	23.2 %	2.8 %	0.9 %	0.9 %
Xanthi	35,912	17,049	47.5	7,811	2,135	372	285	5,404	604	83	62	293
				45.8 %	12.5 %	2.2 %	1.7 %	31.7 %	3.5 %	0.5 %	0.4 %	1.7 %
Serres	29,640	14,049	47.4	4,301	500	393	238	7,214	620	299	71	413
				30.6 %	3.6 %	2.8 %	1.7 %	51.3 %	4.4 %	2.1 %	0.5 %	2.9 %

Source Population census, ELSTAT. Unfortunately, only cities whose population is greater than 10 000 inhabitants are informed

8.5 Change Sequences: Before and Behind the Mirror, Two Worlds Apart

Considered through *obshtina-dimotiki enotita* units (2011), change sequences demonstrate the existence of strictly reverse behaviors between the two countries. Despite the current European dogma of *open spaces* promoting weak frontiers and territorial integration, these documents remind to anybody had forgotten it that the "border effect" of this former section of the Iron Curtain is deeply anchored in geography and history.[22] Some geopolitical myths have to be in touch with reality. Any journey in these confined districts brings evidence to this assessment. With the intent not to produce too many maps, five sequences have been selected (Sect. 8.2.6).

Interpreting Fig. 8.28 is not easy. As we noted above, the period covers the end of the Ottoman era and the perilous times of the Balkan Wars, WWI, the Greek–Turkish conflict, and population exchanges. This map cannot return the complexity

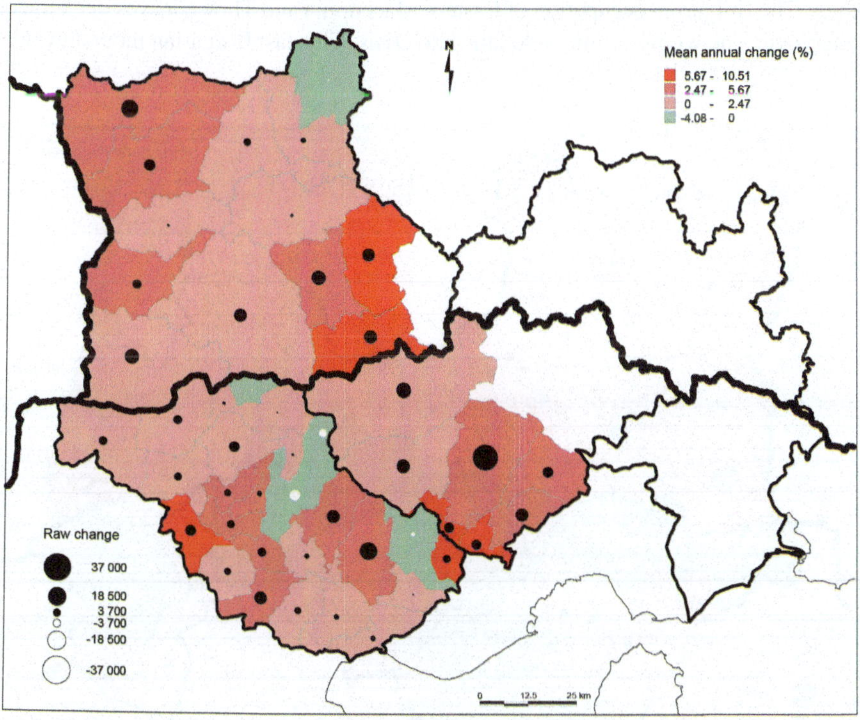

Fig. 8.28 Population change 1905–1934 (BG) and 1905–1940 (GR). *Source* Ottoman counting, 1905. Population censuses (1940-ELSTAT and 1934-NSIB). Data by *dimotiki enotita* & *obshtina*

[22] Beside demography, the same applies to many other statistical and geographic variables.

of so many upheavals that respond to the definitive alteration of a "timeworn Ottoman world."

Against all expectations, a balanced development emerges from this chaos. The first renewal of Gorna Dzhumaya is obvious, whereas Serres suffers from its disconnection with the hinterland. Population exchanges, in general, have promoted the countryside. The growth of the rural areas of Nevrokop-Gotse Delchev and Drama is symptomatic of this. As a consequence, this war and interwar period might be viewed as an adjustment time to new border realities. Mass migrations produce consistent restructurings that take precedence over border effects. In addition, the large areas covered by the *dimotikes enotites-obshtini* generate a graphic smoothing obscuring higher-scale changes. House to house, neighborhood after neighborhood, *divisions,* and further *recombinations* are a matter of individuals and families.

This smoothing effect does not recur afterward. In Fig. 8.29, the WWII and Greek Civil War lay down the foundations for the lasting land division we know today. It becomes clear that the Iron Curtain is not a late reply of Greece against the threat of a "diverging" communist Bulgaria (and vice versa). The partition is early. General Metaxas had already established a legal status regulating the Greek national confines. The Bulgarian occupation of Eastern Macedonia and Thrace leaves deep traces afterward: The whole countryside falls into crisis. This last Bulgarian incursion ends

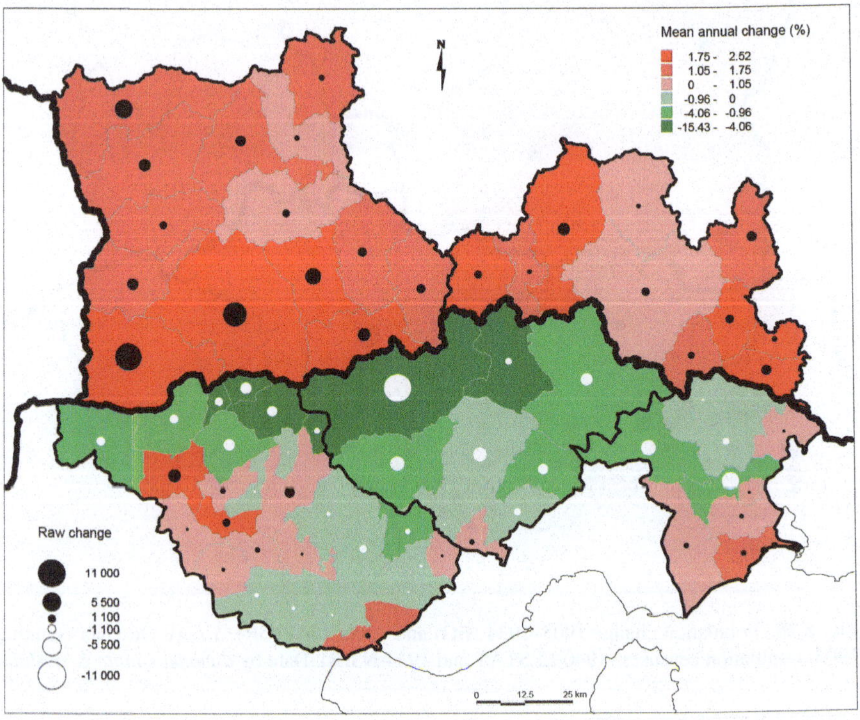

Fig. 8.29 Population change 1934–1946 (BG) and 1940–1951 (GR). *Source* Population censuses (1940–1951 ELSTAT and 1934–1946 NSIB). Data by *dimotiki enotita & obshtina*

up with a brutal backflow bringing away most of the *Slavic Macedonians* who were still living in the uplands.[23] After the Balkan Wars and WWI, this second phase makes the Greek border districts definitively repellent to any human presence. In the opposite scenario, the arrival of ethnic-political refugees in Bulgaria maintains the far-southern lands, newly communist, in a key policy position. The Bulgarian *Drang nach Süden* initiated earlier continues. In this tormented decade, homogeneous areas replace the spatial heterogeneity of the previous period. Be that as it may, the *obshtini* of Petrich, Sandanski, and Nevrokop take advantage of the cumulated events as if the Bulgarian confines' strengthening had been driven by some neighborhood retraction, rather than by national redeployment.

The end of WWII-Civil War allows for the application of dual-track planning policies (Fig. 8.30). From the midterm of the century, the border is no longer seen as temporary, but as a permanent input. More than reinforced, the borderline entrenches itself into the ground, so to speak. From a military perspective, the "hazardous" frontier area is almost emptied, as left on its own by the events. Tens of villages are condemned to ruin and will never recover their former positions.

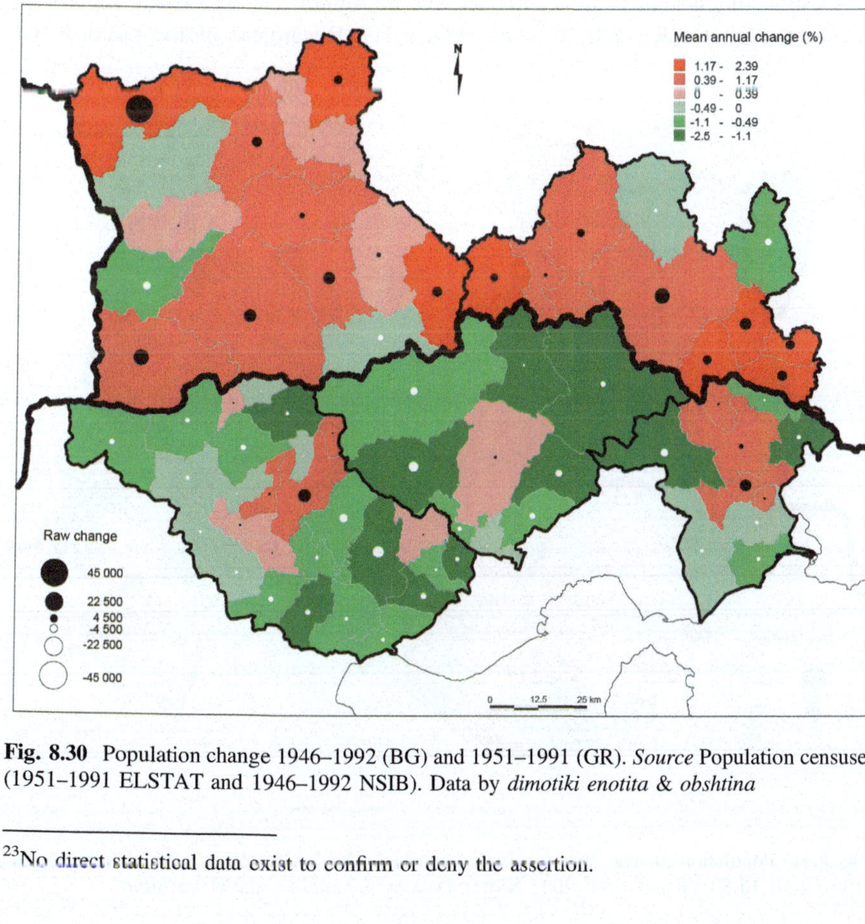

Fig. 8.30 Population change 1946–1992 (BG) and 1951–1991 (GR). *Source* Population censuses (1951–1991 ELSTAT and 1946–1992 NSIB). Data by *dimotiki enotita* & *obshtina*

[23] No direct statistical data exist to confirm or deny the assertion.

From the late 1950s to 1991–1992, the boundary is closed almost completely, without bilateral exchanges or mutual influence. In Bulgaria, the development of remote districts is an official planning purpose. Local Bulgarian communities gain a support that is the envy of many Greek settlements. Development/underdevelopment models are built on well-defined strategies. The agrarian collectivization does not arouse severe depopulation as local head districts receive industrial and infrastructural investments in return (Rey 1996; Billaut 1960; Brossier and Petit 1975). Blagoevgrad and Smolyan emerge as powerful cities.

Quite the contrary, in Greece a strong urban draining leads rural areas to downfall. Slowly but surely, wild vegetation and trees gain ground in the once populated mountains. The creation on both sides of the "surveillance" and "forbidden" belts speeds up the transformation. The process may deserve another ethno-linguistic interpretation. The Bulgarian Rhodope, where Pomak populations prevail, shows an impressive birth dynamic, and this singularity finds an echo in the Pomak sectors of Greece: The greater dynamism of this ethnic group expresses a form of informal *resistance* against the achievement of the border project, which implies a brutal closure of its living space and a deep questioning of its traditional way of life.

Despite official talks, the 1990s achieve no significant operational rapprochement between Greece and Bulgaria (Fig. 8.31). Regardless of the political and

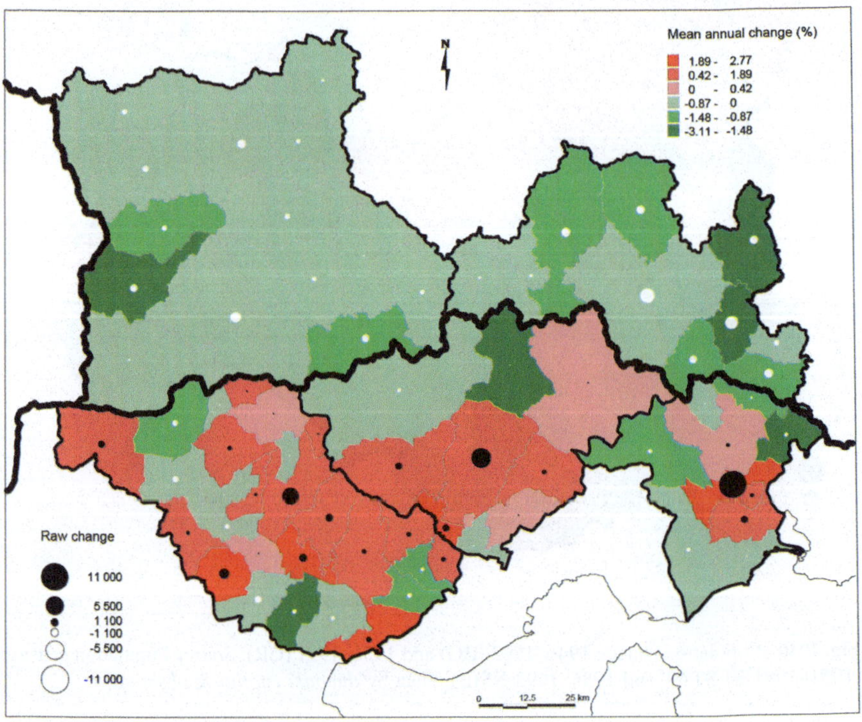

Fig. 8.31 Population change 1992–2001 (BG) and 1991–2001 (GR). *Source* Population censuses (1991–2001 ELSTAT and 1992–2001 NSIB). Data by *dimotiki enotita* & *obshtina*

economic revolution occurring in Bulgaria, the censuses show growing gaps between states. Southern Bulgaria faces its greatest demographic crisis. The "colonization" logic falls down. The general collapse leads to a massive emigration, which sometimes takes the form of close cross-border transfers.

Meanwhile, northern Greece has found some new attractiveness. The country finally opens its doors to migrants. The growth is no longer limited to the suburbs of Salonika and Athens. It also reaches smaller cities. Even a slight rural upturn is evidenced in the Macedonian countryside. The event partially unleashes the Hellenic village communities from their long-term decline and aging trend. For the first time, northern regions appear as remarkable places for the experimentation of new planning strategies.

The perspective of underdevelopment changes sides without destroying the Bulgarian demographic advantage immediately. The Bulgarian emigration has nothing to do with what happens in southern Albania: There, dozens of settlements sometimes lose until 80 % of their inhabitants in a few years (Darques 2004; Boulineau 2004; Deslondes 2004). This "resistance" results from the old national support offered to remote districts—a policy that still bears fruit. The areas dotted with balanced urban networks better withstand the post-communist downfall, at least by temporarily fixing rural populations. The industrial workers of the *oblast* of Blagoevgrad struggle against a general economic suffocation more successfully than the remote and off-centered area of Smolyan (Fig. 8.32).

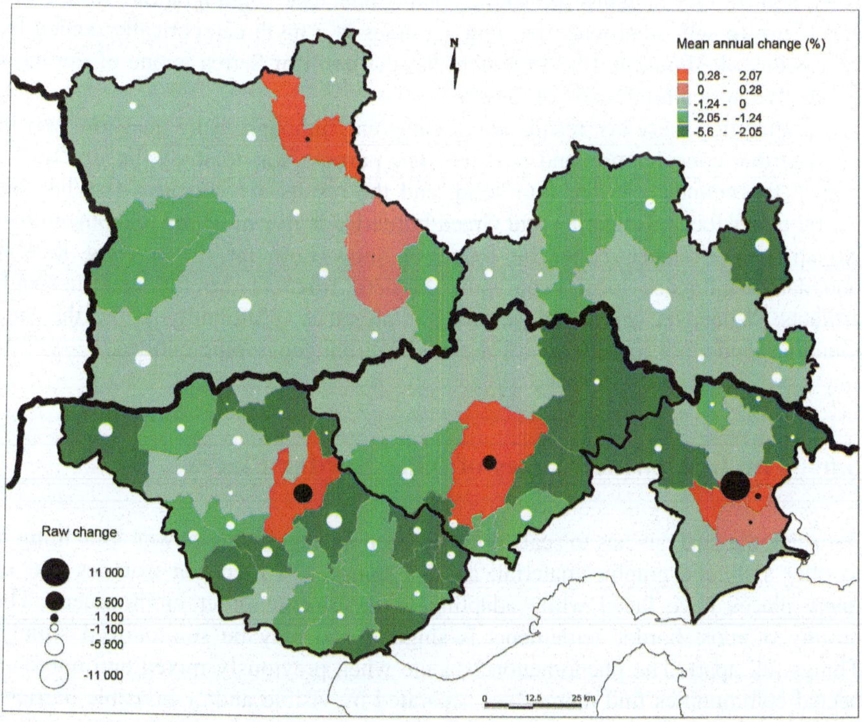

Fig. 8.32 Population change 2001–2011. *Source* Population censuses (ELSTAT and NSIB). Data by *dimotiki enotita & obshtina*

Unfortunately for the Greek countryside, the remission period does not last long. At the end of the process, both European countries unite in a common decay. The price to pay for this "alliance" is a global economic crisis whose causes are many and rooted in macro-economic trends. Bulgaria seems asleep, caught in a lingering recession. Petrich and Sandanski suffer from the downfall of transboundary movements and trade, while Smolyan is doing its best to limit damages. The Bulgarian demographic crisis poses a long-term threat to the clothing microbusinesses and other offshore manufacturing productions in search of a cheap workforce.

Despite the European integration of Bulgaria, the border effect is still there. After a decade of transient "euphoria," the Hellenic countryside faces a major demographic crisis whose whys and wherefores are not completely intelligible at the time of writing of the present work. Serres, Drama, and Xanthi strengthen their position against their Bulgarian competitors. Xanthi gains 10 000 residents in 10 years. Since 1991, the population of this remote city has increased by one-third.

Rural lands tend to have the exact opposite profile. Most importantly, we note the sharpest demographic drop ever registered over the last century in times of peace. The regional unit of Serres loses 25 000 inhabitants in 10 years (−15 %). The downfall reaches impressive rates in tiny villages. The municipal unit of Visaltia, for instance, loses 43 % of its population (8 500 in 2001, 4 900 in 2011). One-third of these *enotites* fall by more than 25 %. Even the communist collapse did not cause such decay across the boundary. The economic crisis has reinforced the weight of cities against the whole countryside. The "outmigration" from cities and return to self-subsistence farming activities is a myth categorically denied by census data. It should not be forgotten that the basin of Serres is one of the most productive agricultural areas of Greece.

Commenting these last results is not our point, although many questions would merit further consideration and research. The purpose was to study the structuring role of the boundary in the long-term, and the results are conclusive: Whatever period under focus, Bulgarian and Greek borderlands live in opposite momentums. No stronger evidence could be found to show how far the moving Balkan boundaries influence local geography. Modern frontiers are full and thorough artifacts: Whenever and wherever they apply, they profoundly affect the surrounding lands, not in half-way. They are *absolute* geographic entities.

8.6 The Intertwined Lives of Nine Border Places

As usual, the analysis has to scale down to keep its convincing power. One thing is to offer synthetic graphic material; another thing is to focus on troubles that so many places have faced while adapting to their new border environment. The destiny of most border settlements is singular and may be summarized simply: Things fall apart. The phenomenon is acute when previously mixed and interconnected communities find themselves separated by visible and/or invisible barriers.

The many anthropological studies addressing the issue suffer from limited spatial perspectives. From one village to the other, there is, in practice, considerable variation. When seized by the "hard border materiality," these places lose most of their monitoring capacities. Former "semi-independent" communities are condemned to the sudden relinquishment of their "sovereignty." The boundary abruptly introduces external players into local *games*. The result varies from one settlement to the other. The way "foreign" actors disrupt the game is decisive. In most cases, the result is clear: The army comes into power for decades, which dooms entire districts to a lasting decline. Borderlands are *suffocating* places: The close contact with a "despised enemy" and the everyday military pressure drive people to get *fresh air* far away.

By way of illustration, nine cases have been chosen on vicinity criteria with the boundary (Fig. 8.2). Observing the wide range of local situations was one of the main thrusts of the analysis. What is the cost to pay for neighboring the boundary in practical terms?

1. ***Ano Vrondu-Gorno Brodi (Άνω Βροντού-Горно Броди)***, "Bulgarian" community on Greek territory (25 km northeast of Serres, 10 km away from the borderline)

Completely destroyed twice, Ano Vrondu/Gorno Brodi is one of those places that have never recovered their former status. With its 4 800–6 800 inhabitants in 1905, it used to be one of most populated places by "Bulgarians." The community lives its late Ottoman years in an atmosphere of fierce competition between Exarchists and Patriarchists (Καφταντζής 1996). The village is directly involved in the Ilinden uprising.[24] Among the Ottoman troops, Andartes, and IMRO Komitadji, it hangs on in the middle of great quarrels and disputed lands.

In 1913, Greek bombings leave no house up, and the "Bulgarian" inhabitants are forced to run away. They settle in Gotse Delchev and Plovdiv, mainly. Few of them —maybe the Patriarchists—shall return home. In 1919, the Treaty of Neuilly introduces a Greek–Bulgarian population exchange—barely applied. The last "Macedonian" families leave Gorno Brodi in 1924, following paramilitary clashes in the neighborhood. At this point in time, some Greek refugees from Asia Minor and Thrace settle in the village. The buildings are once again destroyed during the Civil War, in reprisal to the ELAS activities. Since then, the village was kept isolated, peacefully employed to potato cultivation, surrounded by the Menikio–Orvilos–Vrondu peaks and their invasive forests.

The decade 2001–2011 is, however, a black period for the "municipal unit": It loses 56 % of its population. The community suffers from major operational problems, which include low productivity agriculture, difficulties faced in exporting harvest surplus for monetary reasons, indirect costs attributable to remoteness, without forgetting the impact of various administrative reforms (Table 8.4 and Fig. 8.33).

[24]Famous revolt against the Ottomans (1903).

Table 8.4 Ano Vrondu/Gorno Brodi's population over the century

1905	1913	1920	1928	1940	1951	1961	1971	1981	1991	2001	2011
4 500 Schismatic Bulgarians and 300 Orthodox Greeks[a]	794	1122	887	1118	348	567	450	347	408	452	199
6 500 Exarchist Bulgarians and 250 Patriarchist Bulgarians[b]			Including 245 refugees								

Sources [a](Φιλιππίδης 1912), [b](Brancoff 1905), Greek population censuses, ELSTAT

Fig. 8.33 Ano Vrondu/Gorno Brodi at the beginning of the century. *Source* земи (1919)

2. ***Melnik-Meleniko (Мелник-Μελένικο)***, Greek city on Bulgarian territory (13 km east of Sandanski, 13 km away from the borderline)

Melnik is singular case. This city of medium importance is the only one to have escaped the successful development scenario observed everywhere else in south-west Bulgaria over the twentieth century. The decay originates in the major Greek presence observed at the end of the Ottoman era, at a time when all head villages of the upper Struma and Mesta were in the hands of Muslims. In 1905, the Hellenic share reaches almost three-fourths of the total population, associated with minor Muslim and Bulgarian Exarchist groups. The thriving place, which had founded its prosperity on wine production and export in the eighteenth and early nineteenth centuries, had already lost its old haughtiness. Nonetheless, with around 3 500–4 000 people, it still was viewed as a town.

Melnik loses its Greek community abruptly with the Balkan Wars. The city is partially destroyed and its population thrown out toward Greece. The Patriarchist families move not only to Sidirokastro and Serres, but also to Salonika. Bulgarian

Table 8.5 Melnik/Meleniko's population over the century

1905	1920	1926	1934	1946	1956	1965	1975	1985	1992	2001	2011
2 580 Orthodox Greeks, 500 Muslims and 345 Gypsies[a]	723	722	552	472	522	551	417	329	287	243	217
2 650 Greeks, 950 Turks, 500 Bulgarians, 200 Gypsies and 30 Vlahs[b]											
3 825 Orthodox Greeks, 220 Patriarchist Bulgarians and 40 Exarchist Bulgarians[c]											

Sources [a](Φιλιππίδης 1912), [b](Кънчов 1900), [c](Brancoff 1905), Bulgarian population censuses, NSIB

settlers from the Aegean Macedonia enter deserted houses, but heavy destructions limit the inflows. In 1920, only 700 people stay there. The decline never came to an end. In 1968, the Bulgarian government decides to convert the community into a museum place. Despite its international reputation and tourist attractiveness, Melnik disappears almost completely, as many other places studied onward. The architectural heritage probably prevents the place from a thorough destruction (Table 8.5 and Figs. 8.34, 8.35).

3. ***Lialovo (Лялъово-Λιάλοβον)***, "Greek–Muslim" village on Bulgarian territory (11 km south of Gotse Delchev, 5 km away from the borderline)

The case of Lialovo is specific. It used to be the only place in Bulgaria populated by Greek–speaking Muslims, otherwise called *Valahades*.[25] The 1905 census reports the existence of "1 000 Muslims speaking Greek as a mother tongue and of Greek culture." On the other hand, V. Kanchov mentions 620 "Greek Muslims" in 1900. The origin of the population is much discussed (Hellenized Pomaks, Greeks from Halkidiki converted to Islam, Vlach-Turkish semi-nomads). Their position recalls the Turco-Cretans depicted by E. Kolodny, in a different situation (Kolodny 1995).

Anyhow, with the Balkan Wars the community moves to Greece. Some villagers come back in 1916. In 1920, there are only 360 people left. They finally leave the area in 1928, toward Turkey this time, without return. In 1934, only 146 souls occupy the remaining dwellings. A few "Bulgarian" refugees from the Aegean

[25]They could also be identified as "Grecomans"—a name granted to the Albanian, Vlach, or Bulgarian communities that belonged to the Greek "nation." The inhabitants of Lialovo might be viewed as linguistic/cultural Greeks, Muslims according to their religion, and Bulgarians according to their citizenship after 1912. However, at the beginning of the century, each fighting group or nation could claim that kind of minor communities as attached to them. Their future was sealed in advance.

Fig. 8.34 Melnik in the 1920s. *Source* www.lostbulgaria.com. *At this date, the wealthy city of the eighteenth century is a distant memory. The last destructions and migrations have turned the place into a ghost village*

Macedonia have replaced the former inhabitants, but the new community cannot avoid a severe decline in the first years of the collectivization (migration to Koprivlen, mainly). The locality disappears from the statistics in the 1960s. Ramblers shall only find today the surviving walls of fallen down houses.

4. ***Ano Porroïa-Gorni Poroi (Άνω Πορρόια-Горни Порой)***, mixed Vlach–Bulgarian village on Greek territory (8 km northwest of Lake Kerkini, 3 km away from the borderline)

Fig. 8.35 A view of Melnik today. *Source* R. Darques, 2005. *The relics of the old city have disappeared, mainly under the effects of erosion and forest growth. A "new-age" place has emerged*

Beside its appreciated cold waters, mighty plane trees, living taverns, and stunning views over the Kerkini Lake, Ano Porroïa is a place whose history is complicated. In 1905, the last Ottoman statistics detail a mixed population: 790 Orthodox Greeks (among which 725 Vlach speaking people), 352 "Romanians," 1 487 Schismatic Bulgarians, and 306 Muslims. In other words, the locality reaches about 3 000 people, divided into three main groups: Bulgarians-Macedonians, Vlachs of different political persuasions, and a small Muslim component.

In 1913, the Greek troops burn the village. Most of the Slav dwellers find refuge in Bulgaria where they settle around Petrich, Gorna Dzhumaya, and Plovdiv. Some Vlach families manage to stay. In 1920, 1 750 souls are registered. This number could signal the clearance of any Slavic presence. In fact, the situation is more complex. The settlement of Greek families from the upper Struma is likely at this date—a necessary income for keeping up so many people despite the "Bulgarian" massive outflow.

Afterward, the Greek component gets stronger with the arrival of Asia Minor and Thrace refugees. These are 837 in the village in 1928, but several hundreds of them find other grounds to occupy in the surroundings. From the 1940s onward, the place suffers the outcomes of the rural exodus although managing to keep a consistent population until 2001 (about 1 500 residents). This result is a remarkable achievement for an area locked within the "control zone" of the Iron Curtain for decades. The Vlach presence, favorable sunlight exposure, and fertile lowlands dedicated to tobacco crops have probably limited the rural outflow—without mentioning few tourist activities. Since 2001, however, the fall has reached unexpected levels (−37 %).

5 ***Kolarovo-Kolarevo (Коларово-Κολάροβον)***, Muslim village on Bulgarian territory (8 km west of Petrich, 5 km away from the borderline)

Kolarovo brings us back to a case of entirely renewed society. In 1905, 1 000 Muslims and 125 Schismatic Bulgarians form the population of the village. The site is opposite to Ano Porroïa. Halfway up the Beles slopes, it controls the "Iron Gates" pass, which crosses the mountain range in its central part. Like Ano Porroïa, it takes advantage of lowlands in the Strumeshnitsa Valley but suffers from a poor solar exposure on the north-facing side of the Beles.

Once again, the Greek troops burn the place in 1913. The Muslim community will never come back. To "compensate" for the loss, many "Bulgarian" refugees from the Aegean and Vardar Macedonia enter the standing buildings after the Balkan Wars. They come from Porroïa and Demir Hisar/Sidirokastro, mainly. Back and forth population transfers occur across the new borderline, without interruption. One can hardly imagine the chaos caused by neighboring families commuting their destroyed homes and settling without official permission, apart from the one granted by the international treaties.

In 1920, the locality reaches 850 souls. It will never stop growing after this date. In the 1950s, the community has increased twofold, whereas its Greek counterpart (Ano Porroïa) had only succeeded in keeping its position thanks to the settlement of Asia Minor refugees. According to the last census and despite the "bad years" of communist collapse, Kolarovo succeeds in maintaining about 2 000 people.

The two following cases are also astonishing. They concern a specific region called *Chech (Чеч* or *Чеча)*. At the beginning of the twentieth century, demographic data are missing—they stand outside Salonika's *Vilayet*. The Chech area stretches east of the Nevrokop *kaza* until Dospat and includes today's part of the Gotse Delchev *oblast* and the northern *nomos* of Drama. The Chech "region" was composed of dozens of Pomak localities (from 60 to 120), with few "Turkish" Muslims among them (Srebranov 2006). The homogeneity of the population is remarkable compared with the complex examples studied further west. After the Balkan Wars, the borderline cuts the area into two parts. Let us focus on two Pomak villages.

6 ***Skaloti-Liban-Liubene (Σκαλωτή-Λιμπάν-Либан-Любене)***, former Pomak village on Greek territory (30 km northeast of Drama, 11 km away from the borderline)

This village exclusively populated by Pomaks of the Drama Chech belongs to a complex network of settlements mainly devoted to pasture, small craft industry, and subsistence farming. Within the limits of the *kinotita*, no fewer than eleven hamlets are reported in the early Greek years, which count for 2 300 souls at least. Skaloti is the only place having escaped complete desertion—but the term is excessive. Facing the invasive forest of Elatia-Karadere, Skaloti is credited with 84 inhabitants in 2011. The decrease was happened stage by stage.

The *Committee for Western Thrace* points out 936 people in the settlement of Liban itself before 1912 (Comité de la Thrace Occidentale 1922). In front of the Greek troops' advance, many families probably run away during the Balkan Wars, connecting their lot with that of other Turkish-speaking communities. In 1920, 439 people still live in the community. The Treaty of Lausanne validates the decision taken by those who had already left. As composed of Muslim people, Pomak families are forced to move to Turkey, where they settle in the outskirts of Edirne, mainly. In 1928, 125 souls still live in Liban, of whom 113 are Asia Minor and Thrace refugees. A short period of development follows: The scattered refugees gather in the main hamlet, but the Civil War and the Iron Curtain sign the death sentence of the Karadere region. The future is committed to wildlife.

7 ***Kochan-Kochen (Кочан-Кочен)***, Pomak village on Bulgarian territory (25 km east of Gotse Delchev, 6 km away from the borderline)

Across the boundary, in the Nevrokop Chech, another place exclusively populated by Pomaks has faced an exact opposite fate. During the first Balkan War, the Chech villages join to support the Ottoman army until its defeat. Afterward, they take part in the *Gümülcine Republic*. On this occasion, they commit to helping Turkish–Muslim fighting groups. In 1913, when Bulgaria confirms its rights over northern Chech, many families fearing reprisals from the Bulgarian authorities decide to seek refuge in Turkey. From its almost 1 300 people of 1905, only 984 souls remain in 1920. In 1923, few Pomak households coming from the Drama Chech settle in the village. The population reaches 1 170 in 1926 and will never stop its growth so far.

In 2011, 2 600 people were dealing with tobacco cultivation as a main activity. Despite the historical autonomy of those Pomak communities, the communist regime successfully managed to achieve its *regular* farming revolution and collectivization. Thanks to high birthrates and through the deployment of infrastructures and microindustries, the authorities succeed in maintaining consistent populations close to the border. Nevertheless, the recent years have pushed hundreds of Kochantsi to emigrate, at least temporarily, to Greece and other foreign countries.

The same scenario could have been expected further east, in other Pomak villages, but Western Thrace has escaped the population exchange of 1923. This exception has created the necessary conditions for disconnecting Bulgaria from the Aegean shore. The Thracian border appears as an antithesis of its Macedonian twin because the old ethnic-confessional balances have not been much disturbed. However, the persistence of a demographic Pomak cross-border area does not mean that former interrelationships could continue to operate. At least since WWII, the interdiction measures have severely weighted on this territory where any transboundary exchange has been banished (Labrianidis 2001, 1997).

8 ***Ehinos/Sahin/Shahin (Εχίνος-Şahin-Шахин)***, Pomak village on Greek territory (16 km northeast of Xanthi, 11 km away from the borderline)

Ehinos is a Thracian place that has learned to live self-sufficiently long ago. As head of the Pomak villages of Xanthi, it has been condemned to confinement for decades. This remoteness has eventually prevented the community from the terrible downward path endured by the Chech residents. The tragic developments of the early twentieth century provide impetus for massive departures toward Turkey. From its 2 200 souls before 1912, only 1 400 still live in the area in 1920. In 1923, the Muslims of Thrace avoid the Greek–Turkish population exchange. The exodus stops. Regardless of official recommendations, the 1928 census indicates that 50 Greek Asia Minor refugees have decided to settle down. From then on, the self-dynamic of the population and its remarkable marketplace position have allowed Ehinos to retain about 2 500 people. This rather favorable fate is paid back by the mountain settlements, whose remoteness has increased. Ehinos, much less than Xanthi itself, has become a local outlet for Pomak people in search of some "urban" comfort.

9 ***Smilyan-Ismilyan (Смилян-Исмилян)***, Pomak village on Bulgarian territory (8 km southeast of Smolyan, 10 km away from the borderline)

Located in a basin of the upper Arda Valley, surrounded by deep forests and grassy pasturelands, Smilyan used to be a major Pomak settlement. According to the *Committee for Western Trace*, 1 900 "Turks" lived here before the Balkan Wars. In December 1912, the place is destroyed by Bulgarian troops who tried to convert the inhabitants to Christianity. In 1920, only 700 people live on site. However, from this date on, the village has become increasingly important.

In 2011, 1 790 people occupy a modern locality provided with administrative buildings and various tourist facilities. Significant efforts have been made during the communist period. Modernization is here much more efficient than a few kilometers southward, on the Greek side of the Rhodope, where few localities have received infrastructure and technical facilities. In Bulgaria, most scattered settlements whose economy was based on subsistence farming are abandoned in the 1940s and 1950s—a movement driven by the collectivization. In return, the surviving communities have become the target of external investments. They were able, in these conditions, to keep up significant activities and maintain a territorial control on large areas.[26] Meanwhile, in the Greek Rhodopian countryside, mountain settlements often had to experience a slow and painful agony. It follows from the above that the urban–rural limit can hardly find a similar definition in Bulgaria and Greece.

[26]This process explains the traditional difference between the areas covered by the *obshtini* (large surfaces) and *kinotites* (probably 1/3rd to 1/4th of their Bulgarian equivalent).

8.7 Perspectives on Confinement: From Demography to Socioeconomics

These examples may help understand the level of complexity reached by local issues. Most Balkan settlements impacted by border fluctuations in the past centuries confront the same thorny problems: Upstream, no measurement tool can help unravel identities and territories; downstream, political-demographic crises lead to disastrous outcomes that escape any survey. In this respect, research projects are necessarily limited. This explains why the Balkans are usually regarded either as an open bazaar or as a black box system.

The mapping exercise validates the idea of fundamental changes induced by the creation of the new Greek–Bulgarian borderline in a mountainous region that, for centuries, had been dependent on Aegean and other far away Ottoman cities. Against such a background, it seems in hindsight that the Bulgarian Rhodope was condemned to inevitable decline. However, as an outcome of serious efforts and proactive planning policies applied during the communist era, a "new political-territorial order" has emerged. For a time, remoteness could no longer be equated to underdevelopment.

Recent border hardening or softening cannot be understood without addressing past legacies. Decades of dual track planning policies have utterly upset the old Ottoman territorial "order." From this it might be deduced that most border areas have fallen "victim" of a great "trauma," whose geographic solution has found different ways: bleeding and freezing on the one side; support and development on the other side. In many places, migration was not only an expedient to tragic events. The mobility of people has been established as a lasting way of life. Balkan societies are well known for their close links with different global diaspora communities. It comes as no surprise that international population flows have become increasingly appealing in the 1990s. Households who have been through forced relocations find it easier to leave temporary asylums for even more promising skies.

Behind the integration praised by Brussels, this work also confirms that a demographic implosion is under way in the southern Balkans. Regardless of official statements, cross-border diverging behaviors have survived the 1990s. Things got even worse rather than better recently. This is what the analysis of the 2011 population censuses emphasizes. Regional imbalances do not evolve toward a uniform resolution. European cohesion programs have been thought as ways of reducing territorial inequalities especially between neighboring countries. A number of crisis-related factors seem instead to lead to enhanced local imbalances these last years, especially at a subnational level, which in turn seriously challenge the efficiency of cross-national integration policies.

The European crisis in the Balkans demonstrates that addressing territorial discrepancies with the intent to establish stable and balanced relationships between local community "partners" may be more difficult than expected, for known reasons: Lands and territories are not stable abstract entities. They evolve continuously under the pressure of external and internal inputs. Crisis times simply take ordinary processes to higher unexpected levels.

Legal provisions and policies that were supposed to bring further harmonization between Greece and Bulgaria before and after 2007 have not achieved the required territorial benefits. Such mixed results must be seen against the background of long-term inter-regional relationships. Throughout the communist period, the two *models* had cumulated so great mutual ignoring capacities that the contemporary border "disaggregation" might be viewed as a process of discharging forces that had remained under control for decades (overdevelopment vs. underdevelopment). Implementing coordinated cross-border development policies in this context is obviously risky. The post-2007 crisis serves as a catalyst.

Not surprisingly, the statistical analysis has been subject to various impediments. Collecting long-term data was a real challenge. The greatest obstacle is the *interoperability* of available information between neighboring countries (Blasius and Thiessen 2006). Against all expectations, this work demonstrates that the official transnational mappings edited by Brussels have to be considered with great caution. Data harmonization and scale agglomeration produce obvious mind-numbing effects and obliterate the varied characteristics of local territories. And this is without considering that, historically, most statistical indicators have been thought and collected *not* to be compared from one country to the other.

Of course, the application of *converging* algorithms shall always produce "comparable" recombined or agglomerated metadata. Broader scales may even bring a solution to high-scale "heterogenous" raw data. What can we expect from such leveled inputs? Nothing significant. Most GIS operators underestimate the importance of these issues when developing geodatabases. Only few *specialists* really master data accuracy assessment techniques. Have you ever seen thematic maps that mention estimated error rates? Should these rates reach 20–40 %, would not you reject the output for its irrelevance? Countless maps arising from improperly mastered GIS databases are, in fact, meaningless. Similarly, comparing data issued from statistical institutes with those computed from satellite imagery, for instance about land occupation, often leads to unexpected and divergent results. The *soft uncertainty* of quantitative analyses is much more easily accepted by the public than other *hard uncertainties*, such as the geography of international boundaries!

The demographic dimension of this final chapter is not contingent. A long time has been spent trying to find comparable indicators in other socioeconomic fields. Most attempts have ended up in stalemate, despite our efforts. Between national statistical offices, most data mismatch: Indicators have different definitions, definitions vary across time, data availability is uncertain at the right scale, collection methods differ, the simultaneity of acquisitions is not fulfilled, and repeated administrative changes are usually ignored. Against the opinion of many officials who praise the increased coherence of various indicators, looking closer at the issue demonstrates that figures strongly resist.

Of course, mapping potato productions or irrigated acreages is still possible. What about comparing company numbers and characteristics, unemployment rates, migration flows, or standards of living? Such statistics are supposed to be meaningful at national and international levels. In our experience, at local scales, statistical biases pose virtually insoluble problems. Was the great scandal of Greece's truncated statistics a real surprise when made public in 2008? Only in the eyes of poorly aware observers (Darques 2002).

Obviously, cross-border studies still await their grail. International comparisons are made possible through data interpolations and adjustments that reduce their conclusive power. Here, as in other fields, national specificities are difficult to overcome, but the statement is of poor visibility considering that virtually all geographic studies are carried out within national frameworks.

A series of successive tests have been carried out on seemingly comparable socioeconomic variables. These confirm that the acute Greek–Bulgarian disparities pointed out earlier have equivalents in fields as diverse as agriculture, population structure, dwellings, tourism, forestry, and industry. The hypothesis of unachieved *transitions* is validated: Between Greece and Bulgaria, additional time will be needed before restoring some land and human "continuity."

References

Ancel J (1930) La Macédoine, son évolution contemporaine. Delagrave, Paris

Andréadès A (1928) Les effets économiques et sociaux de la guerre en Grèce. PUF, Paris

Behar C (1998a) Qui compte? Recensements et statistiques démographiques dans l'Empire ottoman, du XVIe au XXe siècle. Histoire et mesure 13(1–2):135–145

Behar C (1998b) Sources pour la démographie historique de l'empire ottoman: les tahrirs (dénombrements) de 1885 et 1907. Population 53(1–2):161–177

Bieber F (2000) Muslim identity in the Balkans before the establishment of nation states. Muslim Minorities Balkans 28:13–28

Billaut M (1960) La collectivisation agraire en Bulgarie: l'exemple du village de Petartch. Annales de Géographie 69(375):484–492

Blasius J, Thiessen V (2006) Assessing data quality and construct comparability in cross-national surveys. Eur Sociol Rev 22(3):229–242

Boulineau E (2004) Le déclin démographique bulgare. Bilan d'une décennie de transition post-socialiste. Espace Populations sociétés 3:545–558

Brancoff D (1905) La Macédoine et sa population chrétienne. Plon, Paris

British War Office (1908) Salonika. The Balkans. Turkey, London

Brossier J, Petit M (1975) Un exemple d'agriculture socialiste: L'agriculture bulgare. Etudes Rurales 60:55–72

Carte des écoles chrétiennes de la Macédoine (1905) Erhard, Paris

Chirot D (1989) The origins of backwardness in Eastern Europe. University of California Press, Berkeley-Los Angeles-Oxford

Comité de la Thrace Occidentale (1922) Les statistiques de la population en Thrace occidentale. Constantinople

Cook L (2004) The quality and qualities of population statistics, and the place of the census. Area 36(2):111–123

Cowan JK (2008) Fixing national subjects in the 1920s southern Balkans. American Ethnologist 35(2):338–356

Dalègre J (1997) La Thrace grecque: populations et territoire. L'Harmattan, Paris

Darques R (2000) Salonique au XXe siècle: de la cité ottomane à la métropole grecque. Espaces et milieux, CNRS Editions, Paris
Darques R (2002) La Grèce et l'Europe: une intégration paradoxale. CEMOTI 33:177–194
Darques R (2004) L'Albanie en transition: mutations démographiques et recomposition territoriale (1989–2001). Espace, Populations, sociétés 3:559–575
Darques R, Deslondes O (2008) Bulgarie-Roumanie, les nouveaux horizons de l'Union européenne. Méditerranée 110:136
Deslondes O (1997) Les fourreurs de Kastoria entre la Macédoine et l'Occident. CNRS Editions, Paris
Deslondes O (2004) La population des Balkans depuis 1990: aspects géographiques de la crise. Espace populations sociétés 3:487–498
Desrosières A (2009) How to be real and conventional: a discussion of the quality criteria of official statistics. Minerva: Rev Sci, Learn Policy 47(3):307–322
Dimitropoulos D (2008) Limites intérieures dans l'espace grec au temps de la domination ottomane. L'aspect géographique des administrations politiques et ecclésiastiques. Historical Review-La Revue Historique 5:239–253
Dragostinova T (2011) Between two motherlands: nationality and emigration among the Greeks of Bulgaria, 1900–1949. Cornell University Press, New York
Foody GM (2007) Map comparison in GIS. Prog Phys Geogr 31(4):439–445
Gartner GF, Ortag F (eds) (2010) Cartography in central and Eastern Europe. Springer, Berlin
Glenny M (2012) The Balkans. Nationalism, war, and the Great Powers, 1804–2011. Penguin Books, New-York
Gregory IN (2002) The accuracy of areal interpolation techniques: standardizing 19th and 20th century census data to allow long-term comparisons. Comput Environ Urban Syst 26(4): 293–314
Gregory IN, Ell PS (2006) Error sensitive historical GIS: Identifying areal interpolation errors in time series data. Int J Geogr Inf Sci 20. (Taylor & Francis Ltd)
Hall RC (2011) The modern Balkans. University of Chicago Press, Chicago, A history
Karpat KH (1978) Ottoman population records and census of 1881-82-1893. Int J Middle East Stud 9(3):237–274
Karpat KH (1985) Ottoman population 1830–1914. The University of Wisconsin Press, Madison, London, Demographic and social characteristics
Kasaba R (2010) A moveable empire: ottoman nomads, migrants and refugees. University of Washington Press, Seattle-London
Kolodny E (1995) Des musulmans dans une île grecque. Les "Turcocrétois". Mediterranean World 14:1–16
Koulov B (1992) Tendencies in the administrative territorial development of bulgaria (1878–1990). Tijdschrift Voor Economische En Sociale Geografie 83(5):390–401
Καφταντζής Β (1996) Ιστορία της πόλεως Σερρών, vol 3. Thessaloniki
Кънчов В (1900) Македония. Етнография и статистика. Darzhavna Pechatnitsa, Sofia
Labrianidis L (1997) The impact of the Greek military surveillance zone on the Greek side of the Bulgarian—Greek borderlands. Bound Secur Bull 7(2):82–93
Labrianidis L (2001) 'Internal Frontiers' as a hindrance to development. Eur Plan Stud 9(1): 85–103
Ladas SP (1932) The exchange of minorities. Macmillan, New-York, Bulgaria, Greece and Turkey
Lampe JR, Jackson MR (1982) Balkan economic history, 1550–1950. Indiana University Press, Bloomington
Lixin L, Revesz P, Egenhofer MJ, Mark DM (2002) A comparison of spatio-temporal interpolation methods. In: Egenhofer MJ, Mark DM (eds) GIScience 2002. Proceedings (Lecture Notes in Computer Science vol 2478), Berlin, Springer
Mazower M (2000) The Balkans. A short history. Weidenfeld & Nicolson, London
Mentzel P (2000) Conclusion: millets, states, and national identities. Nationalities Papers 28(1):199–204

Michailidis ID (1998) The war of statistics: traditional recipes for the preparation of the macedonian salad. East European Quarterly 32(1):9

Mouzelis N (1976) Greek and bulgarian peasants—aspects of their sociopolitical situation during interwar period. Comp Stud Soc Hist 18(1):85–105

Neuburger M (2000) Pomak borderlands: muslims on the edge of nations. Nationalities Pap 28 (1):181–198

Μπαρτζούδης Γ (1995) Τα εγγειοβελτιωτικά έργα στη πεδιάδα Σερρών. Soil Amendment Directorate, Serres

Panzac D (1992) La population de la Macédoine au XIXe siècle (1820-1912). Revue du Monde Musulman et de la Méditerranée 66:113–134

Panzac D (1993) La population de l'Empire Ottoman. Cinquante ans (1941–1990) de publications et de recherches. Travaux et documents de l'IREMAM. CNRS, Aix-en-Provence

Peace CEfI (ed) (1914) Report of the international commission to inquire into the causes and conduct of the balkan wars. Washington

Péchoux P-Y (1968) Les paysans de la rive orientale du bas Nestos (Thrace grecque). Etudes rurales 29:5–61

Péchoux P-Y (1977) Formation du réseau administratif et intégration du territoire dans la Grèce moderne. In: IEDES (ed) Etat, pouvoir et espace dans le Tiers Monde. Paris, pp 187–212

Prévélakis G (1994) Les Balkans. Cultures et géopolitique, Nathan, Paris

Rey V (1996) Les nouvelles campagnes de l'Europe centre orientale. CNRS editions, Paris

Schindler S, Curado N, Nikolov SC, Kret E, Carcamo B, Catsadorakis G, Poirazidis K, Wrbka T, Kati V (2011) From research to implementation. Nature conservation in the Eastern Rhodopes moutains (Greece and Bulgaria), European Green Belt. J Nat Conserv 19(4):193–201

Shaw SJ (1978) The ottoman census system and population, 1831–1914. Int J Middle East Stud 9 (3):325–338

Sivignon M (2009) Les Balkans. Une géopolitique de la violence, Belin, Paris

Srebranov R (2006) Bulgarian muslims from the chech region and their linguistic self-identification. Int J Sociol Lang 2006(179):131–143

St-Hilaire M, Moldofsky B, Richard L, Beaudry M (2007) Geocoding and mapping historical census data—The geographical component of the Canadian Century Research Infrastructure. Hist Methods 40(2):76–91

Χουλιαράκης Μ (1973–1976) Γεωγραφική, διοικητική και πληθυσμιακή εξέλιξις της Ελλάδος, 1821–1971, vol 1–4. Athens

Yosmaoğlu İK (2006) counting bodies, shaping souls: the 1903 census and national identity in ottoman macedonia. Int J Middle East Stud 38(1):55–77

Δημοτική Πινακοθήκη Θεσσαλονίκης (2004) Χαρτογραφώντας τη Μακεδονία. Museum of the Macedonian Struggle, Thessaloniki

Φιλιππίδης Δ (1912) Μακεδονία. Εθνολογική στατιστική των βιλαετίων Θεσσαλονίκης και Μοναστηρίου. Nomikis, Athens

земи Кзинб (1919) Македония в образи Court Printing Office, Sofia

Chapter 9
Conclusion

By some extraordinary paradox, the desire to address the *balkanization* process is as much palpable among people as there is an obvious lack of empirical knowledge of the issue. Political risk and military pressure, through the vacuum effect they create, lead to a form of geographic *oblivion*. Among the few available documents allowing addressing the border issue, old maps are essential propaganda tools before they are testimonies of peace treaties. What do they teach us about borderlands? Almost nothing. They are expressions of political injunctions. The Balkans had never been more so mapped than in the nineteenth century, when the nations were still at a draft stage and dreamed about future events able to achieve their territorial ambitions.

Once delineated and marked, boundaries leave the confines of people's minds and become cartographic *taboo*. Suddenly, entire regions fade away from the earth. How do they switch from the publicly discussed issues to official secrets? Deprived of the right to be mapped and studied, whole districts have been "put at a distance" for decades, as lands in the back of beyond. It comes as no surprise for us now that vast areas able to broadcast their influence and special status far in the backcountry could no longer be swept under the rug.

Maps have an incredible capacity: They can make anybody believe that, beyond certain limits, nothing else exists than abysmal void (Agnew 1998, 1994, 2010). Communities separated by a few hundred meters can thus spend decades without any contact, even formal. Maps teach us that real barriers are mental. Sometimes, graphic representations take precedence over reality. When ordinary tourists crossed the Bulgarian or Albanian boundaries in the 1960s, they actually went from one world to another. The distance from a Greek border village to its Albanian equivalent was *meant* to be the same as the earth–moon distance. And no one ever saw a map figuring the Sea of Tranquility and the Mediterranean Sea on the same sheet.

This world of complete information blackout now belongs to the past. GIS data repositories are built today to give details on whatever place of the globe, with an

© Springer International Publishing Switzerland 2017

R. Darques, *Mapping Versatile Boundaries*,
Springer Geography, DOI 10.1007/978-3-319-40925-2_9

excellent degree of precision. However, despite terabytes of digital information, many areas remain out of the spotlight (Pickles 1995). The present work aimed at filling in the gaps while restoring in some way the tarnished reputation of the Balkan area.

Nowadays, most works in geography promote a practice of the discipline where maps are assigned little importance. Ethnographic and sociological methods prevail. Such mapping rejection is questionable. In these writings, the creation of various maps and the use of remote sensing imagery were an opposite, although non-exclusive, choice. Graphically picturing the area was essential in bringing the study to a successful conclusion. Without applied cartography, the approach would have failed: Unlikely Balkan confines would have remained by the wayside. Leaving apart the few road sheets edited by major commercial companies, who has ever handled a pleasing map of Kosovo? What about Dobrudja, Syrmia, Rhodope, or Bosnia? Contested places are often fantasized; they are randomly explored and badly pictured.

9.1 Deep Times and Unlikely Territories

The Balkan area is anachronistic land. The nearest historical equivalent to the current European "open space" goes back to the Ottoman era. For most Europeans and North Americans, these *Turkish* ages are lost in the depths of time. In the eyes of local individuals, however, the end of the Ottoman Empire is a lively past, strongly established in all minds. We therefore should not be surprised to observe that mixed feelings among people accompany the European enlargement and border opening. There is no doubt that the fall of capitalist/communist fences has brought relief to people who used to live in silence and resignation until recently. The new era guarantees freedom of expression and restores the mobility of goods and people after decades of mutual blockade. Despite these all, a space "without borders" is also perceived by Balkan peoples with anxiety as a general setback supported by nation-states that had fiercely fought against the "Turks" to impose inviolable sovereign perimeters.

The Ottoman past itself is seen with more duplicitous feelings than the official historical rejection would suggest. Dwarf states are proud of their independence, although self-reliance has various drawbacks: Because of their limited weight, they are always at the risk of losing the post-modern interstate competition. Byzantine, Ottoman, and other Slavic medieval Empires promoted at least some idea of universality and established powerful dominions without definite frontiers. They provided perspectives. Quite the contrary, border-based nation-states would entail a general waiver arising from their restricted operational capabilities. Supporting the European venture especially in view of the current economic downturn is a way of maintaining some external projection capabilities against the negative consequences of past legacies–backwardness and remoteness.

Borders and boundaries play a much greater role in shaping the whole Balkan geography than imagined. Balkan countries are young beings haunted by distant memories. Borderlands are open-air museums, not only through the many historical monuments scattered in backcountries, but also through the built mental models. They easily wear wrong size clothes imposed by remote authorities. Between the actual and imagined attributes of borderlands, bridges had to be built.

Some may say that the study of past frontiers would have led us to focus on a great many *ghost borders*. A *ghost border* is usually defined as the landmark left by a boundary, once moved away or suppressed. Even though the concept may apply in some areas, it does not seem to us operative in the Balkans, unless we make it a universal model. Balkan "ghosts" are everywhere because frontiers live many lives. Once they got rid of their political character and they have become internal interfaces, they may even gain importance. Ghostly visions fail as they only consider two states of being and ignore the many reincarnation forms of each borderline. Frontiers are translations of geographic discontinuities; as such, they cannot be given a single political meaning. The distinction between external and internal limits thus loses significance.

Besides, the initial premise of research is that sovereign boundaries are limits among others. Some internal interfaces, with or without administrative recognition, are sometimes more significant and often less studied than state confines. International boundaries are not the alpha and omega of everything in the Balkans. They reveal the complexity of many domestic situations and demonstrate how difficult it is to govern fragmented lands. Without complying with the established patterns, the Balkan borders adopt kaleidoscopic forms. Their complexity may thus be depicted through a combination of paradoxes as it is often said about people: Everyone is the sum of his self contradictions (Fig. 9.1).

The nature of borderlands is difficult to determine, because of constantly changing conditions (Paasi and Zimmerbauer 2016). Standard mappings will not help us overcome the problem. Everything is wrong in applied cartography. Lines are used in place of areas, points represent composite entities, and plain colors replace infinitely nuanced hues. All pictured data are the rough expressions of a complex reality. The complexity of each situation is perhaps the reason why conventional boundary models and clean linear symbols achieve great success: Too much difficulty leads to enhanced simplification. The focus of interest on border issues may thus be seen as a reverse tendency: Graphic symbols seize minds retroactively, in such a way that their pre-eminence is self-confirmed.

The recent development of computer tools and GIS databases did not solve the problem; it made it worse. People are now haunted not only by clear linear features, but also by unlimited accuracy requirements. The combined advances in photogrammetry, geodesy, and both material/software informatics create a cocktail of parameters enforcing the symbolic power of maps to an utmost never experienced before. Are "real" people and territories ready to accept a geometric *diktat*? How far is a symbol able to constrain reality (Painter 2008)?

Uncertainty is a concept this study has enhanced many times as a basic application of geography and land planning, although the human mind seems reluctant to

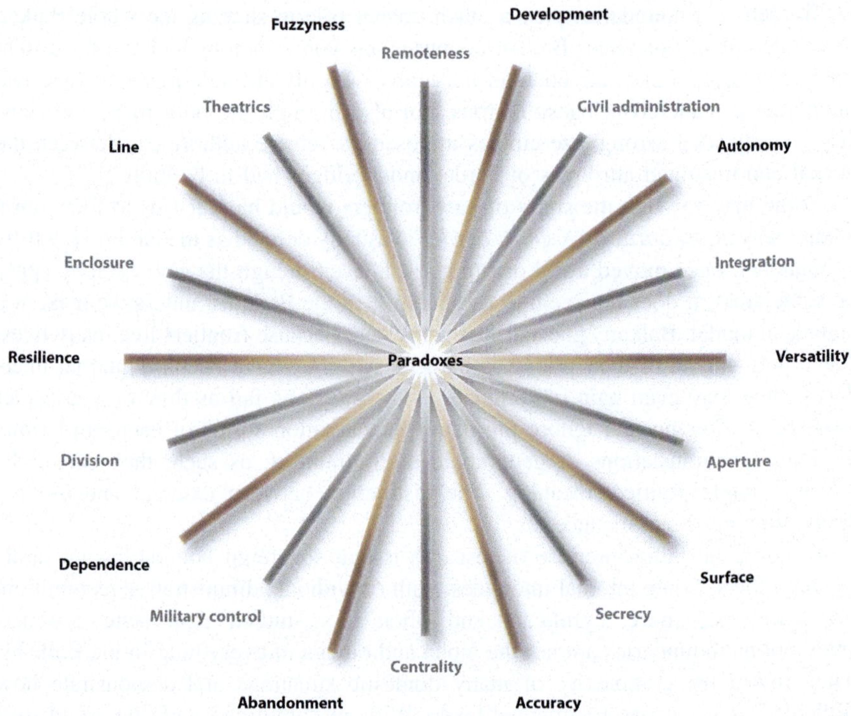

Fig. 9.1 Frontier paradoxes. *Source* R. Darques, 2015

such idea and prefers sharp edges to help identify objects. Applied geography opposes a strong resistance to political claims over land through the necessary analysis of complex territorial systems. Digital image processing techniques can help understand uncertainty as a principle pertaining to any spatial body. In remote sensing, confusion results from the inability to classify image pixels through "pure" electromagnetic signals. This has led to the creation of an interesting neologism: "mixel." A *mixel* is a pixel made of mixed electromagnetic signatures. "Identifying" a pixel and assigning it to one or more landuse classes imply to think the smallest image unit in terms of statistical probability. Spectral mixing teaches us that any classified land surface without associated standard error is nonsense. Mixels disclaim the dictatorship of "clean" thematic mappings. Remote sensing techniques and classical cartography usually live in worlds apart. One of the key challenges of this work was to bring these two *divided parties* together.

Many examples can be found to demonstrate that the uncertainty principle prevails (Merry 1995; Brown 1995; Schofield and Gallego 2011). Let us remember the "neutral" areas left aside by various international peace treaties, the deliberate decision to leave unsettled the boundaries of some states, the inability to map land with accuracy, the mobility of any natural landmark, and the incongruity of a linear boundary in the eyes of expanding states or nomadic societies (Ebel 2008). Despite its claim to universality, territorial *steadiness* is a geographic aberration and a lasting social myth altogether. Maybe because of its negative connotation, some analysts prefer replacing the word *uncertainty* by *risk*. A risk is measurable. It brings some security feeling in front of the unknown. However, the pejorative sense of the word *uncertainty* is of limited scope in geography. Our goal was not to define a supposed political or military risk associated with borderlands, but to help understand the conditions that lead to the emergence of such negative preconceptions.

9.2 Adakale: The Devil (and the Angels) Are in the Details

Doubtlessly, recent history has introduced high degrees of centralization few early states or Empires have been able to achieve. The process of increasing the power of central authorities is double: Improved means allow control of land confines that used to escape any superintendence before; small nation states mechanically solve distance problems and reduce internal disparities.

Since the nineteenth century, land fragmentation has allowed the emergence of small cities scattered in the countryside. The downside of this is the creation of human deserts and green corridors along multiple boundaries. Jacques Ancel would recognize here the paradoxical resurrection of ancient times, when little peasant communities measured their territories through forest clearing abilities. The restoration of the *gora* (deep forest) across the Iron Curtain is a remarkable event of these days (Vahdat 2013; Terry et al. 2006).

The conversion of many frontiers into *easy* nature parks shows that modern borders are not expanding, but shrinking limits. In other words, the fall of Yugoslavia was an implosion, not an explosion. The human and natural void left by the last wars is typical of communities that build their independence on secured rather than conquering territorial projects. In general, national "reconstructions" are narcissistic movements. Should we consider that endogenous spaces are essential properties of emerging nations? In fact, it goes beyond the expected.

Historians argue back about the events that have led the early-conquering Ottoman Empire to become a stable state. How did it move away from its *Ghazi* spirit? How did it manage to replace a world-conquering envy with a territorial management of the acquired? When did it renounce its steppe origin to become an Empire of peasants? Endless disputes between specialists (Kasaba 2001). For an expanding political unit, the idea of boundary is nonsense. Anything settled is an impediment to the project.

The emergence of nation-states has been taken as the starting point of the present study. Some may say this was not the right choice. In 1699, the Treaty of Karlowitz brings an end to a 14-year war between the Ottoman Empire and the Habsburg Monarchy. Peace negotiations start a revolution (Abou-el-Haj 1969). For the first time, both opponents agree to accept a clear demarcation line, which breaks with the fuzzy frontier zones of previous times. The *marchlands* had no fixed status, paid double taxes, and remained under the constant threat of smugglers and soldiers. With the new agreement, the division becomes persistent and exclusive. Adopting *territorial integrity* as a principle constitutes a historic precedent.

The idea of a permanent peace that would replace a semi-permanent war leads to a rooted territoriality, i.e., to the definition of sovereign areas. Borderland controls will be enforced subsequently. The unprecedented act signed by the Ottomans and Austrians introduces a new classification. The soldiers who used to quarrel across land confines become legally limited in their movements. To carry on their raids, they have to change into rebels and outlaws.

The implications of such a decision are huge and affect the whole contact zone of the Ottomans with the rest of the Christian world. The nomadic peoples who faced the Austrian, Polish, and Russian territories are suddenly denied their old way of life and urged to settle down. The "border revolution" announces such a deep internal turmoil that the Ottoman Empire probably never recovered from the loss.

In fact, fixed boundaries are pure products of Western diplomacy imposed to a weakened Empire. With the agreement, the Porte is convinced to convert itself to patterns of land management that contradicts its old governing principles. The application of the rule meets stiff internal resistance. For these human groups living under the *Pax Ottomanica*, settling boundaries was a threat to their way of life. Beside the hushed atmosphere surrounding negotiation tables, a "wild" world would continue to exist, denying any legitimacy to this Western land control apparatus. Semi-nomadic groups shall carry on ignoring the newly established boundaries. They account for a large part of the Balkan population. Furthermore, an early nomadic heritage will continue to soak the existing institutions and regulatory frameworks of the Ottoman Empire. The mistrust against sedentary lifestyle has been pervading the dominant culture until the leading class converted itself to Western standards.

In the Balkans, the spatial mobility of people finds deep historical origins. J. Lampe and M. Jackson suggest that this might be an outcome of the early colonization policy of the conquerors (Lampe and Jackson 1982). According to the authors, soon after the abolition of the *Timar* system, the establishment of *çiftliks* has caused the emergence of a land-owning class whose policy has fostered the development of an unproductive agricultural sector. Many workers would have moved to other mountain spaces for their own safety, to escape heavy imperial taxation or to run away from the çiftliks where they were maintained in a serfdom-like status. These would have built "free" villages and learned to live mainly from livestock grazing. Mountain regions held by *zadruga* communities and lower *çiftlik* areas become "worlds apart".[1]

[1]Both terms are employed here in a wide figurative sense, as the *çiftlik* tenure system and the *zadruga* family structure penetrate the Balkan districts at different degrees according to the periods and regions.

The price to pay for this Ottoman land management system is heavy. The Balkan countryside would have lost five million residents between the sixteenth and the mid-eighteenth centuries—from 8 000 000 to 3 000 000 people (Lampe and Jackson 1982). Recurring wars and diseases follow the same line. The rise of nationalist movements is rooted in rural communities that have learned to live by their own means and rules a long time ago. Renewed interest in agricultural development was an essential dimension of the Tanzimat reforms, which intended to bring back some competitiveness to the Ottoman economy.

Linking the economic and demographic downturn of the Ottoman Empire to some "backwardness" is obviously tempting (Chirot 1989; Janos 2000). The idea is highly disputable, although often forwarded in many historical works about the causes of the downfall (McCarthy 2001). Even today, *modernity* is an idea typically associated with tight territorial controls. However, observations should not lead to wrong inferences. There is no Ottoman *backwardness* in the absence or weak practice of any mapping tool. This "default" expresses an early assumed political choice summed up in three words: *low border territoriality*.[2] This land management dogma has a counterpart: It produces different levels of *extra territoriality* (Bruneau 2004).

Not surprisingly, since the eighteenth century more and more accurate Austrian maps have been produced. For their part, the Ottoman officials never resigned themselves to map the Empire. In the late nineteenth century, when changes could no longer be avoided, they decided to seek the advice of Western cartographers.

Once deprived of its external goal, the Porte could not avoid being the target of armed banditry. There is an old saying that "the fish always stinks from the head downwards"; in former Ottoman lands, territorial decomposition has begun by the end of the tail onward. Such reasoning leads to refute the Oriental image associated with the Balkans: The border obsession of all modern nation-states expresses a raging desire to uphold the Western standards of territoriality. The balkanization process is nothing like the bloody reminiscence of eastern instincts.

In his novel dated 1872 *The man with the golden touch* (Jokai 1963), the Hungarian writer Mor Jokai accounts a story full of developments about an anonymous strip of land located in the middle of the Danube, south of Orşova. Without any clear date indication, the narration takes place in the first half of the nineteenth century. Downstream the Iron Gates, the Danube forms a long boundary line dividing the Ottoman and Austrian Empires. The border area is depicted as a land of all traffics. Smugglers move almost freely; they cross the river despite custom houses, defense garrisons, fortified places, controlled paths, and escape quarantine measures. Each village of the border zone has to supply watchmen;

---

[2] *Borderless* is a term to avoid as it directly relates to the widespread theories of "borderless worlds", see (Kahler and Walter 2006; Newman 2006). The use of this word would be unnecessarily anachronistic. In the Ottoman and Balkan thought, the concept of boundary is not denied. Both internal and external frontiers are simply deprived of any *dictatorial* power. This does not mean that the relationship between land and space is *deterritorialized*: Territoriality adopts different forms.

simultaneously, local communities produce their own smugglers. In the area, everyone speaks two or more languages: Romanian, Serbian, and Hungarian. Educated people also practice Greek, Turkish, and German. Catholics, Calvinists, Protestants, Orthodox, and Muslims cohabit and sometimes change faith.

The Iron Gates are not seen as a north–south, but as an east–west barrier:

> "The Danube rushed on beneath like a fiery Phlegethon, in its midst an island with towers and massive buildings, all glowing as if part of a huge furnace through which every creature coming from the pestilential east to the frontier of the healthy west, must pass as through purgatory" (p. 39).

The boundary is legally positioned in the middle of the river. Ships charged with goods and people navigate along the stream and cross one side of the river to the other, always playing with police controls. The Danube is an unpredictable companion. During months in winter, its surface freezes. With the ice thawing, spring floods often alter the landscapes. Across hundreds of kilometers, shores disappear, other massive alluvial deposits erect new barriers, and flows cover extensive fields. They bring the valuable fertilizing elements for further promising crops. Some pieces of land appear; others are eaten away by the erosion. Natural landscapes are versatile, and the vegetal environment is luxuriant. Fixing a boundary in such conditions is a real challenge. The linear artifacts mentioned on most maps shall remain a symbolic fable.

Mor Jokai captures the human vanity of a linear boundary and a mastered river. For this reason, he invents a hidden paradise in the middle of the Danube. Near Ostrova, a spring flood creates a new island referenced on no map, soon colonized by plentiful vegetation. The "nameless island possessed by no one" is a garden of delights planted with fruitful trees and delicious flowers. The inhabitants banish guns and money to preserve their own security. The island is the lowest piece of land of the earth. Rocky mountains are its antimodel. To the sterile and unbreakable heights, the islet opposes a tenuous existence left to the destructive vagaries of the Danube.

The metaphor is clear and meaningful: Border territories have a chance to survive when they escape the scrutiny of their protecting powers. They need anonymity and total deregulation to live on their own. The 30-day distance to Vienna and 40-day distance to Istanbul are not enough to provide security. To achieve this, cartography has to fail: Without mapping mistakes, the magnificent garden would change into a swampland.

> "Every year the Danube retreats a little further from its banks and helps on the good work of extending the boundaries of the woods, and because of this the huts built twenty-five years ago on the shore, now have a position much farther inland. And anyone wanting to cross the river without a passport can find in the young brushwood an entirely neutral territory" (p. 261). "Smuggling, on the banks of this wooded river, was a regular profession, with its own constitution, its own schools, its secret regulations, an autonomous state" (p. 266).

> "-I [the visitor, supposed to be Mor Jokai himself] informed him [Deodatus] that Hungary was now united to Austria by the word "and".He blew a smoke of his pipe, and the significance of that cloud of smoke was, 'My island has nothing to do with that'.I told him of our heavy taxes, and the smoke replied, 'We have no taxes here.'I described to him the fearful wars which had been waged in our kingdom and all over the world, and again the silent answer was, 'We wage war here with no one.'There was at that time a great panic on the exchanges, the oldest firms had failed; and this too I explained to him. Only his pipe's steady puffs seemed to say, 'Thank God, we have no money here.'I described to him the bitter struggle between parties, the ambitious strife between religions and nations.The old man shook the ashes out of his pipe, as if to say, 'We have neither bishops, electors, nor ministers here'. (p.399)."This is the present state of Nobody's Island. The privilege granted by two kingdoms, that this speck of ground should be excluded from any map, will last for fifty years more.Fifty years! Who knows what will have become of the world then?" (p. 400).

At the end of his book, Mor Jokai clearly delivers his message through the figure of Deodatus answering the questions of a visitor:

Intentionally, Mor Jokai creates the antithesis of the corrupt border condition which was weighing on extensive spaces at that time. The renowned author is so inspired that reality surpasses fiction a few years after the publication of his book.

The narrow island of Adakale has a weird destiny. Just next to the Iron Gates, its strategic fortress surveys an "end of the world" for ages. During the seventeenth and eighteenth centuries, the island ceaselessly switches between Ottoman and Austrian hands, and according to the wars, the two mastodons wage against each other. Alas! the semi-independence gained by the Principalities of Wallachia and Serbia at the beginning of the nineteenth century weakens its military importance. As a symbol of forgotten fights, the Austrian fortress is left to decay. At the end of the Russo-Turkish War of 1877–1878, the Treaty of Berlin defines the conditions for a considerable withdrawal of the Ottomans from the northern Balkans. Mor Jokai's world falls down. However, the treaty "forgets" to settle the fate of Adakale. Between 1878 and 1923, the Danubian island nominally remains the property of the Sultan, although 150 km away from the nearest boundary of his Empire.

What exactly is Adakale? A *nowhere land* left at the corner of a table like a breadcrumb? Another Balkan oddity? For fifty years, the island has been living in extra-territorial conditions: The Ottoman sovereignty remained theoretical, the place was under effective Austrian control, and it enjoyed fiscal exemption. In 1885, it was declared an open territory. In 1923, the few hundred inhabitants choose to be part of Romania. The dreadful old fortress turns into a delightful cruise stop for tourists. The territorial anomaly reveals its advantages for a while. Unfortunately, the new Đerdap dam erected in the 1970 a few kilometers downstream creates an enlarged Danubian reservoir that soon submerges the island. No visible trace of its past existence is left. Many Muslims leave their homes to join Turkey, while other residents reach the main Romanian cities and Dobrudja. As a compensation for the major public works achieved when building the electric station, two national parks on both Serbian and Romanian sides of the river convert the Iron Gates into an environmental sanctuary.

Adakale is often presented as a "lost paradise" with nostalgia, as opposed to painful memories always attached to other Balkan borderlands. Indeed, the peaceful resolution of Adakale's problem is remarkable when compared with other mixed areas whose fate was decided by military means. Some may say the case is anecdotal. However, Balkan borderlands have been affected by so many troubles that Adakale might be considered, with other examples of this study, as a symbolic "revenge" of territory/geography against some official–political "black-and-white" vision of land.

9.3 Questioned Centralities

Borders and boundaries play a key role in understanding the Balkans, but the existence of borderlands supposes that corelands prevail over them. And precisely, the permanent mobility of all kinds of frontiers, their varied legal definitions and administrative interlocking capacity, and the land fragmentation resulting from fluctuating political entities are all essential elements challenging the influence of central places. Tiny states are not expected to find balanced regional powers able to tackle the growing weight of their capital cities. Balkan countries are made of heads out of proportion attached to weakened bodies. Until recently, rural depopulation played a central role in securing most national borders, in a much more efficient way than would be the case with heavy military contingents. To protect sovereign perimeters, all forms of regional selfgovernment must be limited or kept at a formal level. The oversized weight of Athens, Istanbul, Sofia, Tirana, Belgrade, Zagreb, Skopje, Ljubljana, Podgorica, Sarajevo, and maybe tomorrow Prishtinë[3] will have trouble finding opposite forces. The development of abnormally enlarged national *bodies* and the elimination of any internal resistance are issues still to be studied in many countries.

The defeat of any subnational "identity" is thus a necessary stage in the historical process of the establishment and growth of the states. The "hostility" to all expressed or latent forms of selfgovernment is not just about land confines; it also applies to the core civilian districts themselves, directly submitted to the "dictatorship"/prevalence of their respective capitals. When confronted with the Balkan mess, most media usually find it easier to focus on "ethnic" conflicts, but in this case, their approach is biased as they consider consequences rather than causes. Balkan states are first obsessed by their viability. Only "later,"[4] in front of uncertain futures, will they fix their attention on border issues; hence, the incapacity often assumed to distinguish conventional interstate wars from civil conflicts (Newman 2014).

[3]Some of these cities even go as far as to gather half of the national population. The syndrome is long known: Balkan capitals are villages quickly transformed into sprawling urban centers. This explains most of their serious planning problems.

[4]This refers to the general reasoning, not to a chronological order.

The struggle of borderlands against centers was typical of the nineteenth century in Ottoman Rumelia. The battle is closely associated with other national uprisings against "foreign" occupiers. Building a nation-state is thus a process of borderlands becoming central places—at a certain scale level at least. To do so, new confines are needed. Their "invention" enables self-proclaimed communities to set up or reinforce necessary central places. Independence is scale transfer.

Istanbul had already jeopardized the core–periphery balance in the eighteenth and nineteenth centuries. Most historians believe that the end of the Ottoman Empire was planned by the Tanzimat reforms, which aimed at limiting the growing power of the *ayans*. These local "notables" were recipients of many privileges granted by the Porte. Officially, they remained subjects of the Sultan—in fact, their loyalty was fictitious (Karpat and Zens 2004; Lampe and Jackson 1982). The autonomy claimed by such early leaders adopted prenational characteristics, promoting land appropriation, tax exemptions or transfers, and military self-governance. However, they never defined themselves on pure ethnic attributes against the dominant powers.[5]

A widespread schizophrenia thus permeated all levels of the Ottoman administrative hierarchy. On the one hand, official and legal provisions referred to the high sovereignty of the Sultan over the lands placed under his dependence. By contrast, the everyday life denied any application of this principle and allowed communities to enjoy a practical semi-independence. The *millet* system itself was a form of power transfer among others. This schizophrenia, which leads to distrust against central public authorities, is maybe the most common legacy found nowadays in the "post-Ottoman" countries—with other cooking recipes (Hajdarpašić 2008). Account should also be taken of the distrust against the state in former communist countries. All these lead us to understand the serious problems all Balkan states have to manage when trying to govern territories and populations. Constantly "inventing" borders not only creates land fragmentation, but also causes tough disorders to handle afterward.

The resonance of different events that have seized the Peninsula in the 1990s reminds us how much spatial balance and frontier issues are closely related. All evidences show that "local" forces are not completely eradicated and may still find ways to express their resistance against distrusted external "giants": districts against nations and nations against international *patronage*. Reinforced centralization is often the answer adopted to limit centrifugal forces with many subsequent risks. The Tanzimat reforms had led in their time to an increased mutiny of the borderlands, a reverse result to the expected one.

The mafias—seen as trading systems, as well as social organizations—which are deciding forces today in the area are nothing more than local counterpowers seeking to challenge weakened nation-states. They become as much invasive as boundaries thwart the necessary mobility of men and goods. They partially derive their force

[5]Official records usually transform these spontaneous local uprisings into purely national/ethnic movements afterward

from the denied territoriality of borderlands. They develop as much as local populations recognize border systems as a threat against their activities.

Clientelism is presented these days as a general plague in the Balkans although it is nothing more than a contractual arrangement between parties, the territorial rooting of which is well documented. Some corrupt practices derive from land fragmentation as tiny spaces make close acquaintanceships easier. It is acknowledged, for instance, that most early national *leaders* come from anonymous rural places lost in the middle of nowhere. Once smugglers and bandits, these "commanders" build their reputation through any form of "local" resistance and fighting epic. They run for respectability afterward, when new "democratic" parties confirm their official position. The relationship between bribery, black economy, local privileges granted to the clan *versus* clientelist group the, political instability, and mafias competing with central governments is outcome of a specific geographic framework. Such words echo the current situations of Kosovo and other former Yugoslav provinces, but these are far from being the exclusive supporters of such practices.

Isn't it what Milan Kundera says about tiny states? The notorious Czech-born writer explains that the very essence of small nations does not lie in their limited surface, but in their precarious destiny (Goodby 2014; Kundera 1984). This principle would generate powerful nationalist movements and foster community resilience, whose evidence is embodied in a fierce opposition to the interchangeability of people. In practice, this means that horizontality is a main feature characterizing Balkan societies—often perceived externally as engulfed in some structural "light anarchy." Combining this with the enhanced geographic centrality of capital cities creates the conditions for lasting political tensions and territorial imbalances.

Despite this, the concept of balkanization has lost some of its attractiveness these last years. With the end of the wars, the successful post-communist *transition*, and the promised or enacted EU integration of many candidate countries (Solioz and Stubbs 2009), the ghosts of the past seem to have been exorcised, at least partially. The Peninsula would have entered the new millennium hoping for a better future. The post-WWII generation that had lived in an oppressive atmosphere over decades now passes the baton to younger people who live as fully as possible in the globalized world. Everyone has noted the remarkable progress achieved by states that have made every effort to improve their economic condition. It is no secret that this renewal is also possible through the emergence of modern political circles trained in the very best universities worldwide. Much has been made.

As a consequence, border issues appeared until recently to have faded away from people's minds all the more easily as the young generation displays relative indifference to and ignorance of past litigations. Nowadays, public attention is drawn on economic and social questions. *Crisis* has become a much trendier word than *balkanization*. Some may say that this new inclusive view qualifies a certain defeat of geography in front of what really matters today. Global-scale economic conflicts would have taken precedence over outdated nation-based interests. In a time when world cities seem to have achieved global domination, borderlands are no more than insignificant places whose marginality provides the opportunity to develop environmentally friendly services.

Unfortunately, this position is denied by a number of recent events (Branka and Janczak 2015). The refugee crisis and the ongoing Greek collapse seem to have brought back the materiality principle into the European policy arena. Both events demonstrate that the increased inequality arising from the *transition/crisis* cannot be without any consequences particularly in view of border management and policy. The new walls and razor wires erected along the borders of Greece, (FY) Macedonia, Hungary, Serbia, Croatia, Slovenia, Austria, and Bulgaria have a particular emotional sting. They evidence that certain territorial processes cannot be ignored and that long-lasting spatial differentiation models still weigh in international relations despite cross-border cooperation programs whose successful completion is challenged by ambiguous ambitions.

Milan Kundera argues that "small nations have their own vision of the world, a vision based on a deep distrust of history." One might be tempted to add that, reversely, big nations and powers show a blatant disregard for geography. This stigma is nowhere more evident than in border areas. Against all odds, geography still matters.

References

Abou-el-Haj RA (1969) The formal closure of the Ottoman frontier in Europe: 1699–1703. J Am Orient Soc 89(3):467–475

Agnew J (1994) The territorial trap: the geographical assumptions of international relations theory. Rev Int Polit Econ 1(1):53–80

Agnew J (1998) Geopolitics. Re-visioning World Politics, Routledge, London

Agnew J (2010) Still trapped in territory? Geopolitics 15(4):779–784

Branka T, Janczak J (eds) (2015) Boundaries revisited: conceptual turn in European border practices. Logos Verlag, Berlin, Thematicon

Brown C (1995) Serpents in the sand: essays in the nonlinear nature of politics and human destiny. University of Michigan Press, Ann Arbor

Bruneau M (2004) Diasporas et espaces transnationaux. Anthropos, Paris

Chirot D (1989) The origins of backwardness in Eastern Europe. University of California Press, Berkeley-Los Angeles-Oxford

Ebel KA (2008) Representations of the frontier in Ottoman town views of the sixteenth century (English). Les représentations de la frontière dans les vues des villes ottomanes du XVIe siècle (French) 60(1):1–22

Goodby JE (2014) The survival strategies of small nations. Surv: Glob Polit Strateg 56 (5):31-39

Hajdarpašić E (2008) Out of the ruins of the Ottoman empire: reflections on the Ottoman legacy in South–Eastern Europe. Middle East Stud 44 (5)

Janos AC (2000) East Central Europe in the modern world: the politics of the borderlands from pre- to postcommunism. Stanford University Press, Stanford, Calif

Jokai M (1963) The man with the golden touch. Corvina, Budapest

Kahler M, Walter B (2006) Territoriality and conflict in an era of globalization. Cambridge University Press, Cambridge-New York

Karpat KH, Zens RW (2004) Ottoman borderlands: issues, personalities and political changes. University of Wisconsin, Madison

Kasaba R (2001) L'Empire ottoman, ses nomades et ses frontières. Aux XVIIIe et XIXe siècles. Critique internationale 12:111–127

Kundera M (1984) The tragedy of Central Europe. N Y Rev Books 31(7):33–38
Lampe JR, Jackson MR (1982) Balkan economic history, 1550–1950. Indiana University Press, Bloomington
McCarthy J (2001) The Ottoman peoples and the end of empire. Arnold-Oxford University, London–New York
Merry U (1995) Coping with uncertainty: insights from the new sciences of chaos, self-organization, and complexity. Praeger, Westport, Conn
Newman D (2006) The lines that continue to separate us: Borders in our 'borderless' world. Prog Hum Geogr 30(2):143–161
Newman E (2014) Understanding civil wars: continuity and change in intrastate conflict. Routledge, London, New York
Paasi A, Zimmerbauer K (2016) Penumbral borders and planning paradoxes: relational thinking and the question of borders in spatial planning. Environ Plan A 48(1):75–93
Painter J (2008) Cartographic anxiety and the search for regionality. Environ Plan A 40(2):342–361
Pickles J (1995) ground truth: the social implications of geographic information systems. Guilford Press, New York
Schofield N, Gallego M (2011) Leadership or chaos: the heart and soul of politics. Springer, Heidelberg; New York, Berlin
Solioz C, Stubbs P (2009) Emergent regional co-operation in South East Europe: towards 'open regionalism'? Southeast Eur Black Sea Stud 9(1–2):1–16
Terry A, Ullrich K, Riecken U (2006) Green belt of Europe. From Vision to Reality Gland, Cambridge
Vahdat V (2013) Das Grüne, vol Europas. Diplomica Verlag, Hamburg, Im Spannungsfeld von Naturschutz und Tourismus

Glossary

Andartis, αντάρτης Greek revolutionary, guerilla, irregular fighter

Armatole, αρματολός Member of the Greek militias charged by the Ottoman authorities with the task to maintain order in the remote districts of the Empire

ASPRS American Society for Photogrammetry and Remote Sensing

Ashkalis, Ашкалије (Evgjit) Albanian-speaking community, often identified as "Egyptians" in the official surveys and confused with the Romani

Ayan Ottoman local leaders, wealthy people in reference to the property-owning class

Bayrak, bajrak Albanian political entity of customary law based on clan organization(s). The leader of a bayrak is called the Bayraktar

Banovina First-level administrative district in First Yugoslavia (1929–1941)

Bashi Buzuk, Başıbozuk Irregular soldier of the Ottoman army, mercenary

Bey Ottoman dignitary, usually owner of estates in the countryside

Blato Slavic name for a swampland

Brovey transformation In remote sensing, resolution merge techniques that make it possible to combine multispectral and panchromatic data. It produces highly contrasted images

Bunjevci, Буњевци Catholic South Slavic minority originating from Herzegovina and Dalmatia, living today in Southern Hungary and Vojvodina

Chams, Çam Former inhabitants of Chameria- Çamëria-Τσαμουριά (today Thesprotia). Muslim people ethnically referenced as Albanians, although they mainly belonged to Greek-speaking communities in the old times

Cheta, чета, četa Generic term for an armed gang, military, or paramilitary unit (Serbia, Macedonia, Bulgaria)

Chetnik, четник Member of a cheta

© Springer International Publishing Switzerland 2017
R. Darques, *Mapping Versatile Boundaries*,
Springer Geography, DOI 10.1007/978-3-319-40925-2

Chiftlik, Çiftlik Big agricultural estate, Ottoman system of land management

CRS, Coordinate Reference System Set of data describing a full georeferencing system. This includes projection and datum with all related data

DEM, Digital Elevation Model or DTM, Digital Terrain Model Raster or vector digital file used for building 3D views of the earth

Dimos, δήμος Fourth-level administrative unit in Greece (see Sect. 9.2.1 in Chap. 9). Urban community

ELAS, Ελληνικός Λαϊκος Απελευθερωτικός Στρατός Greek People's Liberation Army

ELSTAT, Ελληνική Στατιστική Αρχή Greek statistical office (formerly ΕΣΥΕ-NSSG)

Eparhia, επαρχία In Greece, intermediate administrative unit between nomos and dimos-kinotita (see Sect. 9.2.1 in Chap. 9)

Exarchist Supporter of Bulgarian national Orthodox church in front of the Ecumenical Patriarchate of Constantinople

(FY)ROM (Former Yugoslav) Republic of Macedonia

GADM, Global Administrative Areas GIS database for the location of the world's administrative areas

Geodesy Science of measuring the shape of the earth and its dimensions, especially through gravity field measures

Ghazi Muslim conquerors fighting in the name of Holy War

GIJNA, Geografski Institut Jugoslovenske Narodne Armije Geographic Institute of the Yugoslav People's Army

Gora, гора Wooded mountain or, sometimes, wooded area

Gorani, Горани Slav-speaking Muslims of Kosovo, Eastern Albania, and Macedonia

Grecomans, Γραικομάνοι, Гъркомани, Grekomanë Hellenophile minorities of the Balkans, sometimes confused with Patriarchist citizens outside Greece

Hajduk, хайдут(ин), κλέφτης Bandit or highwayman. The thief fighting against wealthy "Ottomans" could rely on a positive image in people's mind

HPF, High Pass Filter Image fusion technique applying a spatial filter that emphasizes high spatial frequencies and reduces the weight of spectral informations

HSI, Hue Saturation Intensity Image display method that fits to the human visual perception. It implies a color transfer from the standard RGB three-dimensional coordinates. When including a panchromatic band, it may be used as a fusion technique

IEBL, Inter-Entity Boundary Line "Border" dividing the two constitutive Entities of Bosnia–Herzegovina

IFOR, Implementation Force NATO peacekeeping forces of Bosnia–Herzegovina (1995–96)

IGME, Ινστιτούτο Γεωλογικών και Μεταλλευτικών Ερευνών Greek Institute of Geology and Mineral Exploration

IMRO, Вътрешна Македонска Революционна Организация (ВМРО) Internal Macedonian Revolutionary Organization

INSTAT, Instituti i Statistikave Albanian Statistical Institute

Kachak Albanian guerilla group

Kambos, κάμπος Flat plain, often repellent, as opposed to the mountain

Kaza Third-level administrative unit in the Ottoman Empire (after 1864)

KFOR, Kosovo Force NATO peacekeeping forces of Kosovo (from 1999 onward)

Kinotita, κοινότητα Fourth-level administrative unit in Greece (see Sect. 9.2.1 in Chap. 9). Rural community

Komitadji, Комити Members of revolutionary militias fighting for the national liberation (mainly in Macedonia and Bulgaria), in the late nineteenth and early twentieth centuries

Krashovani, Caraşoveni, Krašovani, Крашовени South Slavic ethnic group originating from Caraş-Severin (present Romania), sometimes registered as Bulgarians

Kriging Geostatistical method allowing the spatial interpolation of a variable

KTEL, Κοινά Ταμεία Εισπράξεων Λεωφορείων Main public bus transport service in Greece

Millet Legal system acknowledging the existence and organizing the control of religious communities in the Ottoman Empire

Nahija Fourth-level administrative unit in the Ottoman Empire (after 1864)

NDVI, Normalized Difference Vegetation Index In remote sensing, technique that aims at measuring the major fluctuations in vegetation activity

Nomos, νομός Former second-level administrative unit in Greece. Replaced by the Regional Units after 2010 (see Sect. 9.2.1 in Chap. 9)

NSIB, Национален Статистически Институт Bulgarian National Statistical Institute

Obshtina, općina, opština, община Second-level administrative unit in Bulgaria, (FY)ROM, Slovenia and Croatia
Third-level administrative unit in Serbia and Bosnia–Herzegovina
First-level administrative unit in Montenegro

Oblast, област First-level administrative unit in Bulgaria

Okrug, округ, окръг Second-level administrative unit in Serbia
First-level administrative unit in Bulgaria (until 1987)

Pashalık, paşalık Ottoman administrative territory headed by a pasha

Patriarchist Supporter of a unified Orthodox church under the guardianship of the Ecumenical Patriarchate of Constantinople

Photogrammetry Science whose purpose is to study the form, dimensions, and position of an object through its optical perspectives acquired by remote sensing means

Planina, планина Grassy mountain dedicated to pasture

Plemena, племена (or bratstva, братства) Traditional Slavic tribes and clans (brotherhoods). These are social as well as territorial units

Pomaks, Πομάκοι, Помаци Bulgarian-speaking Muslims of the Rhodope

Pondi, Πόντιοι Greeks originating from the Black Sea and the USSR

Raya Ottoman lower-class subject paying the Haraç tax

Rreth Second-level administrative unit in Albania

Rumelia, Ρωμουλία Name of the Balkan–European part of the Ottoman Empire

Rusyns Slavic people, otherwise called Ruthenes, Carpatho-Russians or Ukrainians

SORS, Републички Завод за Статистику Statistical Office of the Republic of Serbia

Sandjak Second-level administrative unit in the Ottoman Empire (after 1864)

Sarakatsani, каракачани, Σαρακατσάνοι Greek-speaking nomad people of the Southern Balkans living from pastoralim, mainly

SFOR, Stabilization Force NATO peacekeeping forces of Bosnia–Herzegovina (1996–2004)

SHIK, Shërbimi Informativ Kombëtar National Intelligence Service of Albania, successor of the much-dreaded Sigurimi (Directorate of State Security) after 1991. Replaced by the SHISH in 1998

SLC, Scan Line Corrector Landsat motion corrector instrument. The permanent failure in 2003 leaded to degraded image quality afterward

Šokci, Шокци Slavic-Croatian minority of the Danube and Sava. They profess Catholicism as a religion

SRTM, Shuttle Radar Topography Mission NASA space shuttle mission launched in 2000 for acquiring a global DTM of the earth

Székely Subnational Hungarian group in charge of keeping the Transylvanian frontier under the Habsburgs

Tanzimat Series of deep reforms undertaken after 1839 aiming at modernizing the Ottoman Empire

Tasseled cap transformation This remote sensing technique converts the original bands of a satellite image into composite values of brightness, wetness, and greenness. Mainly used in the frame of vegetation studies

Timar Early Ottoman land management system. Revocable allotment of estates as a compensation for military service

Torbeshi, Торбеши Slav-speaking Muslims of Macedonia

Turcocratia, Τουρκοκρατια Pejorative term referring to the era of Ottoman rule

Turco-Cretans, Τουρκοκρήτες, Girit Türkleri Greek-speaking Muslims of Crete

UÇK, Ushtria Çlirimtare e Kosovës Kosovo Liberation Army

UÇPMB, Ushtria Çlirimtare e Preshevës, Medvedgjës dhe Bujanocit Liberation Army of Preševo, Medvedja and Bujanovac

UNMIK United Nations Interim Administration Mission in Kosovo (from 1999 onwards)

USGS United States Geological Survey

Valahades, Βαλαχάδες Islamized Greek-speaking communities of the Southern Balkans

Vilayet First-level administrative unit in the Ottoman Empire (after 1864)

Vladika Name of the Prince-Bishop heading Montenegro until 1851. Theocratic political system

Vlahs, Власи, Βλάχοι, Vllehët Otherwise called Aromanians. These pastoral nomads of the Balkans speak a Latin language and profess Orthodoxy as a religion

Yuruks, Yörüks, Κοναριώτες, Юруци Muslim Turkish-speaking nomads, sometimes called Turkmen by extension

Zadruga, задруга Extended family structure of the upland villages, which formed armed clans placed under the guidance of patriarchs

Index

© Springer International Publishing Switzerland 2017
R. Darques, *Mapping Versatile Boundaries*,
Springer Geography, DOI 10.1007/978-3-319-40925-2

The manufacturer's authorised representative in the EU is Springer Nature Customer Service Centre GmbH, Europaplatz 3, 69115 Heidelberg, Germany. If you have any concerns regarding our products, please contact ProductSafety@springernature.com

Printed and bound by CPI Group (UK) Ltd, Croydon, CR0 4YY
30/09/2026
02225571-0010